THE ORGANIZATIONAL FRONTIERS SERIES

The Organizational Frontiers Series is sponsored by the Society for Industrial and Organizational Psychology (SIOP). Launched in 1983 to make scientific contributions to the field, the series has attempted to publish books that are on the cutting edge of theory, research, and theory-driven practice in industrial/organizational psychology and related organizational science disciplines.

Our overall objective is to inform and to stimulate research for SIOP members (students, practitioners, and researchers) and people in related disciplines, including the other subdisciplines of psychology, organizational behavior, human resource management, and labor and industrial relations. The volumes in the Organizational Frontiers Series have the following goals:

1. Focus on research and theory in organizational science, and the implications for practice
2. Inform readers of significant advances in theory and research in psychology and related disciplines that are relevant to our research and practice
3. Challenge the research and practice community to develop and adapt new ideas and to conduct research on these developments
4. Promote the use of scientific knowledge in the solution of public policy issues and increased organizational effectiveness

The volumes originated in the hope that they would facilitate continuous learning and a continuing research curiosity about organizational phenomena on the part of both scientists and practitioners.

Previous Frontiers Series volumes, all published by Jossey-Bass, include:

Personality and Work
Murray Barrick, Ann Marie Ryan, Editors

Managing Knowledge for Sustained Competitive Advantage
Susan E. Jackson, Michael A. Hitt, Angelo S. DeNisi, Editors

Work Careers
Daniel C. Feldman, Editor

Emotions in the Workplace
Robert G. Lord, Richard J. Klimoski, Ruth Kanfer, Editors

Measuring and Analyzing Behavior in Organizations
Fritz Drasgow, Neal Schmitt, Editors

The Nature of Organizational Leadership
Stephen J. Zaccaro, Richard J. Klimoski, Editors

Compensation in Organizations
Sara L. Rynes, Barry Gerhart, Editors

Multilevel Theory, Research, and Methods in Organizations
Katherine J. Klein, Steve W. J. Kozlowski, Editors

The Changing Nature of Performance
Daniel R. Ilgen, Elaine D. Pulakos, Editors

New Perspectives on International Industrial/Organizational Psychology
P. Christopher Earley and Miriam Erez, Editors

Individual Differences and Behavior in Organizations
Kevin R. Murphy, Editor

The Changing Nature of Work
Ann Howard, Editor

Team Effectiveness and Decision Making in Organizations
Richard A. Guzzo, Eduardo Salas, and Associates

Personnel Selection in Organizations
Neal Schmitt, Walter C. Borman, and Associates

Work, Families, and Organizations
Sheldon Zedeck, Editor

Organizational Climate and Culture
Benjamin Schneider, Editor

Training and Development in Organizations
Irwin L. Goldstein and Associates

Productivity in Organizations
John P. Campbell, Richard J. Campbell, and Associates

Career Development in Organizations
Douglas T. Hall and Associates

Health and Safety in Organizations

A Multilevel Perspective

David A. Hofmann

Lois E. Tetrick

Editors

Foreword by Neal Schmitt

JOSSEY-BASS
A Wiley Imprint
www.josseybass.com

Published by Jossey-Bass
A Wiley Imprint
989 Market Street, San Francisco, CA 94103-1741   www.josseybass.com

Jossey-Bass books and products are available through most bookstores. To contact
Jossey-Bass directly call our Customer Care Department within the U.S. at 800-956-7739,
outside the U.S. at 317-572-3986 or fax 317-572-4002.

Jossey-Bass also publishes its books in a variety of electronic formats. Some content that
appears in print may not be available in electronic books.

Library of Congress Cataloging-in-Publication Data

Health and safety in organizations : a multilevel perspective / by David A.
Hofmann, Lois E. Tetrick, editors : foreword by Neal Schmitt—1st ed.
        p. cm.—(The organizational frontiers series)
Includes bibliographical references and index.
    ISBN 0-7879-5846-8
  1. Industrial safety.   2. Industrial hygiene.   I. Hofmann, David A.
II. Tetrick, Lois E.   III. Series.
    T55.H385   2003
    658.3'82—dc21

                                                      2002155633

Printed in the United States of America
FIRST EDITION
HB Printing   10 9 8 7 6 5 4 3 2 1

Contents

Foreword

This is the twentieth book in a series of books published by Jossey-Bass and initiated by the Society for Industrial and Organizational Psychology (SIOP) in 1983. Originally publishing this as the Frontiers Series, the SIOP executive committee voted in 2000 to change the series name to the Organizational Frontiers Series in an attempt to enhance its identity and visibility. The purpose of the publication of these volumes in a general sense was to promote the scientific status of the field. Ray Katzell first edited the series. He was followed by Irwin Goldstein and Sheldon Zedeck. The editorial board chooses the topics of the volumes and the volume editors. The series editor and the editorial board then work with the volume editor in planning the volume and occasionally in suggesting and selecting chapter authors and content. During the writing of the volume, the series editor often works with the editor and the publisher to bring the manuscript to completion.

The success of the series is evident in the high number of sales (now over 45,000). Volumes have also received excellent reviews, and individual chapters as well as volumes have been cited very frequently. A recent symposium at the SIOP annual meeting examined the impact of the series on research and theory in industrial and organizational (I/O) psychology. Although such influence is difficult to track and volumes varied in intent and perceived centrality to the discipline, most participants concluded that the volumes have exerted a significant impact on research and theory in the field and regard them as being representative of the best the field has to offer. Our first volume, edited by Tim Hall (1986), *Career Development in Organizations,* was our most successful in terms of sales. The current volume reflects new thinking and research on health and safety in organizations that we hope will redirect and reenergize researchers in this important area of our discipline and stimulate new ideas about how to produce healthy organizations among practitioners.

This volume, edited by David A. Hofmann and Lois E. Tetrick, represents a significant and successful effort to consider health and safety at various levels, including that of the individual, the group or team, and the organization. Hofmann and Tetrick introduce the book with a discussion of various definitions of health. Their discussion of the amazing number of different ways researchers have defined this topic and the framework they devise to organize these definitions are good precursors to the variety of perspectives one encounters in this book. In Part One, the authors deal with topics that I/O psychologists have frequently addressed in their research and practice. The new spin, however, is to explicitly consider how these topics affect health and safety in organizations, thereby forging new ground and, we hope, encouraging a better integration of health and safety issues into the mainstream of our field in the future. Another important aspect of Part One is its breadth with respect to levels of analysis. For example, the various chapters include a focus on individuals (Chapters Two and Three), work design (Chapter Four), as well as groups and organizations (Chapter Five; Chapter Six; Chapter Seven).

Part Two is likely to be novel to many in our field. For example, it includes chapters discussing the impact of human resource management strategies on health and safety (Chapter Eight), an evaluation of the degree to which workers' compensation laws and insurance actually promote health (Chapter Twelve), and the role that regulation in various countries of the world plays in the promotion of health (Chapter Thirteen). These topics as well as others represent areas about which few I/O psychologists are likely to be informed. Even some of the topics that some members of our discipline have addressed, such as the interface between work and family (Chapter Ten) and shiftwork and telework (Chapter Nine) are areas in which most of us remain relatively uninformed. I have mentioned some chapters, but I believe all of this book will prove to be highly informative and useful to all who are concerned about the welfare of the people with whom we work and interact as well as those we study.

Our target audiences include graduate students in I/O psychology and organizational behavior as well as doctoral-level researchers and practitioners who want to gain knowledge on the most up-to-date data and theory regarding the health and well-being of individuals and organizations. I believe that this book represents a significant departure in the way that most I/O psychologists have viewed health and safety issues. It should make accessible to our discipline research and theory that will stimulate a concern and appreciation for the impact of the broader environment and the ways we can manipulate that environment to produce more health-sustaining organizations and healthy individuals. The degree to which it does so is the degree to which it will meet the primary goal of the Organizational Frontiers Series as stated earlier in this Foreword.

I also believe that this book will stimulate and promote a much broader perspective on health and safety and promote investigation of a wider array of interesting and important questions. Finally, I think it will be of interest to all of us as individuals as we evaluate the environments in which we work or those that we create for the individuals in our organizations or supervisory units. The chapter authors deserve our gratitude for pursuing the goal of clearly communicating the nature, application, and implications of the theory and research they describe in this book. Production of a volume such as this involves the hard work and cooperative effort of many individuals. The chapter authors and the editorial board all played important roles in this endeavor. Because all royalties from the series volumes are used to help support SIOP financially, none of the chapter authors received any remuneration. They deserve our appreciation for engaging in a difficult task for the sole purpose of furthering our understanding of organizational science. We also express our sincere gratitude to Cedric Crocker, Julianna Gustafson, Matt Davis, and the entire staff at Jossey-Bass. Over many years and several volumes, they have provided support during the planning, development, and production of the series.

February 2003

NEAL SCHMITT
Michigan State University
Series Editor, 1998–2003

Preface

Webster's Dictionary defines *health* as the absence of illness and as the state of well-being (i.e., flourishing). This definition, however, actually constitutes two different criteria: the absence of illness and the presence of health. For example, a person may not have any signs of illness but still have a cholesterol level that is too high, not exercise sufficiently, and have a few extra pounds to lose. Thus, that person's fundamental internal systems work well (i.e., show no illness or disease), yet he or she may not show signs of being healthy if we consider health to be long-term well-being and flourishing. When thinking about healthy organizations, we often make the same distinction. For example, we could think about organizations lacking illness (e.g., not harming employees or the environment) as well as possessing the presence of health (e.g., long-term adaptability and thriving). The overall objective of this volume is, therefore, to consider a broader definition of health and safety in organizations—one that encompasses both the absence of illness and the presence of health—and the implications it has for industrial/organizational (I/O) psychology.

In light of this broader definition of health, we believe that within I/O psychology specifically and applied psychology generally, a great deal of research into traditional research domains bears on individual and organizational health but without explicitly considering these linkages nor integrating them into a coherent framework. Thus, the first purpose of this volume is to consider the impact of certain more traditional areas of inquiry within our field in the context of health and safety outcomes. In addition to these more traditional areas, a number of areas of research have either not received high visibility in our field or are new and emerging areas that also affect health and safety in organizations. Thus, the second purpose of this volume is to provide a mechanism not only to introduce these areas to the larger field but also to explore possible new frontiers of research for I/O psychologists.

Structure of the Book and Overview of Contents

I/O psychology professionals increasingly recognize the multilevel nature of the phenomenon that we study. In light of this, we sought to explicitly address the multilevel nature of health and safety across the various contributions to the volume. Part One (Individual and Organizational Effects) starts, for example, at the individual level of analysis and then moves on to consider higher levels of analysis such as groups, organizational climate, and leadership. Part Two (Strategy and Policy), although not as explicitly multilevel, does consider how different strategies and policies affect both individuals within the organization and the health of the organization as a whole.

Prior to Part One, however, we dive more deeply into the meaning of individual and organizational health. In Chapter One we review a number of different definitions of health (both medical and organizational) and develop a framework that integrates these various definitions both in terms of content and level of analysis. We asked the authors to consider the implications of this framework for their contributions, and therefore we hope to have provided a common theme cutting across the various contributions.

After Chapter One we begin Part One of the book, Individual and Organizational Effects. This begins at the individual level in Chapter Two by Spector, on the implications of individual differences for health and well-being in organizations. Spector views individuals as interacting with their environment such that individual reactions to external environments—in terms of the experience of strain—can be viewed only in light of individual differences. In Chapter Three Burke and Sarpy consider how to affect employee health and safety through interventions, with a primary focus on safety training in organizations. Parker, Turner, and Griffin then discuss the notion of designing healthy work in Chapter Four. For example, is it possible to design work such that one can jointly optimize performance and health? Chapter Five by Tesluk and Quigley considers how groups, and the norms that develop within these groups, can influence health and safety. Specifically, Tesluk and Quigley use the familiar input-process-output model of group performance to consider how various aspects of this model can directly and indirectly affect health and safety. Next, Duffy, O'Leary-Kelly, and Ganster (Chapter Six) investigate antisocial work behavior and how this behavior can affect both organizational and individual health. Concluding Part One, Zohar (Chapter Seven) considers how leadership and organizational climate affect organizational health and safety, with a particular focus on safety climate research.

Part Two of the book, Strategy and Policy, introduces a number of areas of research that are either new and emerging or that have not received a great deal of visibility in the field of I/O psychology. Leading off is Chapter Eight by Shaw and Delery, considering how strategic human resource management affects organizational health. Although a number of recent studies have documented the relationship between human resource management strategies and organizational performance, none has explicitly considered how these policies influence health-related outcomes. Shaw and Delery take up this task. Next, Smith, Sulsky, and Ormond (Chapter Nine) consider the impacts of various work arrangements and their implications for health in organizations. Chapter Ten by Perrewé, Treadway, and Hall continues the focus on work policies by considering the health implications of work-family policies. Griffiths and Munir (Chapter Eleven) shift the focus away from organizational policies to health promotion efforts taken up by organizations in an effort to improve the health of their workforce. Griffith and Munir focus most extensively on smoking cessation efforts. The next two chapters consider the regulatory environment within which organizations reside and the potential implications of these regulations for health in organizations. Roberts (Chapter Twelve) focuses on workers' compensation laws and insurance and how these regulations affect health outcomes. Brotherton (Chapter Thirteen) reviews developments in health and safety policies across a number of countries and discusses their implications for health outcomes.

We wrote the concluding chapter, Chapter Fourteen, in which we take the framework developed in Chapter One and revisit it in the context of the various contributions of the volume. Given the diversity of the authors' contributions, we attempt to provide some integration across the chapters as well as discuss potential synergies that future research could exploit.

Acknowledgments

First and foremost, we would like to thank each of our chapter authors for their patience with us as we managed the process of pulling this volume together as well as for their informative, insightful, and high quality chapters. In addition, each of the authors was extremely responsive to our feedback in revisions. For this we are truly grateful.

We also need to acknowledge the support of Neal Schmitt and the rest of the board of the SIOP Organizational Frontiers Series. We greatly appreciate their encouragement throughout the process. Also, all the fine folks at Jossey-Bass deserve special recognition for being patient with us as we fell "a little" behind schedule.

Finally, we conclude our comments on a sad note. Carlla Smith (coauthor of Chapter Nine) lost her courageous fight with cancer while we were working on this volume. Carlla will be remembered by all in the field of I/O psychology not only for her many scientific contributions but also for her courageous spirit, gentle soul, and warm heart. It is to Carlla Smith that we dedicate this volume.

February 2003

DAVID A. HOFMANN
Chapel Hill, North Carolina

LOIS E. TETRICK
Houston, Texas

The Contributors

David A. Hofmann is currently associate professor of Management at the Kenan-Flagler Business School at the University of North Carolina at Chapel Hill. He received his B.A. degree from Furman University (1986), his M.S. degree from the University of Central Florida (1988), and his Ph.D. in Industrial and Organizational Psychology from The Pennsylvania State University (1992). Within the health and safety domain, his research has focused on how individual, group, and leadership factors relate to safety problems, safety role definitions, the interpretation of accident causes, and accident occurrences. Other research interests include multilevel analysis, organizational climate and leadership, organizational change, and group performance and ineffectiveness. His research appears in the *Academy of Management Journal, Academy of Management Review, Educational and Psychological Measurement, Human Performance, Journal of Applied Psychology, Journal of Applied Social Psychology, Journal of Management, Journal of Safety Research, Organizational Behavior and Human Decision Process, Personnel Psychology,* and *Leadership Quarterly.* He has also coauthored a number of book chapters and presented numerous papers/workshops at professional conferences. In 1992, he was awarded the Yoder-Heneman Personnel Research award by the Society for Human Resource Management.

Lois E. Tetrick received her doctorate in Industrial and Organizational Psychology from Georgia Institute of Technology in 1983. She was on the faculty for twelve years at Wayne State University in Detroit where she was program director of the Industrial and Organizational Psychology Doctoral program and served as the interim director of the Masters of Arts in Industrial Relations program. She served as the primary mentor to one of the first postdoctoral fellows in Occupational Health Psychology funded by the American

Psychological Association and the National Institute for Occupational Safety and Health. Dr. Tetrick moved to the University of Houston in 1995, where she is director of the Industrial and Organizational Psychology Program and co-director of the Occupational Health Psychology minor within the Department of Psychology. Professor Tetrick has served as associate editor of the *Journal of Applied Psychology* and is on the editorial boards of the *Journal of Organizational Behavior* and the *Journal of Occupational Health Psychology*. Dr. Tetrick's research has focused primarily on individuals' perceptions of the employment relationship and their reactions to these perceptions including issues of occupational health and safety, occupational stress, and organizational/union commitment. She is co-editor of the *Handbook of Occupational Psychology* with Jim Quick and *Changing Employment Relations* with Julian Barling. She is active in the Society for Industrial and Organizational Psychology and was recently elected to represent the Society on the American Psychological Association Council. She also is active in the Academy of Management and is past-chair of the Human Resources Division. Dr. Tetrick is a fellow of the Society for Industrial and Organizational Psychology, the American Psychological Association, and the American Psychological Society.

Chris Brotherton is professor of applied psychology at Heriot Watt University in Edinburgh. He was director of the Centre for Occupational Health and Safety, which provides postgraduate education to Health and Safety Executive inspectors, until August 2002. Brotherton has been chair of both the section and the division of Occupational Psychology of the British Psychological Society. He currently chairs the Faculty of Psychology and Management for the Chartered Institute of Personnel and Development. He has produced about 120 scientific papers and four books, including most recently *Social Psychology and Management: Issues for a Changing Society* (Open University Press, 1999).

Michael J. Burke is the A. B. Freeman Distinguished Professor of Organizational Behavior at Tulane University. He will serve as the president of the Society for Industrial and Organizational Psychology in 2003–2004. His current research focuses on the meaning of employee perceptions of work environment characteristics (psychological and organizational climate); the meaning of job performance, with emphasis on worker safety performance; the efficacy of worker health and safety training; and procedures for estimating interrater agreement. He has authored numerous articles and book chapters on these subjects and others, and he has served on the editorial boards of many leading journals, such as *Journal of Applied Psychology, Personnel Psychology, Journal of Management,* and *Journal of Occupational Health Psychology.* He received his Ph.D. from the Illinois Institute of Technology in 1982.

John E. Delery is an associate professor in the Sam M. Walton College of Business Administration at the University of Arkansas. He received his Ph.D. from Texas A&M University. His research interests include the strategic management of human resources, the structure of human resource management systems, and selection interviews.

Michelle K. Duffy is an assistant professor in the Gatton College of Business and Economics at the University of Kentucky. She received her Ph.D. from the University of Arkansas. Her current research interests include the investigation of antisocial behavior at work (social undermining and moral disengagement processes) and diversity. Funded by the Society for Human Resource Management, her research has appeared in *Academy of Management Journal, Journal of Applied Psychology,* the *Journal of Management, Journal of Organizational Behavior,* and *Research in Personnel and Human Resource Management.*

Daniel C. Ganster is professor of management and holds the Charles C. Fichtner Endowed Chair in Management at the University of Arkansas, where he is also department chairman. He received his Ph.D. from the Krannert Graduate School of Management at Purdue University. Funded by the National Institute for Occupational Safety and Health and the National Institute of Mental Health, his research on job stress and employee health has been published in many journal articles and book chapters.

Mark A. Griffin is Principal Research Fellow in the School of Management, Queensland University of Technology. He received his

Ph.D. in industrial/organizational psychology from the Pennsylvania State University, and his research has been published in journals such as *Academy of Management Journal, Journal of Management, Journal of Applied Psychology, Personnel Psychology,* and *Journal of Organizational Behavior.* Griffin has consulted widely with organizations in Australia and internationally to assess and improve work effectiveness. His primary research focus is the link between employee well-being and organizational performance. The integration of multiple data sources is a key theme in the investigation of this link.

Amanda Griffiths is professor of occupational health psychology at the Institute of Work Health and Organisations at the University of Nottingham. She received a Ph.D. in developmental psychology from the University of Nottingham. Her recent research concerns the effects of the management, design, and organization of work on individual and organizational health. She is a founder member of the European Academy of Occupational Health Psychology and visiting professor at the Universities of Lund and Kristianstad in Sweden. She is associate editor of the international scientific journal *Work and Stress* and book series editor for *Issues in Occupational Health* (Taylor & Francis).

Angela T. Hall is a Ph. D. candidate at Florida State University in human resource management and organizational behavior. She received a J.D. from Florida State University in 1993. Hall's research interests include organizational politics, accountability, and workplace accommodations. She has a chapter forthcoming in *The Dark Side of Organizational Behavior.*

Fehmidah Munir is a lecturer in applied psychology and director of an M.Sc. in workplace health promotion at the Institute of Work, Health and Organisations at the University of Nottingham. She gained a Ph.D. in neuropsychology at the University of Nottingham in 1999. Her recent research concerns the impact of chronic illness upon working life and workplace health promotion. Munir is a chartered psychologist, a graduate member of the British Psychological Society, a member of European Academy of Occupational Health Psychology and a member of the Task Force 6 Health Promotion Activity Group of the World Health Organisation.

Anne M. O'Leary-Kelly is a professor in the department of management at the University of Arkansas. She received her Ph.D. from Michigan State University and has taught at Texas A & M University and the University of Dayton. Her research interests include the study of aggressive work behavior and organizational attachments. She has published in *Academy of Management Review, Academy of Management Journal, Journal of Applied Psychology, Journal of Management, Journal of Organizational Behavior,* and *Journal of Management Inquiry.* She has received both the Outstanding Publication in Organizational Behavior Award and the Dorothy Harlow Outstanding Paper Award from the Academy of Management.

Wayne E. Ormond holds an M.S. in industrial/organizational psychology from the University of Calgary, awarded in 1998, and is currently a 2000–2002 Killam Scholar. He has been twice awarded the Province of Alberta Graduate Scholarship and was nominated for the Governor General's Gold Medal Award for outstanding master's research. He has also served as student representative for the Canadian Society for Industrial and Organizational Psychology and is a member of two research units at the University of Calgary: the Workplace Learning Research Unit and Creating Organizational Excellence. His primary areas of research interest and specialization include employee performance monitoring and leadership development.

Sharon K. Parker is associate professor in organizational behavior at the Australian Graduate School of Management. She obtained her doctorate at the University of Sheffield. Her research interests include work design and its effects on employee health, well-being, safety, and organizational performance; the role of work design in mediating the effects of organizational changes and structures; perspective taking and its application to organizations; and women in the workplace. She has published in journals such as *Journal of Applied Psychology* and *Academy of Management Journal.* She coauthored *Job and Work Design: Organizing Work to Promote Well-being and Effectiveness* (Sage, 1998).

Pamela L. Perrewé is the Jim Moran Professor of Management and serves as the associate dean for graduate programs in the College of Business at Florida State University. She received Ph.D. from the University of Nebraska. Perrewé's research focuses on job stress, organizational politics, and personality. She has published widely in books and journals, including *Journal of Applied Psychology, Journal of Management, Journal of Organizational Behavior, Journal of Vocational Behavior, Journal of Occupational Health Psychology, Human Relations Research in Personnel and Human Resources Management,* and *Journal of Applied Social Psychology.* She is coeditor for *Research in Occupational Stress and Well Being.*

Narda R. Quigley is a Ph.D. candidate in organizational behavior in the department of management and organization at the Robert H. Smith School of Business, University of Maryland. Her dissertation examines the effects of core self-evaluations and transformational leadership on team efficacy in a high-pressure environment. Her research interests include group and team influences on individuals in organizations, team efficacy, cohesion, and processes over time, as well as personality as a compositional factor in work teams. She is also interested in the interface between organizational culture and leadership.

Karen Roberts is professor in the School of Labor and Industrial Relations at Michigan State University. She received her Ph.D. from the Department of Urban Studies at M.I.T. Prior to joining the university, she was an economist at the Workers' Compensation Research Institute in Cambridge, Massachusetts, and developed simulation models of income replacement. She now researches topics related to workplace disability and workers' compensation. Her work on disability has been published in *Industrial Relations, Journal of Risk and Insurance, Human Resource Management Journal, Journal of Labor Research, Journal of Occupational and Environmental Medicine, Negotiation Journal,* and *Journal of Occupational Health Psychology.*

Sue Ann Sarpy is clinical assistant professor of environmental health sciences and evaluation coordinator for the Center for Applied Environmental Public Health in the School of Public Health and Tropical Medicine at Tulane University. She has served as principal scientist for various research projects that evaluate the effectiveness of health and safety interventions including labor and minority worker safety training, exposure control in the automotive industry, and community-based lead prevention programs. Sarpy has written technical reports and spoken at conferences worldwide concerning these evaluation efforts. Her research interests include intervention effectiveness in reducing health risks, occupational health and safety management, and public health workforce preparedness.

Jason D. Shaw is associate professor and Clark Material Handling Company Professor of Management in the Gatton College of Business and Economics at the University of Kentucky. He received his Ph.D. from the University of Arkansas. His research interests include the effects of compensation systems on individual and organizational outcomes, organizational consequences of voluntary and involuntary turnover, and personality-environment congruence.

Carlla S. Smith was professor of psychology at Bowling Green State University. She served as director of the graduate program in industrial/organizational psychology at Bowling Green and was instrumental in launching a graduate training program in occupational health. Smith was an editorial board member for scholarly journals, including *Journal of Occupational Health Psychology*. Her research focused on the effects of shift work and alternative work schedules on worker health, as well as the effects of work stress on employee health and performance. She consulted with several manufacturing and service organizations, including Fortune 500 companies, in the areas of work stress and employee health.

Paul E. Spector is professor and director of the industrial/organizational psychology program at the University of South Florida. His interests include both content (counterproductive work behavior, occupational health psychology, occupational stress, and personality) and research methodology. He has published in many of the field's leading journals and has served as editor for several journals, including *Journal of Occupational Health Psychology*. In 1991 the Institute for Scientific Information listed him as one of the 50 highest impact contemporary researchers (out of over 102,000) in psychology worldwide.

Lorne M. Sulsky received his Ph.D. in industrial/organizational psychology from Bowling Green State University in 1988. Associate professor of psychology at the University of Calgary, his areas of research include performance management, employee training, and occupational stress management. Sulsky served as president of the Canadian Society for Industrial and Organizational Psychology in 1999. He is currently director of human resources for a nonprofit agency concerned with rehabilitation from brain injuries, and he is editor-designate of *Canadian Journal of Behavior Science,* published by the Canadian Psychological Association. Recently, he edited a special issue of the journal that examined industrial/organizational psychology within the Canadian Armed Forces.

Paul Tesluk is assistant professor in management and organization at University of Maryland and associate director of the Center for Human Capital, Innovation, and Technology at the Robert H. Smith School of Business. He received his Ph.D. from the Pennsylvania State University in industrial/organizational psychology. His research concerns the design and implementation of high involvement workplace systems, work team performance, and employee and managerial development. He has published in journals such as *Personnel Psychology, Academy of Management Journal,* and *Journal of Applied Psychology,* and he has received awards from the Society for Industrial and Organizational Psychology for his research on team effectiveness and work experience.

Darren C. Treadway is a Ph. D. candidate at Florida State University in human resource management and organizational behavior. His dissertation is titled *The Role of Age and Influence in the Performance Evaluation Process: Test of a Process Model Incorporating Age Norms and Political Skill.* He received an M.B.A. from Virginia Tech. Treadway's research interests include the role of age in the workplace, reactions of diverse groups to organizational politics, multilevel processes in organizations, and work-family conflict. Before his doctoral studies, Darren worked for eight years as a human resource and operations manager.

Nick Turner is research associate at the Institute of Work Psychology, University of Sheffield, and adjunct assistant professor,

Queen's School of Business, Queen's University, Canada. His research explores workplace safety from a variety of approaches and different levels of analysis. His work has been published in a number of books and journals, including *Journal of Applied Psychology* and *Journal of Occupational Health Psychology*. Turner has also served as a consultant to a number of public and private organizations in the United Kingdom, dealing mainly with the implementation of teamworking and its effect on employee health and safety.

Dov Zohar earned his Ph.D. at the University of Maryland. He is associate professor of industrial/organizational psychology at the Faculty of Management, Technion—Israel Institute of Technology. Zohar has published articles on safety climate, occupational stress, and burnout. His current climate research includes the study of factors influencing climate formation and strength, climate measurement methodology, and climate-based intervention methods. Current stress-research projects include event-level measurement of negative affect and interruptions at work.

Health and Safety in Organizations

CHAPTER 1

The Etiology of the Concept of Health

Implications for "Organizing" Individual and Organizational Health

David A. Hofmann
Lois E. Tetrick

When we were originally approached about editing this volume, the proposed focus was stress and safety in organizations. Given our respective backgrounds in these lines of research (Hofmann, Jacobs, & Landy, 1995; Hofmann & Morgeson, 1999, in press; Hofmann, Morgeson, & Gerras, in press; Hofmann & Stetzer, 1996, 1998; Stetzer & Hofmann, 1996; Quick & Tetrick, 2002; Tetrick & LaRocco, 1987; Tetrick, 1992), this seemed to make a great deal of sense. In response to this initial salvo, however, we proposed broadening the focus to include both safety and health. Broadening the focus to include health resulted in the asking and answering of somewhat different questions. For example, it encouraged questions that moved beyond a focus on the reduction of accidents and stress or strain to questions about how one can go about creating health, both for individuals and organizations; that is, it shifts the focus from reducing illness to creating the presence of health. In fact, some of our own research as well as other recent research suggests that many of the organizational characteristics that lead to effective performance (both in-role and extra-role) have also been shown to lead to increased health.

Given the added focus on health, the first objective of the volume—and therefore the purpose of this chapter—is to figure out exactly what the term *health* means. On the surface this seems pretty straightforward. For example, one could simply use the dictionary definition: (1) the condition of being sound in body, mind, or spirit, especially the freedom from physical disease or pain; and (2) a flourishing condition (Merriam-Webster Dictionary, 1997). Alternatively, one could refer to the oft-cited World Health Organization (WHO) definition of health as a "state of complete physical, mental, and social well-being and not merely the absence of disease or infirmity" (WHO, 1948). The beauty of this approach is, of course, that these definitions both center on the same themes, namely, that the definition of health encompasses more than merely the absence of illness. Rather, health seems to engender notions of flourishing and well-being as well.

Further inspection of these definitions, however, reveals that there is more meat on those bones to chew on than first appears. On the one hand, the aspect of the definition focusing on the absence of disease seems pretty straightforward and, at least at the individual level, could be assessed empirically (medically). Other aspects of the definition, however, introduce a number of questions that keep resurfacing for further consideration:

- What does *flourishing* mean?
- From whose perspective is it assessed?
- How does one define *well-being* in the physical, mental, and social domains?

These questions prove to be a bit vexing because when one moves from discussions of disease (or the absence thereof) into discussions of health, a strange thing happens. Specifically, even though the segue between the two obviously related questions may appear seamless, the reality is that one has crossed a rather large chasm. Resting in the middle of this chasm is the notion of values, goals, expectations, and the like. Health, well-being, and flourishing by definition involve questions of value: Health from whose perspective? In relation to what goals and values? In relation to what type of functioning? In order to get a better handle on both the conceptions of illness and health as well as the reasons that health will likely prove to be a construct with a fair degree of plasticity, we first turn to fields that have been grappling with these notions much longer than psychologists, the interconnected fields of medicine and philosophy.

Perspectives on Health

Philosophy and medicine have been interconnected since the time of Socrates and Plato (Lidz, 1995), and this interconnection has led to a number of theoretical debates regarding the distinctions among disease, illness, and health. The centrality of this debate, it seems, centers on the degree to which each of these terms—*disease, illness,* and *health*—is a value-laden or value-free term. On the value-free side of the equation, Boorse (1975) attempted to argue that one could define disease and illness (and for that matter health) without invoking value-laden statements. By opting to use strictly biologically based definitions, Boorse argued that (1) disease subsumes the notion of illness and (2) an entity is diseased if there is deviation from the natural functioning of entities in that class (i.e., a deviation from the statistical norm). By defining disease in terms of normal functioning for that species, he attempted to skirt the more value-laden discussions of health. In other words, he argued that one can talk about normal functioning in relation to the design of the entity without making a value judgment regarding the usefulness of this functioning (Is the functioning good or bad in terms of some end goal or state?). Thus, according to Boorse, a disease becomes an illness when it is (1) undesirable for its bearer, (2) a title to special treatment, and (3) a valid excuse for normally criticizable behavior.

Although we do not need to delve into the rationale underlying these conditions, the value-free claim of Boorse's arguments (1975) seem to be on thin ice given the subjectivity involved in defining what normative reference to use (What is the statistical sample upon which one computes the norm?) as well as the definition of *undesirable,* both of which will involve value judgments.

Since these original arguments, several other authors have attempted to reconcile the desire to achieve value-free definitions of disease, illness, and well-being and the inherent value-laden assessments that seem to emerge in these discussions (e.g., Lennox, 1995; Wakefield, 1995).

DeVito (2000) recently concluded after critically reviewing each of these arguments that although it is possible to ascertain the presence of a medical malfunction in value-free terms, this does not determine whether one is diseased or ill. In fact, he concludes that

> more than life and reproduction are important to the concepts of health and disease, and *what* is important will depend upon the interests of people. In the end, because making people healthy really turns out to be making people better (and the "better" is a subjective and value-laden matter), the concepts of health and disease must be intrinsically value-laden. (p. 563)

Any discussion of illness and health therefore will require value-laden statements because one will have to consider the vantage point from which one is assessing these outcomes.

This is not to say that all definitions or perspectives on health and illness will be equally value-laden. For example, as one more narrowly defines health as the absence of illness (or disease), the definition can be better captured by biologically based empirical assessments. As one moves toward broader definitions of health, the concept not only becomes more complex and fuzzy but also involves more value-laden concepts. This range of various approaches to health is probably best reflected in Larson's four schools of thought (1999) regarding the conceptualization of health as discussed hereafter.

The medical model emerges from early machine-based conceptualizations of the human body (e.g., Descartes). These conceptualizations resulted in health definitions that focused on the absence of disease; that is, illness occurs when the body's structure or function is disrupted. This conceptualization has served the medical community well and has resulted in a highly productive advancement of the medical sciences (Wood, 1986). Despite these benefits, however, the medical model has been criticized due to its deemphasis on preventative medicine as well as through the increasing recognition that disease and illness involve biological as well as social and economic factors. These latter two factors are generally excluded from this perspective. In other words, its strength—namely, a narrow focus on the causes of and cures for disease—is also the model's weakness.

The second model discussed by Larson (1999) is the WHO model, reflected by the definition in the WHO's constitution (1998) cited earlier (i.e., health is a state of complete physical, mental, and social well-being). Essentially, this model defines health as emerging from a tripartite confluence of physical, mental, and social aspects of one's life. We should mention, however, that several attempts at operationalizing measures of this definition of health have been more successful at assessing the physical and mental as opposed to the social aspects of health (Berkman & Breslow, 1983; Ware, Brook, Davies, & Lohr, 1981). Although considered idealistic and unrealistic when initially introduced, the WHO definition of health currently is the most popular (Larson, 1999).

Larson (1999) termed the third health viewpoint the wellness model. In our view the focus of this perspective overlaps considerably with the WHO approach to health. From this perspective health is defined as the strength and ability to overcome illness (i.e., having a reserve of health), making progress toward higher-level functioning, and having an optimistic view of one's future and potential. In sum, the wellness model suggests that health moves past merely the absence of illness toward a discussion of things such as well-being, energy, ability to work, and efficiency (Schroeder, 1983; Larson, 1999). Again, we believe this is quite similar to WHO's definition (1998) of physical, mental, and social well-being.

The environmental model, the final of Larson's health perspectives (1999), emphasizes individuals' adaptation to their environment, where environment is broadly defined to include physical, social, and other environments. Parsons (1972) more precisely defines health from this perspective as "the state of optimum capacity of an individual for the effective performance of the roles and tasks for which he [she] has been socialized" (cited in Navarro, 1977, p. 14). Similar to the Merriam-Webster definition we cited earlier, health under this perspective exists "when an organism works with its environment successfully and is able to grow, function, and thrive" (Larson, 1999, p. 131; see also Abanobi, 1986). One criticism of this perspective is that it does not include an explicit consideration of the healthiness of the environment.

Going back to our discussion regarding the inherent value-laden nature of health, we can see that as one moves from the medical to the wellness and finally to the environmental perspective, the concept of health becomes increasingly value-laden. Both in the medical and organizational sciences, the absence of illness is likely to represent more objective indicators that will likely be agreed upon across situations and times. For example, having cancer or heart disease, we suspect, are pretty well agreed upon as bad outcomes and represent illness. Furthermore, they can be rather objectively defined across situations and times. Similarly, an organization with extraordinarily high accident rates or with a production process that contributes to employee disease (e.g., asbestos) is likely to be perceived rather uniformly as an ill organization.

The WHO, wellness, and environmental models, however, cannot escape more value-laden propositions. Wellness involves individual assessments of a person's life in broad terms such as satisfaction, optimism, and the like. Obviously, these subjective assessments will necessitate the inclusion of both short- and long-term goals as well as other personal values and objectives. The environmental perspective adds an additional set of value-laden propositions: those dealing with successful performance of roles and tasks. The introduction of successful performance puts into the mix both the expectations of external constituents (i.e., role senders) as well as an appraisal of how successful the organism is at performing these roles. As a result, this perspective introduces the values and demands of outside observers or stakeholders and the extent to which these are satisfied. The other two perspectives on health did not consider these external stakeholders.

Thus, assuming that the WHO and the wellness models share enough in common to fold into one definition (hereafter referred to as the wellness model), we are left with three different models of health: (1) the medical model, (2) the wellness model, and (3) the environmental model. In general terms these definitions vary with respect to the breadth of their definition of health. At the most narrow level is the medical model, with its definition of health as the absence of illness. The second level would not only include the absence of illness but also expand the health criteria to include discussions of well-being, higher-level functioning, and a positive or optimistic view of the future. The third perspective not only involves the absence of illness and individual well-being but adds the additional criterion of effective performance of tasks and roles demanded by the external environment. With these three emerging viewpoints of health as a backdrop, we now turn to a review of how the term *health* has been used in organizational behavior and applied psychology.

Health in the Organizational Sciences

The notion of health has been discussed, either implicitly or explicitly, within the organizational sciences for quite some time. In fact, a number of authors reached the conclusion, as did the medical profession, that discussing the notion of health within the bounds of organizations is a difficult thing to do. Argyris (1958, p. 115), for example, concluded after an intensive case study in a single organization that "perhaps the most important [lesson] is that organizational health turns out to be a very complex idea." He determined that a number of traditional measures of effectiveness (e.g., low absenteeism, turnover) may not be indicative of health, because these indicators represent the outcomes of the system. In the organization under investigation, Argyris found that even though the outcomes of the system indicated a healthy organization, a deeper investigation of the system itself indicated a number of indicators pointing to an unhealthy organization. A few years later, Argyris (1964) tackled the notion of trying to integrate individual and organizational goals such that both individual and organizational outcomes could be optimized. In the introduction to his treatise, he noted:

> We are interested in developing neither an overpowering manipulative organization nor organizations that will "keep people happy." Happiness, morale, satisfaction are not going to be highly relevant guides in our discussion. Individual competence, commitment, self-responsibility, fully functioning individuals, and active, viable, vital organizations will be the kinds of criteria that we will keep foremost in our minds. (p. 4)

With that as the introduction, Argyris (1964) continued to examine the interrelationships between individuals and organizations. He concluded the volume by suggesting that on a theoretical level the concepts of effective organizations and individual positive mental health are congruent, yet there is a lack of empirical evidence. One of the purposes of this volume is to examine the empirical evidence that has materialized since Argyris's initial theoretical consideration.

About this same time, McGregor's theory X and Y (1960) emerged on the scene; he was concerned with how to maximize individual and organizational health. Similarly to Argyris, McGregor (1960, p. 46) rejected the notion that satisfaction is the sine qua non of organizational health:

> It has become clear that many of the initial strategic interpretations accompanying the "human relations approach" were . . . naive. . . . We have now discovered that there is no answer in the simple removal of control—that abdication is not a workable alternative to authoritarianism. We have learned that there is no direct correlation between employee satisfaction and productivity. We recognize today that "industrial democracy" cannot consist in permitting everyone to decide everything, that industrial health does not flow automatically from the elimination of dissatisfaction, disagreement, or even open conflict. Peace is not synonymous with organizational health; socially responsible management is not coextensive with permissive management.

Paralleling Argyris's integration (1964), the core of McGregor's theory Y argument (1960, p. 49) boiled down to integration, namely, "the creation of conditions such that the members of the organization can achieve their own goals *best* by directing their efforts toward the success of the enterprise."

Integration is a theme that continues through to more recent definitions of health in organizations. Our search of more recent literature discovered a number of different definitions of organizational health that are presented in Table 1.1. As we will show, these definitions as well as other organizational research provide clear evidence for each of the health perspectives emerging from the medical domain as well as evidence of the integrative nature of health.

Table 1.1. Definitions of Organizational Health.

Authors	Definition of Organizational Health
Adkins, Quick, & Moe (2000)	The authors adopt four health principles: (1) health is more than the absence of illness (it ranges from disease to well-being); (2) health is a process that requires continual attention and effort; (3) health involves interrelated parts working together in balance; and (4) health can come about only through positive collaborative relationships. The authors adopt a view that focuses on both individual health and goal accomplishment (organizational health).
Bennis (1962)	"The basic features of organization rely on adequate methods for solving problems. These methods stem from the elements of what has been called the scientific attitude. From these ingredients have been fashioned three criteria or organizational mechanisms, which fulfill the prerequisites of health. These criteria are in accord with what mental-health specialists call health in the individual. . . . Nonetheless, it has been asserted that the processes of problem-solving—of adaptability—stand out as the single most important determinant of organizational health and that this adaptability depends on a valid identity and valid reality-testing." (p. 278)
Cox (1992)	Occupational health should logically refer to the dynamic interaction between work and work-related processes and at least physical, psychological, and social health. Cox proposes a three-factor model of health concerns: work, organization, and health. This model would pose three challenges: (1) question the healthiness of work and work environments within their psychosocial and organizational contexts; (2) question the healthiness of organizations as well as of their employees; and (3) question the relationship between the healthiness of the organization and that of its individuals.
Danna & Griffin (1999)	"The term *health* generally appears to encompass both physiological and psychological symptomology within a more medical context (e.g., reported symptomology or diagnosis of illness or disease); therefore, we suggest the term *health* as applied to organizational settings be used when specific physiological or psychological indicators or indexes are of interest and concern. Following from Warr (1987, 1990), on the other hand, well-being tends to be a more broad and encompassing concept that takes into consideration the 'whole person.' Beyond specific physical and/or psychological symptoms or diagnoses related to health, therefore, *well-being* should be used as appropriate to include context-free measures of life-experiences (i.e., life satisfaction, happiness), and within the organizational research realm to include both generalized job-related experiences (e.g., job satisfaction, job attachment), as well as more facet-specific dimensions (e.g., satisfaction with pay or coworkers)." (p. 364)

Table 1.1. Definitions of Organizational Health, Cont'd.

Authors	Definition of Organizational Health
Griffin, Hart, & Wilson-Evered (2000)	The organizational health perspective differs from many of the traditional approaches to reducing occupational stress in two important ways: (1) it emphasizes the need to focus on *both* employee well-being and the organization's "bottom-line" performance; (2) it recognizes that employee well-being and organizational performance is determined by both individual *and* organizational factors.
Hoy & Feldman (1987)	The authors use Miles (1965) for their definition.
Ilgen (1990)	The primary focus is on the health of individuals within the organizational context. Health here not only refers to the reducing signs of illness (e.g., safety) but also to the creation of more positive indicators of health (e.g., health promotion, wellness programs).
Jaffe (1995)	"Organizational health implies an expanded notion of organizational effectiveness. Traditionally, effectiveness was defined as meeting profit, production, service, and continuity goals. Organizational health, as I define it here, adds a further dimension, raising several questions: How well does the organization treat its people? What are the connections between traditional measures of effectiveness and the health and well-being of people working in the organization? Do effective organizations also support the growth and developmental needs of their employees? Another factor is the morale, level of satisfaction, growth and development, and motivation of employees. This concern can be even broader, if the needs of customers, suppliers, owners, and community members that are touched by the organization are considered as well. A healthy organization, I suggest, will create health for its employees and the people and communities it touches." (p. 13)

Katz & Kahn (1978)	"We are accustomed to studying the extent to which . . . [organizational and person role systems] contribute to efficiency, productivity, growth, and other criteria of organizational effectiveness. Has the individual performed his or her role with energy, skill, regularity, and judgment that are sufficient for the continuing success of the organization? It is equally appropriate to ask the complementary questions: Does the enactment of the organizational role enhance or reduce the well-being of the individual? Does it enlarge or reduce the well-being of the individual? Does it enlarge or reduce the person's valued skills and abilities? Does it increase or restrict the person's opportunity and capacity for other valued role enactments?" (p. 580)
Lindström, Schrey, Ahonen, & Kaleva (2000)	"Organizational health implies that an organization is able to optimize its effectiveness and the well-being of its employees, and to cope effectively with both internal and external changes." (p. 83)
Miles (1965)	Miles formally defines organization health as the systems ability not only to function effectively but to develop and grow into a more fully functioning system. "A healthy organization [is one that] not only survives in its environment, but continues to cope adequately over the long haul, and continuously develops and extends its surviving and coping abilities" (p. 378). Miles then develops a representation of health including three broad areas and 10 dimensions. The first broad area concerns organizational goals: (1) goal focus, (2) communication adequacy, and (3) optimal power equalization. The second broad area focuses on the internal state of the system: (4) resource use, (5) cohesiveness, and (6) morale. The third broad area focuses on growth and change: (7) innovativeness, (8) autonomy, (9) adaptation, and (10) problem-solving adequacy.
Murphy & Cooper (2000)	The terms *healthy companies*, *healthy work organizations*, and *organizational health* all refer to the notion that worker well-being and organizational effectiveness can be fostered by a common set of job and organizational design characteristics.
Quick (1999)	The focus of occupational health psychology is *healthy* workplaces, defined as ones in which people may produce, serve, grow and be valued. Specifically, healthy workplaces are ones in which people use their talents and gifts to achieve high performance, high satisfaction, and well-being.

Table 1.1. Definitions of Organizational Health, Cont'd.

Authors	Definition of Organizational Health
Sauter & Hurrell (1999)	Their definition of occupational health psychology focuses on organizational risk factors (e.g., supervisory practices, production processes, and their influence on how work is performed) for illness and injury.
Schmidt, Welch, & Wilson (2000)	"The healthy work organization concept centers on the premise that organizations that foster employee health and well-being are also profitable and competitive in the marketplace. The concept recognizes that work can have a significant effect on employee commitment, satisfaction, and health which, in turn, impact productivity and the effectiveness of the organization." (p. 133)
Vicenzi & Adkins (2000)	"An organization's ability to compete successfully in this emerging 'information-based' economy demands creative thinking, innovation, and rapid adaptation to relevant information generated in the marketplace. Core competencies must continually be upgraded and enhanced through application and internal sharing, and never be allowed to congeal into 'core rigidities.' Linkages and communication with related resources involving critical capabilities are required. The adaptive abilities of the organization will be heavily dependent upon the use of and the value it places on the expertise, talents, skills, and experiences of its members. The critical components of today's products and services are not raw materials and energy, but the ability of people throughout the organization to anticipate and respond to complex and shifting customer requirements. . . . The successful firms in the knowledge economy will be those who effectively manage the consistent creation, sharing, harvesting, and leveraging of their people's abilities and knowledge into intellectual capital." (p. 102)
Williams (1994)	Based on the World Health Organization and International Labor Organization views of occupational health, the four elements of organizational health are (1) environmental factors, (2) physical health, (3) mental (psychological) health, and (4) social health.
World Health Organization (1998)	Health is a "state of complete physical, mental, and social well-being and not merely the absence of disease or infirmity."

Evidence of Medical-Based Definitions

The medical model, although clearly present in the organizational sciences and applied psychology, has typically not fallen under the nomenclature of organizational health. This is not particularly surprising, given the individual orientation of the medical model. This does not mean, however, that no applied research is investigating health within organizations using the medical model. Research investigating the impact of work stress on employee health has often investigated the presence or absence of signs of illness. For example, one of the earliest investigations of the effects of work stress investigated the relationship between accountants' serum cholesterol and the tax deadline (Friedman, Rosenman, & Carroll, 1958). More recently, Schaubroeck, Ganster, and Kemmerer (1994) investigated the combined effects of job complexity and type A behavior on cardiovascular disorder. Interestingly, they found that jobs that would be considered as enriched from a job design perspective were more highly associated with cardiovascular disorder. Other research has investigated the relationship between work and individual characteristics and respiratory illness and immune system functioning (Schaubroek, Jones, & Xie, 2001), blood pressure (Schaubroeck & Merritt, 1997), and health care costs (Ganster, Fox, & Dwyer, 2001). The focus of these investigations was on the presence of dysfunctional physiological reactions, that is, the presence versus absence of indicators of physiological illness.

In addition to the physiological studies, a large number of studies have investigated more psychosocial indicators of illness (e.g., perceptions of stress and strain). Here the focus is on the presence versus absence of negative psychosocial reactions to work (i.e., the absence of illness). For example, Spector (1986) in a meta-analytic review, found that work-related control variables were significantly associated with self-reported physical symptoms as well as emotional distress. Other research focusing on perceived stressors and strains have focused on constructs such as anxiety, frustration, health symptoms, and job stressors (e.g., Spector, Chen, & O'Connell, 2000). In most cases the measures used in these investigations focused on the presence of negative psychosocial factors rather than the presence of health-related factors (e.g., physical symptoms, conflict, job stressors or strains; e.g., Spector & Jex, 1998).

Similarly, much of the research in the safety domain has focused on the occurrence of either unsafe behavior or accidents. Hofmann and Stetzer (1996), for example, investigated organizational correlates of unsafe behavior (e.g., repairing a live system) and accidents (see also Zohar, 2000). Similarly, much of the safety research focused on behavior modification deals with performing actions using accepted safety protocol (Komaki, Barwick, & Scott, 1978; Komaki, Heinzmann, & Lawson, 1980; Krause, 1997). Again, the focus of these measures is either the presence of illness (accidents or injuries) or what we would term the absence of illness, namely, the lack of accidents or injuries or the absence of unsafe behaviors (i.e., actions performed to standard).

Evidence of Wellness-Based Definitions

Several of the definitions of organizational health in Table 1.1 would fall under the wellness model. Danna and Griffin's review (1999) of health in organizations almost exclusively focuses on the health of the individual (both in terms of the absence of illness and the presence of health). The researchers say very little about the degree to which the health relationship in the workplace is reciprocal, with the organization creating health for the individual and vice versa. Similarly, Ilgen (1990) presents ways in which organizations can create health for their employees but says little about either individual performance or how this individual performance relates to the broader performance of the organization (see also Williams, 1994). The emerging field of occupational health psychology also adopts a wellness-based view of health. For example, Sauter and Hurrell (1999) suggested that the main imperative for the field of occupational health psychology was to advance knowledge and expertise regarding the way organizational factors affect worker safety and health, although some scholars recognize the effect of employees' health on organizational health (Adkins, 1999).

Evidence of Environmental-Based Definitions

The majority of the definitions of organizational health in Table 1.1 fall within the environmental health model, because they focus both on individual and organizational performance. In other words, when viewing health from the individual level of analysis, these definitions focus on both the health and well-being of the employee as well as the employee's contribution to the health of the organization through effective task performance (Cox, 1992; Griffin, Hart, & Wilson-Evered, 2000; Hoy & Feldman, 1987; Jaffe, 1995; Katz & Kahn, 1978; Lindström, Schrey, Ahonen, & Kaleva, 2000; Miles, 1965; Murphy & Cooper, 2000; Quick, 1999; Schmidt, Welch, & Wilson, 2000). The definitions proffered by Bennis (1962) and Vicenzi and Adkins (2000) both focus on the organizational level of analysis and clearly concentrate on environmental adaptation. However, they do include other indicators of health (e.g., employee well-being, morale, etc.).

Perhaps the best summary of the environmental model of health from a microperspective is Katz and Kahn's discussion of health (1978). Katz and Kahn used role theory as the foundation of behavior in organizations. Specifically, individuals are exposed to different expectations regarding role performance. In their discussion of health issues, Katz and Kahn noted that organizations are accustomed to assessing the degree to which role performance contributes to the efficiency, productivity, growth, and other aspects of organizational effectiveness. In addition, role-set members often investigate the way in which the focal individual carried out the role (e.g., performed with energy, skill, regularity, etc.). Katz and Kahn suggested it also is reasonable, if less frequently done, to ask (1) if the enactment of role behavior enhances or reduces the well-being of the individual, (2) whether it enlarges or reduces the person's valued skills and abilities, and (3) whether it increases the individual's opportunity and capacity for other valued role enactments. Parallel discussions to Katz and Kahn have recently been occurring within the domain of occupational health (e.g., Cox, 1992; Griffin et al., 2000; Lindström et al., 2000; Quick, 1999; Murphy & Cooper, 2000).

At the organizational level, Jaffe (1995) adopted a similar definition, noting that organizational health is a broader construct than simply organizational effectiveness. According to Jaffe, *effectiveness* has been traditionally defined as meeting profit, production, service, and continuity goals. Organizational health, alternatively, adds an additional set of concerns such as how well the organization treats its employees. This definition raises concerns regarding employee morale, level of satisfaction, growth and development, and motivation. All of these definitions seem to encompass the two questions that Katz and Kahn (1978) posed: (1) Does the employee's role performance successfully affect the organization? (2) Does the performance of this role set affect the employee's health and well-being?

Other organizational-level definitions have focused more on the healthiness of the organization as a whole and its relationship to the external environment. After considering a number of different potential definitions of organizational effectiveness, Bennis (1962, p. 278) concluded that the one criterion that most everyone could agree on was survival: "nonetheless, it has been asserted that the process of problem-solving—of adaptability—stands out as the single most important determinant of organizational health."

Paralleling Bennis (1962), Miles (1965, p. 378) defines organizational health as the system's ability to not only function effectively but develop and grow into a more fully functioning system: "a healthy organization [is one that] not only survives in its environment, but continues to cope adequately over the long haul, and continuously develops and extends its surviving and coping abilities." In addition to this definition, Miles identifies 10 dimensions of organizational health falling under three main categories (see Table 1.1).

From this review of the way organizational health has been operationalized, we see that the organizational sciences contain evidence of the medical, wellness, and environmental models of health. The definitions in Table 1.1 also suggest that organizational health is a multilevel phenomenon. Some perspectives focus more on individual-level outcomes. These wellness definitions typically highlight the interactions between the employee and the immediate organizational context, with the main question being how this immediate context affects employee health and well-being. The environmental definitions expand the focus to include the degree to which the individual contributes to the organizational goals as well (i.e., how well the person performs the work role). The more macroperspective of the environmental model focuses more on the organization as a whole and the degree to which its systems are effective and how these systems influence the employees in aggregate (e.g., the overall morale within the organization).

An Integrative Model

As we consider these three definitions of health more critically, we begin to see several distinguishing characteristics. First, as we noted, they differ with respect to the degree to which they are inherently value-laden constructs, with the medical model being the least value-laden and the wellness and environmental models being more value-laden. But it appears that as more and more value statements are introduced into the health viewpoint, the definition of health increasingly consists of the balancing or integration of competing goals. At the most basic level, the medical model is least value-laden and involves very little competition among goals. For example, this perspective asks if the system currently is functioning as designed or is operating to standards (e.g., as designed, without causing accidents or injuries, without causing dysfunctional physiological outcomes). The introduction of terms in the wellness model such as *well-being, optimism for the future,* and *progress toward higher-level functioning* introduces the balancing of short-term success and long-term goals and values. Within organizations this balancing occurs to the extent that the current work situation is satisfying and helps to develop the capability for growth and development as well as creating an optimistic view of the future. Finally, the environmental model encapsulates the notion of both short- and long-term adaptability to the external demands. This definition again focuses on the degree to which health is created both for the individual and the organization—not only with respect to short-term goals but also in terms of long-term adaptability.

The introduction of balancing of goals brings to the forefront of these health perspectives another dimension upon which they vary, namely, whether the goals are primarily intrinsically or extrinsically oriented. Intrinsic goals are those that the organism pursues in order to satisfy some values or goals internally held, whereas extrinsic goals are those goals that the organism adopts in order to satisfy external stakeholder demands. On this dimension the wellness model typically focuses on the intrinsic goals of the individual: To what extent does the organizational context affect the health of the individual defined in terms of that person's current well-being as well as his or her long-term growth and development? The environmental model, alternatively, is focused more on extrinsic goals held by external stakeholders. Figure 1.1 graphically

**Figure 1.1. An Integrative Framework for
"Organizing" Organizational Health.**

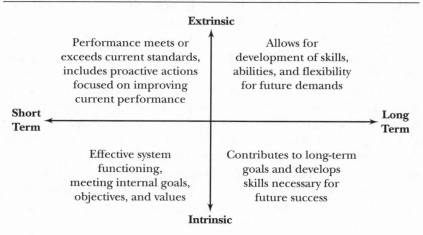

depicts these two dimensions upon which our definitions (particularly the wellness and environmental) vary and a general description of the focus that each quadrant might bring to bear on the question of health. The underlying notion of the model in Figure 1.1 is one of competing goals or objectives. In other words, the question is, How can organizations create environments that jointly optimize short-term intrinsic and extrinsic functioning as well as develop long-term opportunities for growth, development, and external adaptation?

It is important to point out that this notion of jointly optimizing competing goals has been discussed before in the organizational literature. Perhaps the most prevalent model is that of the balanced scorecard for strategic management (Kaplan & Norton, 1992, 1996). The primary focus of the balanced scorecard is to augment traditional, short-term financial metrics with three additional metrics focusing on the customer, internal business processes, and learning and growth. In part, the balanced scorecard system itself could be viewed as a strategic-level model for organizational health, as it discusses how to optimize both the internal health of the organization (e.g., internal processes, long-term learning and growth) as well as the demands of external stakeholders (e.g., customers, financial performance for stockholders). Although individual health typically is not explicitly mentioned to a large degree, employee morale and health would certainly be part of the effective functioning of the internal systems.

Discussions regarding sustainability also have been occurring relatively recently in the strategic management literature (e.g., Hart, 1997; Hart & Milstein, 1999). The concept of sustainability encompasses pursuing and developing goals that balance both short-term and long-term values and objectives (i.e., meeting current needs without compromising the ability of meeting future needs). Although these are couched more in the domain of environmental management, one could draw parallels between this definition of sustainability and Katz and Kahn's discussion (1978) of the impact of role enactment on individuals within organizations (e.g., asking questions such as whether role enactment enhances or reduces the well-being of the individual, whether it enlarges or reduces the person's valued skills and abilities, and whether it increases the individual's opportunity and capacity for other valued role enactments). Essentially, both Katz and Kahn and Hart and colleagues (Hart, 1997; Hart & Milstein, 1999) are talking about sustainability but at quite different levels of analysis. Hart and colleagues discuss it at the organization and environmental level. Katz and Kahn (1978) are essentially talking about designing role sets that are sustainable at the individual level of analysis, that is, work roles that meet current needs of both the organization and individual but do so in a manner that maintains and develops the individual's ability to adapt, grow, and continue to engage in this and future role enactments without negatively affecting the individual's health.

Another more microapplication of this competing-values notion is the leadership model by Quinn and colleagues (e.g., Denison, Hooijberg, & Quinn, 1995; Quinn, 1984, 1988; Quinn & Rohrbaugh, 1983). Initially working from research investigating various indicators of organizational effectiveness, Quinn and Rohrbaugh identified two underlying dimensions (internal/external and structure/flexibility) that are not too different from our dimensions in Figure 1.1. Subsequently, Quinn developed a competing-values model of leadership that suggested that effective leaders are those who can engage in multiple and competing roles. Recent research has supported this notion (e.g., Denison et al., 1995; Hart & Quinn, 1993). Thus, this concept of balancing competing demands appears in several other lines of theory development and research in the organizational sciences as well as occurring implicitly in numerous discussions of organizational health.

With respect to our previous discussions, research on stress and safety has typically focused on the short-term quadrants (see Figure 1.1). For example, performing actions according to safety procedures would largely fall into the extrinsic dimension, whereas research investigating the work context correlates of stress or strain would fall into the intrinsic model. Discussed to a lesser degree in this research are the long-term intrinsic and extrinsic perspectives. This framework and its identification of both short-term and long-term, intrinsic and extrinsic goals brings to mind the question of how research typically focused on one quadrant (e.g., safety research focusing on short-term extrinsic adherence to procedures) relates to the other quadrants in the model. For example, safety climate research has typically been used to predict short-term externally focused behavior (e.g., Hofmann & Stetzer's focus on unsafe behavior, 1996). Could it be, however, that establishing a positive safety climate creates a learning culture within an organization that leads to the long-term development, growth, and improvement in health and safety over time (Hofmann & Stetzer, 1998)? These are the types of questions we hope the framework in Figure 1.1 as well as the contributions of this volume will engender.

We have been deliberately general in our descriptions of what the focus of health might be in each quadrant (Figure 1.1). We do this purposefully, because the final dimension upon which our organizationally focused health definitions vary is across levels of analysis (i.e., from individual to organizational). Essentially, what we have provided as descriptions in Figure 1.1 are the functional outcomes describing health that are not level specific. Take, for example, the short-term, intrinsic health perspective, which focuses on effective internal system functioning and meeting internal goals and values. These functional outcomes (effective system function) are level-free descriptions of health; that is, we could talk about effective system functioning and meeting internal goals and values at the individual, group, or organizational level of analysis. The processes through which these outcomes materialize will likely be fundamentally different across levels of analysis, because the nature of the organism changes from a biologically based system (individual level) to a socially based system (group and organizational levels). Furthermore, the metrics used to assess these functional outcomes at different levels of analysis may vary as well (e.g., Staw, Sandelands, & Dutton, 1981). However, this notion of using functional outcomes as a way to integrate theories across levels, while simultaneously recognizing and exploring differences in the processes through which these outcomes materialize, has been discussed elsewhere as a way to build multilevel theory (Morgeson & Hofmann, 1999). Staw et al. (1981) as well as Fiol, O'Connor, and Aguinis (2001) provide examples of this approach.

The purpose here is to introduce the framework for organizing health (Figure 1.1) so that it can be used as a backdrop against which to view the contributions to this volume. Although most chapters will not specifically deal with each of these descriptions of health, we hope that across chapters each of the descriptions of health included in our model will be fleshed out across the individual, group, and organizational levels of analysis. Furthermore, we believe that the future of investigating organizational health involves research designed to jointly optimize each of the quadrants in Figure 1.1. We will discuss the implications of this model for research and theory in the final chapter in light of each of the contributions to the volume.

References

Abanobi, O. (1986). Content validity in the assessment of health states. *Health Values, 10,* 37–40.

Adkins, J. A. (1999). Promoting organizational health: The evolving practice of occupational health psychology. *Professional Psychology: Research and Practice, 30,* 129–137.

Adkins, J. A., Quick, J. C., & Moe, K. O. (2000). Building world-class performance in changing times. In L. R. Murphy & C. L. Cooper (Eds.), *Healthy and productive work: An international perspective* (pp. 108–131). New York: Taylor & Francis.

Argyris, C. (1958). The organization: What makes it healthy? *Harvard Business Review, 36,* 107–116.

Argyris, C. (1964). *Integrating the individual and organization.* New York: Wiley.

Bennis, W. G. (1962). Towards a "truly" scientific management: The concept of organization health. *General Systems: Yearbook of the Society for General Systems Research, 7,* 269–282.

Berkman, L., & Breslow, L. (1983). *Health and ways of living: The Alameda County study.* New York: Oxford University Press.

Boorse, C. (1975). On the distinction between disease and illness. *Philosophy and Public Affairs, 5,* 49–68.

Cox, T. (1992). Editorial: Occupational health: Past, present and future. *Work and Stress, 6,* 99–102.

Danna, K., & Griffin, R. W. (1999). Health and well-being in the workplace: A review and synthesis of the literature. *Journal of Management, 25,* 357–384.

Denison, D. R., Hooijberg, R., & Quinn, R. E. (1995). Paradox and performance: Toward a theory of behavioral complexity in managerial leadership. *Organization Science, 6,* 524–540.

DeVito, S. (2000). On the value-neutrality of the concepts of health and disease: Unto the breach again. *Journal of Medicine and Philosophy, 25,* 539–567.

Fiol, C. M., O'Connor, E. J., & Aguinis, H. (2001). All for one and one for all? The development and transfer of power across organizational levels. *Academy of Management Review, 26,* 224–242.

Friedman, M. D., Rosenman, R. D., & Carroll, V. (1958). Changes in serum cholesterol and blood clotting time in men subjected to cyclic variation of occupational stress. *Circulation, 17,* 852–861.

Ganster, Fox, & Dwyer (2001). Explaining employees' health care costs: A prospective examination of stressful job demands, personal control, and physiological reactivity. *Journal of Applied Psychology, 86,* 954–964.

Griffin, M. A., Hart, P. M., & Wilson-Evered, E. (2000). Using employee opinion surveys to improve organizational health. In L. R. Murphy & C. L. Cooper (Eds.), *Healthy and productive work: An international perspective* (pp. 15–36). New York: Taylor & Francis.

Hart, S. L. (1997). Beyond greening: Strategies for a sustainable world. *Harvard Business Review, 75*(1), 66–76.

Hart, S. L., & Milstein, M. B. (1999). Global sustainability and the creative destruction of industries. *Sloan Management Review, 41*(1), 23–33.

Hart, S. L., & Quinn, R. E. (1993). Roles executives play: CEOs, behavioral complexity, and firm performance. *Human Relations, 46,* 543–574.

Hofmann, D. A., Jacobs, R., & Landy, F. J. (1995). High reliability process industries: Individual, micro, and macro organizational influences on safety performance. *Journal of Safety Research, 26,* 131–149.

Hofmann, D. A., & Morgeson, F. P. (1999). Safety as a social exchange: The role of leader-member exchange and perceived organizational support. *Journal of Applied Psychology, 83,* 286–296.

Hofmann, D. A., & Morgeson, F. P. (in press). The role of leadership in safety performance. In J. Barling & M. Frone (Eds.), *The psychology of workplace safety.* Washington, DC: American Psychological Association.

Hofmann, D. A., Morgeson, F. P., & Gerras, S. J. (in press). Climate as a moderator of the relationship between LMX and content-specific citizenship behavior: Safety climate as an exemplar. *Journal of Applied Psychology.*

Hofmann, D. A., & Stetzer, A. (1996). A cross-level investigation of factors influencing unsafe behaviors and accidents. *Personnel Psychology, 49,* 307–339.

Hofmann, D. A., & Stetzer, A. (1998). The role of safety climate and communication in accident interpretation: Implications for learning from negative events. *Academy of Management Journal, 41,* 644–657.

Hoy, W. K., & Feldman, J. A. (1987). Organizational health: The concept and its measure. *Journal of Research and Development in Education, 20,* 30–37.

Ilgen, D. R. (1990). Health issues at work: Opportunities for industrial/organizational psychology. *American Psychologist, 45,* 273–283.

Jaffe, D. T. (1995). The healthy company: Research paradigms for personal and organizational health. In S. L. Sauter & L. R. Murphy (Eds.), *Organizational risk factors for job stress* (pp. 13–39). Washington, DC: American Psychological Association.

Kaplan, R. S., & Norton, D. P. (1992). The balanced scorecard: Measures that drive performance. *Harvard Business Review, 70*(1), 71–79.

Kaplan, R. S., & Norton, D. P. (1996). Using the balanced scorecard as a strategic management system. *Harvard Business Review, 74*(1), 75–86.

Katz, D., & Kahn, R.L. (1978). *The social psychology of organizations.* New York: John Wiley & Sons.

Komaki, J., Barwick, K. D., & Scott, L. R. (1978). A behavioral approach to occupational safety: Pinpointing and reinforcing safety performance in a food manufacturing plant. *Journal of Applied Psychology, 63,* 434–445.

Komaki, J., Heinzmann, A. T., & Lawson, L. (1980). Effect of training and feedback: Component analysis of a behavioral safety program. *Journal of Applied Psychology, 67,* 334–340.

Krause, T. R. (1997). *The behavior-based safety process: Managing involvement for an injury–free culture* (2nd ed.). New York: Van Nostrand Reinhold.

Larson, J. S. (1999). The conceptualization of health. *Medical Care Research and Review, 56,* 123–136.

Lennox, J. (1995). Health as an objective value. *Journal of Medicine and Philosophy, 20,* 499–511.

Lidz, J. W. (1995). Medicine as metaphor in Plato. *Journal of Medicine and Philosophy, 20,* 527–541.

Lindström, K., Schrey, K., Ahonen, G., & Kaleva, S. (2000). The effects of promoting organizational health on worker well-being and organizational effectiveness in small and medium-sized enterprises. In L. R. Murphy & C. L. Cooper (Eds.), *Healthy and productive work: An international perspective* (pp. 83–104). New York: Taylor & Francis.

McGregor, D. (1960). *The human side of enterprise.* New York: McGraw-Hill.

Merriam-Webster, Inc. (1997). *Merriam-Webster's collegiate dictionary* (10th ed.). Springfield, MA: Merriam-Webster.

Miles, M. B. (1965). Planned change and organizational health: Figure and ground. In F. D. Carver & T. J. Sergiovanni (Eds.), *Organizations and human behavior: Focus on schools.* New York: McGraw-Hill.

Morgeson, F.P., & Hofmann, D.A. (1999). The structure and function of collective constructs: Implications for research and theory development. *Academy of Management Review, 24,* 249–265.

Murphy, L. R., & Cooper, C. L. (2000). Models of healthy work organizations. In L. R. Murphy & C. L. Cooper (Eds.), *Healthy and productive work: An international perspective* (pp. 1–11). New York: Taylor & Francis.

Navarro, V. (1977). *Health and medical care in the U.S.: A critical analysis.* Amityville, NY: Baywood.

Parsons, T. (1972). In E. G. Jaco (Ed.), *Patients, physicians, and illness* (pp. 107–127). New York: Free Press.

Quick, J. C. (1999). Occupational health psychology: Historical roots and future directions. *Health Psychology, 18,* 82–88.

Quick, J. C., & Tetrick, L. E. (Eds.). (2002). *Handbook of occupational health psychology.* Washington, DC: American Psychological Association.

Quinn, R. E. (1984). Applying the competing values approach to leadership: Toward an integrating framework. In J. G. Hunt, D. M. Hosking, C. A. Schriesheim, & R. Stewart (Eds.), *Leaders and managers: International perspectives on managerial behavior and leadership* (pp. 10–27). New York: Pergamon Press.

Quinn, R. E. (1988). *Beyond rational management: Mastering the paradoxes and competing demands of high performance.* San Francisco: Jossey-Bass.

Quinn, R. E., & Rohrbaugh, J. (1983). A spatial model of effectiveness criteria: Towards a competing values approach to organizational analysis. *Management Science, 29,* 363–377.

Sauter, S. L., & Hurrell, J. J. (1999). Occupational health psychology: Origins, content, and direction. *Professional Psychology: Research and Practice, 30,* 117–122.

Schaubroeck, J., Ganster, D.C.,& Kemmerer, B.E. (1994). Job complexity, "type A" behavior, and cardiovascular disorder: A prospective study. *Academy of Management Journal, 37,* 426–439.

Schaubroeck, J., Jones, J.R., & Xie, J.L. (2001). Individual differences in utilizing control to cope with job demands: Effects on susceptibility to infectious disease. *Journal of Applied Psychology, 86,* 265–278.

Schaubroeck, J., & Merritt, D.E. (1997). Divergent effects of job control on coping with work stressors: The key role of self-efficacy. *Academy of Management Journal, 40,* 738–754.

Schmidt, W. C., Welch, L., & Wilson, M. G. (2000). Individual and organizational activities to build better health. In L. R. Murphy & C. L. Cooper (Eds.), *Healthy and productive work: An international perspective* (pp. 133–147). New York: Taylor & Francis.

Schroeder, E. (1983). Concepts of health and illness. In A. J. Culyer (Ed.), *Health indicators: An international study for the European Science Foundation* (pp. 23–33). New York: St. Martin's Press.

Spector, P. (1986). Perceived control by employees: A meta-analysis of studies concerning autonomy and participation at work. *Human Relations, 39,* 1005–1016.

Spector, P., Chen, P. Y., & O'Connell, B. J. (2000). A longitudinal study of relations between job stressors and job strains while controlling for prior negative affectivity and strains. *Journal of Applied Psychology, 85,* 211–218.

Spector, P.E., & Jex, S.M. (1998). Development of four self-report measures of job stressors and strain: Interpersonal Conflict of Work Scale, Organizational Constraints Scale, Quantitative Workload Inventory, and Physical Symptoms Inventory. *Journal of Occupational Health Psychology, 3,* 356–367.

Staw, B. M., Sandelands, L. E., and Dutton, J. E. (1981). Threat-rigidity effects in organizational behavior: A multilevel analysis. *Administrative Science Quarterly, 26,* 501–524.

Stetzer, A., & Hofmann, D. A. (1996). Risk compensation: Implications for safety interventions. *Organizational Behavior and Human Decision Processes, 66,* 73–88.

Tetrick, L. E. (1992). The mediating effect of perceived role stress: A confirmatory analysis. In J. C. Quick, L. R. Murphy, & J. J. Hurrell (Eds.), *Work and well-being assessments and interventions for occupational mental health.* Washington, DC: American Psychological Association.

Tetrick, L. E., & LaRocco, J. M. (1987). Understanding, prediction, and control as moderators of the relationship between work conditions and well-being. *Journal of Applied Psychology, 72,* 538–543.

Vicenzi, R., & Adkins, G. (2000). A tool for assessing organizational vitality in an era of complexity. *Technological Forecasting and Social Change, 64,* 101–113.

Wakefield, J. C. (1995). Dysfunction as a value-free concept: A reply to Sadler and Agich. *Philosophy, Psychiatry, and Psychology, 2,* 233–246.

Ware, J., Brook, R., Davies, A., & Lohr, K. (1981). Choosing measures of health status for individuals in general populations. *American Journal of Public Health, 71,* 620–625.

Williams, S. (1994). Ways of creating healthy work organizations. In C. L. Cooper & S. Williams (Eds.), *Creating healthy work organizations* (pp. 7–24). New York: Wiley.

Wood, P.H.N. (1986). Health and disease and its importance for models relevant to health research. In B. Z. Nizetic, H. G. Pauli, & P. G. Svensson (Eds.), *Scientific approaches to health and health care* (pp. 57–70). Copenhagen: World Health Organization.

World Health Organization (WHO). (1948). Preamble to the Constitution of the World Health Organization as adopted by the International Health Conference, New York, 19–22 June 1946; signed on 22 July 1946 by the representatives of sixty-one states (*Official Records of the World Health Organization,* no. 2, p. 100) and entered into force on 7 April 1948.

Zohar, D. (2000). A group-level model of safety climate: Testing the effect of group climate on microaccidents in manufacturing jobs. *Journal of Applied Psychology, 85,* 587–596.

PART 1

Individual and Organizational Effects

CHAPTER 2

Individual Differences in Health and Well-Being in Organizations

Paul E. Spector

As we have advanced into the twenty-first century, scholars have put a growing emphasis on issues of employee well-being in organizations. Whereas researchers in the past had a tendency to consider them separately, current researchers see that organizational well-being depends in large measure on employee well-being. The emerging field of job stress has shown clearly that psychosocial features of organizations can affect employee health and well-being, which in turn can affect organizations two ways: (1) directly through increased costs due to absence and health claims and (2) indirectly through employees' reduced effectiveness (Jex, 1998). Complicating matters, however, are the vast individual differences in reactions to organizational conditions, as well as the individual's tendency to expose him- or herself to dangerous or unhealthy situations. We can have a complete understanding of how organizations affect the employees' health and how employees affect organizations' health only when we are clear as to how people differ.

Work and Employee Well-Being

As discussed in Chapter One, well-being is a complex idea that includes elements of both physical and psychological health. On the physical side, well-being is not only the absence of specific illness but the existence of good health, that is, the sound functioning of a person's body systems. On the psychological side, well-being is not just the absence of psychopathology; it includes positive attitudes and feelings about major life domains, including work. Although everyone will experience an occasional bout of illness or distress, positive well-being is indicated by a person who is generally in good health and good spirits.

Work on occupational stress and well-being has been dominated by models linking job conditions (job stressors) to health-related outcomes (job strains). The general idea is that exposure to stressors leads to a variety of strains, but many models are rather mechanistic and fail to fully develop the role of individual differences. Furthermore, making a clear distinction between objective features of the environment or environmental stressors and perceptions of that environment or perceived stressors is essential. Not everyone views situations in the same way, and an individual's perception or appraisal (Lazarus, 1995) of the situation is most relevant for well-being.

Research has identified several work conditions that are classified as job stressors, although Jex and Beehr (1991) noted that a great deal of attention has been given to a small number of workplace stressors, with insufficient attention being paid to some that may be even more important. Strains as indicators of health and well-being can be divided into three categories: behavioral, physical, and psychological (Jex & Beehr, 1991). Psychological strains include both emotional states (e.g., anger or anxiety); attitudes, such as job dissatisfaction; and intentions, such as the intention of quitting the job. Physical strains can be immediate short-term physiological disturbance (e.g., high blood pressure) and somatic symptoms (headache or stomach upset) or long-term illness (heart disease). Behavioral strains are reactions to stressful conditions that can be adaptive or maladaptive. Although they do not necessarily make up well-being itself, these behaviors can be indicators of well-being and in some cases can contribute to positive or negative well-being. For example, absence can be used as a coping method to deal with stressful conditions (i.e., taking a mental health day) but if overdone can exacerbate the problem. Withdrawal from work is not necessarily an adaptive means of coping with a heavy workload if work piles up while the person is away.

A large number of constructs and scales have been developed to assess strains, although most tend to focus on the negative side, reflecting conditions of decreased well-being (Jex & Beehr, 1991). As with stressors, a great deal of attention has been given to a relatively small number of strains. For example, job satisfaction has been extensively studied and has been shown to relate to a variety of perceived job stressors. Anxiety has received considerable attention, but other emotions (e.g., anger or depression) have received relatively little. Physiological health symptoms, such as headache and stomach distress, have been studied, but researchers often combine lists of symptoms into checklists rather than studying them alone. In part, this may reflect that individuals differ in their exact physical responses, with one person responding to stressors with tension headache and another with stomach problems.

A Model of Well-Being at Work

The model shown in Figure 2.1 (see also Spector, 1998) suggests that well-being is a result of an individual's interaction with the work environment. People throughout the day perceive and monitor the

Figure 2.1. Emotion-Centered Model of the Job Stress Process.

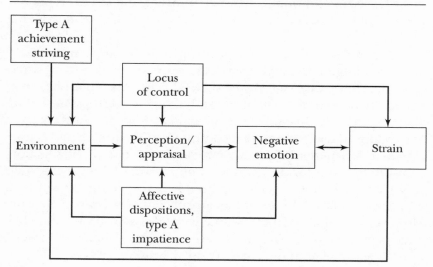

external environment, appraising events and situations that have potential to affect them (Lazarus, 1995). According to the model, stressors are situations that an individual sees as a provocation or threat that induce negative emotions of varying intensity. Negative emotions (e.g., anger or anxiety), themselves a form of strain, will lead to other strains. This model distinguishes emotions from other forms of psychological strains that are typically more stable, such as job satisfaction, but that are likely influenced by negative emotional responses to work (Weiss, in press).

The model shows the usual progression from environment to reaction to strain, but we should not assume that individuals are passive responders as opposed to active agents in their environments. The model shows arrows from emotion to perception/appraisal and from strain to emotion, as well as from strain to environment. This implies that a person's emotional state can color his or her perception/appraisal of the environment. A person who is anxious is more likely to see the environment as threatening than is a person who is calm. Further, strains can affect emotions; for example, a person who is experiencing physical symptoms might find this raises feelings of anxiety. Finally, strains can affect the environment itself. An individual who is experiencing physical symptoms is likely to be less efficient and may perform in a way that supervisors notice. This may result in added pressure to perform, which the individual may perceive as an additional stressor. In this model emotion plays a central role as both cause and effect of perceptions/appraisals and strains.

From an evolutionary perspective, emotions arose to serve an adaptive function in helping a person respond to events and situations that have implications for survival (Plutchik, 1989) by energizing the person physiologically and inducing appropriate action (Simon, 1967). Evaluation of the situation and physiological changes are important components that contribute to motivation and action (Wallbott & Scherer, 1989). Cognitive interpretation of the situation provides the basis for the specific emotion experienced (Schachter & Singer, 1962), and specific emotions help determine the kinds of reactions displayed; for example, anxiety induces escape behavior. However, emotions don't tend to elicit reflexive or immediate behaviors but are more likely to induce action tendencies and intentions to act at a later time (Bies, Tripp, & Kramer, 1997).

Emotions can lead to other strains through several mechanisms. First, emotions themselves produce action tendencies designed to enhance positive and reduce negative affect. This can result in actions designed to constructively deal with the job stressor itself (e.g., staying late to finish a project when workload is high), which is likely when the individual feels he or she both is capable of being effective (self-efficacy) and has control in the situation (Spector, 1998). If these conditions are not met and the person experiences negative emotion, he or she might engage in counterproductive acts against the organization (e.g., sabotage the work), escape the stressor (e.g., call in sick to avoid work), or reduce the emotion itself (e.g., drink alcohol) (Spector & Fox, in press). Second, emotions have physiological components that can affect health. Negative emotions and distress at work have been linked to both heart disease (Greenglass, 1996) and to suppression of the immune system (O'Leary, 1990). Situations people find distressing (i.e., that induce negative emotion) have been shown to increase production of both catecholamines (e.g., adrenaline) and cortisol (Frankenhaeuser & Lundberg, 1982), which may be in part the mediators of long-term health. Chronic exposure to cortisol, for example, has been implicated in development of heart disease (Johansson, 1989).

Individual differences are important at all stages of the model, and they can have effects through a variety of mechanisms, as a number of authors have discussed in relation to the single personality trait of negative affectivity (NA) (e.g., Moyle, 1995; Spector, Zapf, Chen, & Frese, 2000). Researchers have observed (as I will discuss in detail later) that people vary in their tendency to appraise a given situation as a stressor, to experience emotions, and to exhibit various strains. Most of this work has been linked to personality variables as opposed to other possible individual difference variables.

At least two classes of personality variables are relevant for the stress–well-being process. First are affective dispositions, personality variables that reflect an individual's tendency to experience emotions across situations. Examples include trait anxiety (the most thoroughly investigated), trait anger, and the more general construct of NA. Second are control-related variables that reflect an individual's beliefs about his or her own capabilities and capacities to be effective and control situations. The most frequently studied in relation to health and well-being is locus of control, belief in the ability to control reinforcements.

The Role of Individual Differences: Personality

Individual difference variables in general and personality in particular can act through a variety of mechanisms (see Spector, Zapf, et al., 2000), four of which this chapter will discuss:

- Perception/appraisal
- Stressor creation
- Emotion threshold
- Strain sensitivity

The perception/appraisal mechanism suggests that people vary in how they view the world, with some more likely than others to perceive a given situation as a job stressor. This is particularly true for job stressors that involve aspects of the more abstract social environment rather than the more concrete and objective physical environment. For example, Spector, Dwyer, and Jex (1988) investigated convergence between incumbents and their supervisors on seven job stressors. Those concerning concrete aspects of workload showed much higher correlations between sources ($r = .49$ to $.83$) than stressors that were more abstract, such as interpersonal conflict, organizational constraints, and role ambiguity ($r = .08$ to $.31$). These latter job stressor measures left far more variance attributable to individual differences than did the former, suggesting that incumbents and supervisors were viewing the situation quite differently. Taking this a step further, we would expect that individual difference variables should relate to perceptions of various job stressors differentially.

The stressor-creation mechanism suggests that people are not randomly assigned to job stressors, but through their own direct and indirect actions, they create the environments in which they find themselves. This can occur via selection into more or less stressful situations or via actions that enhance or inhibit job stressor development. Selection would occur if people, either by their own intentional actions or because of their characteristics, are placed into situations that are more or less stressful. For example, type A individuals are described as being hard driving and achievement oriented, seeking out situations that allow them to fulfill these tendencies. This would likely lead such people into high-pressure work environments with heavy workloads and short deadlines. On the other hand, it is also likely that certain types of people will be selected into certain environments. For example, Caldwell and Burger (1998) showed that personality traits were related to interview success for new college graduates. Certain types of individuals might find themselves with more job choices and thus would be better able to avoid unfavorable job conditions.

On the other hand, others have argued that people can be the cause of their own stressors (e.g., Dohrenwend, Dohrenwend, Dodson, & Shrout, 1984). Given the same initial work environment, some individuals would be more likely than others to be a cause of job stressors. For example, some people might be overly aggressive and create conflicts with others. Other people might exert little effort to keep up with their workloads, thus causing supervisors to put pressure on them for performance.

The emotion-threshold mechanism suggests that some individuals are more emotionally reactive to the environment than others and will tend to experience negative emotions more frequently. Strain sensitivity says that individuals vary in their tendencies to experience or exhibit strains, either in response to emotions or more directly in response to job stressors. Clearly, the correlations of job stressors and emotions with strains are in most cases quite modest, leaving a great deal of room for individual differences. The perception/appraisal, emotion-threshold, and strain-sensitivity mechanisms imply reactivity of people to the environment. Not only should we expect to find correlations of individual difference variables with job stressors, emotions, and strains, but we should find moderator effects as well. Specifically, we would expect to find that individual differences moderate the relation between the environment and perceptions of that environment. Given the same situation, certain types of people will be more likely to perceive it as a job stressor. Furthermore, given the same level of perceived stressor, some individuals will respond emotionally, and others will not. Unfortunately, there have not been many good tests of moderator hypotheses. Finally, given the same emotional response, some people will exhibit strains, while others will not. Taken together, the four mechanisms suggest a complex interplay of individuals with the job environment, with people both reacting to and causing job stressors.

Although there are many possible individual difference variables that one might consider, few have been given much attention in the employee health domain. In this chapter we will discuss four personality variables that have been studied in the workplace: trait anxiety–NA, trait anger, locus of control, and type A behavior. The reader should keep in mind that not every personality trait works via all four mechanisms and that the way in which the mechanisms operate can differ.

Trait Anxiety–NA

Anxiety is an emotion characterized by both physiological arousal and the immediate feeling of apprehension, fear, nervousness, and worry (Spielberger, Ritterband, Sydeman, Reheiser, & Unger, 1995). Emotion researchers have distinguished the experience or state anxiety from the personality characteristic of trait anxiety. Spielberger (1979) defines *trait anxiety* as anxiety proneness or the tendency to perceive a wide range of situations as dangerous or threatening. Spielberger (1972) notes that this proneness to anxiety tends to be limited to the psychosocial environment and not to situations that might be physically dangerous. Thus, it would be particularly relevant to the commonly studied job stressors that tend to have elements of interpersonal interaction, such as conflict with others, and not to stressors that involve nonsocial aspects of the environment, such as workload.

Watson and Clark (1984) broadened the trait anxiety construct to encompass a wider range of negative emotional experience. Their construct of NA suggests that certain individuals are predisposed to experience not only anxiety but other negative emotional states as well, including fear, hostility, scorn, and disgust. In the organizational domain, researchers have adopted the NA perspective, although many studies really assess trait (or state) anxiety. However, scales of a variety of negative emotions tend to correlate quite strongly with one another and may lack discriminant validity. Much of this is because the most popular scales include items that get at the core of anxiety, asking about feeling ill at ease, nervous, tense, and uncomfortable. Thus, results with scales to assess related constructs of trait anxiety, NA, and neuroticism are often thrown together.

Whereas some researchers have combined a wide variety of negative emotions into the single metatrait of NA, Spielberger and colleagues have explored the distinctiveness of anger and anxiety (e.g., Spielberger, Krasner, & Solomon, 1988), which although related, can have different correlates. This distinction is supported by studies using the trait anger and anxiety scales from his State-Trait Personality Inventory (Spielberger, 1979) that show their correlation is quite modest (e.g., .32 in Fox & Spector, 1999). I will discuss trait anger separately but will combine NA with trait anxiety and other related constructs such as neuroticism, consistent with the workplace literature.

The four mechanisms provide ways in which trait anxiety–NA (and trait anger) might affect job stressors, emotions, and strains. Individuals who rate high on these characteristics have developed a view of the world through their life experiences that results in a tendency to interpret the world as threatening. Even relatively benign situations might induce anxiety as they are seen as a threat to self-esteem and well-being. Trait anxiety–NA is especially relevant to the emotion-threshold mechanism. By definition these traits are concerned with the tendency for individuals to be high or low in emotional reactivity. It is worth repeating that trait anxiety does not bring on a universal hyperreactivity, but rather high trait-anxious individuals are reactive to social situations and stressors (Spielberger, 1972). It is also likely that individuals high in trait anxiety–NA have an enhanced strain sensitivity, although whether this happens directly or is mediated by their greater emotional response is not clear. Finally, individuals high in trait anxiety–NA will tend to create their own stressors via the stressor-creation mechanisms. Some evidence exists that these affective traits relate to characteristics of the jobs people hold. Trait anxiety is associated with several characteristics of people's jobs, assessed with objective methods (Spector, Fox, & Van Katwyk, 1999). Indirect evidence is provided by studies that show individuals low in trait anxiety–NA perform better and are more successful in employment interviews (Caldwell & Burger, 1998; Cook, Vance, & Spector, 2000). Presumably, those low in trait anxiety–NA are able to land the better jobs.

The implications of this model and the specific mechanisms are that trait anxiety–NA should be correlated with job stressors, negative emotional states, and strains. Individuals high on trait anxiety–NA should be more likely than their low trait anxiety–NA counterparts to perceive job stressors, experience state anxiety and other negative emotions, and experience strains. To summarize evidence to support these propositions, I began with the meta-analysis provided by Spector and Jex (1998) of studies relating measures of trait anxiety, as well as NA and trait anxiety, to three job stressors and their scale of somatic symptoms. I then added an additional 11 studies relating NA and trait anxiety to additional job stressors and strains, including Chen and Spector (1991); Fox and Spector (1999); Fox, Spector, and Miles (1999); Frone (1998); Heinisch and Jex (1997); Iverson and Erwin (1997); Jex and Spector (1996); Moyle (1995); Schaubroeck, Judge, and Taylor (1998); Spector, Chen, and O'Connell (2000); and Spector et al. (1999). Mean correlations were computed for each variable across those studies that contained it. Spector and Jex (1998) found that the mean correlation was .30 (three studies) with organizational constraints (conditions in organizations that prevent the individual from performing the job well), .33 (six studies) with interpersonal conflict with others at work, and .13 (five studies) with workload. I found that trait anxiety–NA had mean correlations of .17 (six studies) with role ambiguity, .23 (four studies) with role conflict, and -.28 (one study) with procedural justice. Thus, individuals who are high on trait anxiety–NA tend to perceive more conflict with others, organizational constraints, workload, role ambiguity, role conflict, and injustice. Furthermore, as expected based on Spielberger's argument (1972) that trait anxiety is relevant only to psychosocial stressors, there was a stronger relation with those stressors that had interpersonal components than with those that did not. For example, the mean correlation with interpersonal conflict was almost double that with workload.

On the emotion and strain side, we also find relations with trait anxiety–NA. Spector and Jex (1998) found a mean correlation of .40 (four studies) with somatic symptoms, such as headache and stomach distress. I found mean correlations of trait anxiety–NA with job satisfaction (-.28; seven studies), frustration at work (.32; four studies), anger at work (.41; one study), doctor visits (.23; one study), state anxiety at work (.43; one study), and counterproductive behavior at work (.33; two studies). Note that the rather modest .43 correlation with state anxiety at work clearly suggests that these are distinct variables with good discriminant validity. Looking at research outside of the organizational realm, Booth-Kewley and Friedman (1987) provided evidence to link trait anxiety to cardiovascular disease. NA has been shown to relate to general subjective well-being and physical health complaints (e.g., McLennan, Gotts, & Omodei, 1988; Watson & Pennebaker, 1989).

The workplace studies I have summarized have all used cross-sectional designs with incumbent reports as the data collection method. It is conceivable that the measures of trait anxiety–NA were influenced by the job stressors and strains or that all variables were influenced by mood or other transitory conditions (Spector, Chen, & O'Connell, 2000). A longitudinal study by Spector and O'Connell (1994) casts doubts about such possibilities. The researchers administered a battery of personality scales, including trait anxiety, to a group of graduating college seniors, and in a follow-up a year and a half later, they administered scales of job stressors and strains. Trait anxiety correlated significantly with role ambiguity (.20), role conflict (.24), organizational constraints (.40), interpersonal conflict (.38), state anxiety at work (.32), and physical health symptoms (.23), which are not much different from the mean correlations I noted previously with cross-sectional studies.

So far, I have argued that according to the perception/appraisal mechanism, there will be a relation between trait anxiety–NA and job stressors due to the greater tendency for the highly anxious to see the world as threatening and stressful. However, this mechanism would suggest a more complex role of personality. It is not that the highly anxious see everything as threatening but rather that they have a greater tendency to appraise situations in that way. Thus, there should be a moderator effect of trait anxiety–NA on the relation between aspects of the objective work environment and perceived job stressors. Unfortunately, little if any research has attempted to see if this moderator effect occurs. Rather, the few studies that have looked at moderator effects have seen if personality moderates the relation between perceived job stressors and emotions or strains, which is more of a test of the emotion threshold and strain-sensitivity mechanisms.

Parkes (1990) showed that a measure of NA moderated the relation between perceived job stressors and job strains. These results were replicated by Heinisch and Jex (1997), although more strongly for women than men. Fox, Spector, and Miles (2001) tested for moderator effects of trait anxiety on relations between job stressors and counterproductive behavior as a behavioral strain. They found significant moderators for both interpersonal conflict and organizational constraints. In all three studies, as expected, trait anxiety–NA acted to enhance the relation between job stressors and strains, showing a steeper slope for those high on the personality characteristic. This suggests that the high trait anxiety–NA people are more sensitive to stressors and respond more strongly to them.

Trait Anger

As with anxiety, one can distinguish state from trait anger. *State anger* is an emotional state that can range in intensity from mild irritation to intense rage, accompanied by autonomic arousal (Tanzer, Sim, & Spielberger, 1996). Trait anger is the tendency to perceive a wide range of situations as annoying or frustrating and to respond to those situations with anger. In the work domain, relatively little anger research has been conducted, and the few studies that have been done have focused on frustration (a mild form of anger).

There are many parallels between anger and anxiety-NA in relation to job stress. Although they are different emotions and tend to be triggered by different events (although anxiety can follow an experience of anger), the basic mechanisms and process are quite similar. The perception/appraisal mechanism says that individuals high in trait anger are likely to perceive the world as frustrating and provocative, but the emotion this induces is anger rather than anxiety. Furthermore, many of the same stressors can trigger anger or anxiety in different people, as one person might respond to criticism by a supervisor with anger and another with anxiety. The stressor-creation mechanism suggests that individuals high in trait anger are likely to create their own stressors, either by selection or by their behavior. It seems almost self-evident that the high trait-angry person will create interpersonal conflict by expressing anger to others. The emotion threshold suggests that not only will the trait-angry person tend to perceive frustrating situations, he or she will have a lower anger threshold. Finally, strain sensitivity says those high in trait anger will be more likely to respond to anger with strains.

These mechanisms suggest that trait anger will be related to job stressors, emotions, and strains. Unfortunately, there have been very few studies in the workplace of anger and even fewer of trait anger. Fox and Spector (1999) and Fox et al. (2000) found that trait anger related significantly to interpersonal conflict ($r = .29$), organizational constraints (mean $r = .30$; two studies), procedural justice ($r = -.18$), job satisfaction (-.21), frustration at work (.32), and counterproductive work behavior (mean $r = .48$; two studies). Note that the relation between trait anger and frustration at work (a mild form of anger) was not particularly high, suggesting that the high trait-angry individuals might be more likely to experience anger without necessarily being chronically angry at work, nor is the low trait-angry person never angry. Research outside of the workplace has linked trait anger to heart disease (Booth-Kewley & Friedman, 1987; Greenglass, 1996).

Locus of Control

People have been shown to differ in their tendency to feel in control. Locus of control has been the most studied control-related personality variable. It is concerned with beliefs about personal control of rewards and punishments. Rotter (1966) described people as varying along a dimension from feeling they themselves have control (internals) to feeling that control resides outside of themselves in fate, luck, or powerful others (externals). Locus of control can be thought of hierarchically, with individuals having control beliefs in general (i.e., cutting across all domains in life) or beliefs limited to specified domains such as work (Spector, 1988). These beliefs color perceptions of control in various domains of life, with internals being more likely to see themselves in control. Keep in mind, however, that personality reflects tendencies and not certainties and that some situations might be so powerful as to overwhelm personality tendencies.

In theory the role of locus of control is via perceptions of control and the action tendencies this can produce. On the job stressor side, internals should be less likely to interpret situations as stressors because lack of control acts to enhance the power of stressors, producing greater levels of emotional response (Spector, 1998). A situation that is seen as controllable will have less potential for eliciting negative emotions than a situation seen as uncontrollable. In fact, the uncontrollability of a situation that is a potential threat is quite likely to lead to anxiety, due as much to the uncontrollability as to the situation itself. This would lead us to expect to find a relation between locus of control and perceived job stressors, as well as a moderating role of locus of control on the relation between objective environmental conditions and perceived job stressors, whereby externals are far more likely to perceive situations as stressors.

Locus of control also influences reactions to job stressors, mainly via behavior. Some have argued that feelings of uncontrollability contribute to counterproductive and seemingly maladaptive behavior, such as vandalism and other criminal acts (Allen & Greenberger, 1980). Thus, individuals who find themselves subject to job stressors will be more likely to respond with behavioral strains if they believe they have little control. In addition, research has shown that locus of control can be relevant for health. Kobasa (1982) suggested that locus of control is a component of the meta-trait of hardiness. The hardy individual is one who is well able to cope effectively with stress and thus will experience less strain, enhancing health and well-being. Furthermore, research has shown that internals are more proactive in taking care of their own health, which also contributes to less strain. Hurrell and Murphy (1991) summarized evidence showing that internals tend to have better health than externals and that locus of control has been shown to predict subsequent health in fairly long-term prospective studies.

Two of the four mechanisms by which personality might affect job stressors and strains can be relevant to locus of control. As we already noted, perception/appraisal of the environment can differ, with internals being less likely to perceive job stressors. Stressor creation can occur in two ways. First, the selection idea suggests that internals and externals will be found in different job settings.

Because internals have a tendency to be more proactive, they likely would seek out and persist in pursuing better jobs. Cook et al. (2000), for example, found that internal college students did better in simulated job interviews and had better actual interview outcomes than did their external counterparts. Furthermore, through their greater likelihood to respond to job stressors counterproductively, externals may produce even more stressors for themselves.

All this suggests that we should find locus of control relating to perceived job stressors, emotions at work, and strains. Evidence exists to support these hypotheses. On the job stressor side, studies have shown that externals tend to report higher levels of job stressors, such as role ambiguity and role conflict (e.g., Holder & Vaux, 1998; Newton & Keenan, 1990) and organizational constraints (Fox & Spector, 1999); they report their jobs as more stressful (Kyriacou & Sutcliffe, 1979). Interestingly, it is internals and not externals who tend to perceive greater workload and work demands (Moyle, 1995; Newton & Keenan, 1990). This may reflect that they are more motivated and exert more effort at work (Spector, 1982; Shukla & Upadhyaya, 1986). Locus of control is also associated with emotions, with externals reporting more anger, anxiety, depression, and frustration at work (Cvetanovski & Jex, 1994; Fox & Spector, 1999; Newton & Keenan, 1990). Locus of control has been linked to a number of strains, with externals reporting less job satisfaction (e.g., Evers, Frese, & Cooper, 2000; Fox & Spector, 1999; Moyle, 1995; Newton & Keenan, 1990; Shukla & Upadhyaya, 1986; Holder & Vaux, 1998), mental strain (Evers et al., 2000), physical health symptoms (Cvetanovski & Jex, 1994; Evers et al., 2000), and counterproductive behavior (Fox & Spector, 1999). In addition, externality has been linked to cardiovascular health (Brousseau & Mallinger, 1981).

Spector and O'Connell (1994) included locus of control in their longitudinal study and found that it correlated significantly with role ambiguity, role conflict, interpersonal conflict, job satisfaction, and state anxiety at work. In all cases externals reported greater job stressors and strains. In addition, as expected, internals were more likely to perceive autonomy at work ($r = -.31$). The longitudinal nature of this design suggests some causal influence of personality on perceptions and presumably on experience of job stressors and strains.

Type A Behavior

Perhaps no personality variable has been more strongly linked to health and well-being than has type A behavior. This construct was originally developed by Friedman and Rosenman (1974) to describe the characteristics of the typical coronary heart disease patient: competitive, hostile, impatient, and time urgent. Original research focused on global type A, but subsequent work has isolated several individual subdimensions, some of which are only slightly related and have different correlates (Edwards, Baglioni, & Cooper, 1990; Lee, Jamieson, & Earley, 1996; Spence, Helmreich, & Pred, 1987). In their meta-analysis Booth-Kewley and Friedman (1987) showed that type A has a small relation with cardiovascular disease, but they found considerable variation in results dependent upon the specific type A measure used, as well as the specific component of type A.

It now seems likely that type A relates to job stressors and strains through perception/appraisal and emotion-threshold mechanisms. Glass and Carver (1980) argued that type A individuals have a high need for control and overreact to uncontrollable stressors. Ganster (1986) also concluded that type A individuals are overly sensitive to job stressors. Perhaps the best evidence for this comes from a study by Kirmeyer (1988), who assessed type A, objective workload, and subjective workload in a sample of police dispatchers. Type A individuals reported greater workloads despite the fact that there was no relation between objective workload and type A. Thus, the high type A individuals perceived greater workloads that apparently did not objectively exist. In addition, it seems likely that the hostility and impatience components of type A reflect a low threshold for anger and frustration.

Some researchers have used global type A measures and have found a link between type A and general job demands and physical health symptoms (e.g., Moyle & Parkes, 1999). As we noted earlier, however, type A is composed of relatively independent components that have different correlates, and studies that have used component measures have provided more precise findings. In their longitudinal study, Spector and O'Connell (1994) found a relation of role ambiguity, role conflict, and workload with type A achievement striving but not with type A impatience, and a relation of organizational constraints and interpersonal conflict with type A impatience but not with type A achievement striving. Furthermore, only type A impatience was related to physical health symptoms ($r = .35$), which was similar to findings by Lee et al. (1996) with a more complex type A scale. Neither type A scale was related to job satisfaction, work anxiety, or work frustration. These results are consistent with Moyle and Parkes (1999), who also failed to find a relation between global type A and job satisfaction or psychological distress. However, it runs counter to Lee et al. (1996), who found that the anger, hostility, and time pressure scales of their type A measure correlated with anxiety and depression. Keep in mind that all three studies used different type A scales. Thus, additional research will be needed to clarify results with various type A components in their relations with strains.

Potential Confounding Effects of Personality in the Study of Occupational Health

Watson, Pennebaker, and Folger (1987) expressed concern that NA might confound self-report measures of job stressors and well-being, producing spurious correlations among them and thus throwing into question the basic idea that job stressors and strains are related. Concern about the issue has led some researchers to recommend that NA be routinely statistically controlled via partialing or entering into multiple regression analyses (e.g., Brief, Burke, George, Robinson, & Webster, 1988). However, the partialing of an individual difference variable to remove its potential confound is a dangerous practice, likely to result in erroneous conclusions about relations among the variables of interest (Meehl, 1971). Partialing removes shared variance regardless of the reason or underlying mechanism. If NA is a causal factor in the well-being process, then removing it will remove not a confound but the very effect of interest. If we were certain that NA acted as a confound and only a confound, statistical control would be a reasonable approach. In the absence of strong evidence for confounding and against a substantive role in the process, researchers should not conduct partialing.

Spector et al. (2000) reviewed existing evidence for the confounding effect. Brief et al. (1988) reasoned that if the confounding effect occurs, one should observe substantial reductions in partial correlations with NA controlled versus corresponding zero-order correlations with NA not controlled, and that study provided some initial evidence for this using some general job stress measures. Subsequent research using more specific job stressor measures has failed to find the same degree of reduction (e.g., Chen & Spector, 1991; Jex & Spector, 1996; Moyle, 1995), and covariance structure modeling studies have failed to find evidence for a significant confounding effect (Schaubroeck, Ganster, & Fox, 1992; Williams, Gavin, & Williams, 1996).

Although finding the correlation reduction is evidence that is consistent with confounding, it is also consistent with substantive mechanisms such as the four discussed in this chapter. As we noted earlier, there is considerable evidence that trait anxiety–NA has an important role in the job stress and occupational health process. Evidence supports the four mechanisms (perception/appraisal, emotion threshold, strain sensitivity, and stressor creation), suggesting a variety of avenues through which NA has effects. To statistically control NA would distort results, leading to erroneous inferences. The same would be true about removing effects of other personality or individual difference variables.

Statistical control in the job stress–occupational health domain is a useful tool for investigating specific hypotheses when results are compared both with and without the control. It can test hypotheses about the possible role of a variable when used appropriately. For example, Kenny (1979) has provided guidelines for conducting analyses that would be consistent with or counter to a mediator hypothesis. However, statistical controls of individual differences, even demographic variables, should not be used haphazardly in a blind attempt to remove confounds, unless one is certain a variable is in fact a confound (Meehl, 1971).

Suppose for the sake of argument that role conflict is a cause of psychological strain for both men and women equally. Further, suppose women experience more role conflict (and therefore more strain) than men due to greater family responsibilities. In other words, family responsibilities (associated with gender) cause role conflict that causes strain. If we statistically control for gender, it will reduce the relation observed between role conflict and strain because it will be related to both, and we might be tempted to erroneously conclude that there is little or no relation. The partial results compared to the zero order might give us a clue that gender is important, but it will not tell us the role gender plays. For that we will need additional empirical evidence and theory.

Linking Personality to Both Employee and Organizational Health

It seems self-evident that employee well-being leads directly to organizational well-being, although the connection is not always direct. Certainly, a healthier workforce will result in lower costs for health insurance (if the organization provides this benefit) and less absence. Furthermore, if we look at indicators of positive well-being as opposed to illness, we might expect that people who are happy in their jobs and experience positive emotions at work will be more likely to stay on the job and make efforts beyond required task performance. Certainly, people differ in their health and well-being, and as this chapter summarizes, some of these differences are related to personality.

From a very pragmatic point of view, one strategy to enhance organizational health through employee health might be to use personality tests to screen out potentially unhealthy individuals, much as organizations now use integrity tests to eliminate the potentially dishonest (Sackett & DeVore, 2001). However, a number of potential legal and practical issues are involved in this approach. First, using personality tests to screen for potential health conditions might be legally challenged as a violation of the Americans with Disabilities Act. Any test that has items that can be used to detect a mental disability (e.g., depression) or physical disability cannot be used in preemployment selection but could only be used legally as part of a medical screening after the employer has made an offer (Gutman, 2000). Second, the value of using this selection approach depends on the availability of a sufficiently large applicant pool once screening has been done for needed task-related characteristics, such as experience and skill. Third, this approach is based on the assumption that the personality variables being used to predict health are not related to job performance. If they are, it could result in a healthier but less productive workplace, replacing high performing individuals who might have a tendency toward strain with poor performing individuals with less tendency toward strain. This is quite possible, as Blau (1993) found that locus of control related in opposite directions with different aspects of job performance in bank tellers. Externals outperformed internals in an objective measure of productivity. In this case replacing internals with externals would be likely to reduce overall performance of the work unit.

An alternative approach to enhancing health would be to focus on the managing of employee well-being in addition to job performance. This means making employee well-being an important organizational goal. This is not to say that all decisions must favor employees over organizations but rather that the employee should be a consideration in decisions that have implications for stress and well-being. Where possible, organizations should be designed to avoid undue stress on employees, which might be as simple as providing on-site daycare for employees' children or flexible schedules. When this approach is not feasible, the organization should take steps to make assistance available to those who are being exposed to strain-producing stressors. Supervisors can be the first line of defense, and they might be trained to recognize the signs of excessive stress and deal with it appropriately by offering support and encouragement. This can begin with supervisors paying more attention to the emotional climate in their work areas and even encouraging the appropriate sharing of emotional responses (Meyerson, 2000). Flexible schedules and leave benefits can help the overly stressed take short breaks to recover. Organizations can also offer health promotion benefits (e.g., exercise facilities) and stress management training. In the extreme cases, employee assistance programs can be helpful for those whose well-being has suffered enough that it is affecting job performance.

Individual differences should be kept in mind in any interventions. Not all people respond the same way to situations. Managers need to be sensitive to individual differences in employee reactions to stressful job conditions. Those who are sensitive to stress may need extra social support that can be effective in helping people cope with stress (Beehr & Glazer, 2001; O'Driscoll & Dewe, 2001), as well as training in how to better deal with stressful aspects of the job.

Managers sometimes assume that action taken to benefit employee well-being is necessarily at the cost of organizational productivity or effectiveness. The recent concept of the healthy work organization challenges this assumption (Jaffee, 1995; Sauter, Lim, & Murphy, 1996). According to this view, employee and organizational health are interlinked rather than incompatible (Hart & Cooper, 2001). Although some practices favor one over the other, a healthy work organization is one that encourages both simultaneously. The organization is more effective because its employees have a high level of well-being, and employee well-being benefits from the effectiveness of the organization, for example, by providing more job security and rewards. Sauter et al. (1996) conducted a study that identified a number of organizational practices, climate variables, and values that were positively related to both organizational effectiveness and employee well-being. To be fully effective, such practices must be designed with individual differences in mind. For example, enhanced autonomy and control can be an effective approach for some but not all employees (Spector, 1998). Those organizations that fail to consider employee health and well-being as a worthwhile goal and pay attention to individual differences as an important element in reactions to the workplace are unlikely to be as healthy as they could be.

References

Allen, V. L., & Greenberger, D. B. (1980). Destruction and perceived control. In A. Baum & J. E. Singer (Eds.), *Applications of personal control* (pp. 85–109). Hillsdale, NJ: Erlbaum.

Beehr, T. A., & Glazer, S. (2001). A cultural perspective of social support in relation to occupational stress. In P. L. Perrewé & D. C. Ganster (Eds.), *Research in occupational stress and well-being: Vol. 1. Exploring theoretical mechanisms and perspectives* (pp. 97–142). New York: Elsevier.

Bies, R. J., Tripp, T. M., & Kramer, R. M. (1997). At the breaking point: Cognitive and social dynamics of revenge in organizations. In R. A. Giacalone & J. Greenberg (Eds.), *Antisocial behavior in organizations* (pp. 18–36). Thousand Oaks, CA: Sage.

Blau, G. (1993). Testing the relationship of locus of control to different performance dimensions. *Journal of Occupational and Organizational Psychology, 66,* 125–138.

Booth-Kewley, S., & Friedman, H. S. (1987). Psychological predictors of heart disease: A quantitative review. *Psychological Bulletin, 101,* 343–362.

Brief, A. P., Burke, M. J., George, J. M., Robinson, B., & Webster, J. (1988). Should negative affectivity remain an unmeasured variable in the study of job stress? *Journal of Applied Psychology, 73,* 193–198.

Brousseau, K. R., & Mallinger, M. A. (1981). Internal-external locus of control, perceived occupational stress, and cardiovascular health. *Journal of Organizational Behavior, 2,* 65–71.

Caldwell, D. F., & Burger, J. M. (1998). Personality characteristics of job applicants and success in screening interviews. *Personnel Psychology, 51,* 119–136.

Chen, P. Y., & Spector, P. E. (1991). Negative affectivity as the underlying cause of correlations between stressors and strains. *Journal of Applied Psychology, 76,* 398–407.

Cook, K. W., Vance, C. A., & Spector, P. E. (2000). The relation of candidate personality with selection interview outcomes. *Journal of Applied Social Psychology, 30,* 867–885.

Cvetanovski, J., & Jex, S. M. (1994). Locus of control of unemployed people and its relationship to psychological and physical well-being. *Work and Stress, 8,* 60–67.

Dohrenwend, B. S., Dohrenwend, B. P., Dodson, M., & Shrout, P. E. (1984). Symptoms, hassles, social supports, and life events: Problems of confounded measures. *Journal of Abnormal Psychology, 93,* 222–230.

Edwards, J. R., Baglioni, A. J., Jr., & Cooper, C. L. (1990). Examining the relationships among self-report measures of the type A behavior pattern: The effects of dimensionality, measurement error, and differences in underlying constructs. *Journal of Applied Psychology, 75,* 440–454.

Evers, A., Frese, M., & Cooper, C. L. (2000). Revisions and further developments of the occupational stress indicator: LISREL results from four Dutch studies. *Journal of Occupational and Organizational Psychology, 73,* 221–240.

Fox, S., & Spector, P. E. (1999). A model of work frustration-aggression. *Journal of Organizational Behavior, 20,* 915–931.

Fox, S., Spector, P. E., & Miles, D. (1999, April-May). *Counterproductive work behavior (CWB) in response to job stressors and organizational justice: The moderator effect of autonomy and emotion traits.* Paper presented at the annual meeting of the Society for Industrial and Organizational Psychology, Atlanta, GA.

Fox, S., Spector, P.E., & Miles, D. (2001). Counterproductive work behavior (CWB) in response to job stressors and organizational justice: Some mediator and moderator tests for autonomy and emotions. *Journal of Vocational Behavior, 59,* 291–309.

Frankenhaeuser, M., & Lundberg, U. (1982). Psychoneuroendocrine aspects of effort and distress as modified by personal control. In W. Bachmann & I. Udris (Eds.), *Mental load and stress in activity* (pp. 97–103). Amsterdam: North-Holland.

Friedman, M., & Rosenman, R. H. (1974). *Type A behavior and your heart.* New York: Knopf.

Frone, M. R. (1998). Predictors of work injuries among employed adolescents. *Journal of Applied Psychology, 83,* 565–576.

Ganster, D.C. (1986). Type A behavior and occupational stress. *Journal of Organizational Behavior Management, 8,* 61–84.

Glass, D. C., & Carver, C. S. (1980). Environmental stress and the type A response. In A. Baum & J. E. Singer (Eds.), *Applications of personal control* (pp. 59–83). Hillsdale, NJ: Erlbaum.

Greenglass, E. R. (1996). Anger suppression, cynical distrust, and hostility: Implications for coronary heart disease. In C. D. Spielberger, I. G. Sarason, J.M.T. Brebner, E. Greenglass, P. Laungani, & A. M. O'Roark (Eds.), *Stress and emotion: Anxiety, anger, and curiosity* (Vol. 16, pp. 205–225). New York: Taylor & Francis.

Gutman, A. (2000). *EEO law and personnel practices* (2nd ed.). Thousand Oaks, CA: Sage.

Hart, P. M., & Cooper, C. L. (2001). Occupational stress: Toward a more integrated framework. In N. Anderson, D. S. Ones, H. K. Sinangil, & C. Viswesvaran (Eds.), *Handbook of industrial, work, and organizational psychology* (Vol. 2, pp. 93–114). Thousand Oaks, CA: Sage.

Heinisch, D. A., & Jex, S. M. (1997). Negative affectivity and gender as moderators of the relationship between work-related stressors and depressed mood at work. *Work and Stress, 11,* 46–57.

Holder, J. C., & Vaux, A. (1998). African American professionals: Coping with occupational stress in predominantly white work environments. *Journal of Vocational Behavior, 53,* 315–333.

Hurrell, J. J., Jr., & Murphy, L. R. (1991). Locus of control, job demands, and health. In C. L. Cooper & R. Payne (Eds.), *Personality and stress: Individual differences in the stress process* (pp. 133–149). New York: Wiley.

Iverson, R. D., & Erwin, P. J. (1997). Predicting occupational injury: The role of affectivity. *Journal of Occupational and Organizational Psychology, 70,* 113–128.

Jaffee, D. T. (1995). The healthy company: Research paradigms for personal and organizational health. In S. L. Sauter & L. R. Murphy (Eds.), *Organizational risk factors for job stress* (pp. 13–39). Washington, DC: American Psychological Association.

Jex, S. M. (1998). *Stress and job performance: Theory, research, and implications for managerial practice.* Thousand Oaks, CA: Sage.

Jex, S. M., & Beehr, T. A. (1991). Emerging theoretical and methodological issues in the study of work-related stress. *Research in Personnel and Human Resources Management, 9,* 311–365.

Jex, S. M., & Spector, P. E. (1996). The impact of negative affectivity on stressor-strain relations: A replication and extension. *Work and Stress, 10,* 36–45.

Johansson, G. (1989). Stress, autonomy, and the maintenance of skill in supervisory control of automated systems. *Applied Psychology: An International Review, 33,* 45–56.

Kenny, D. A. (1979). *Correlation and causality.* New York: Wiley.

Kirmeyer, S. L. (1988). Coping with competing demands: Interruption and the type A pattern. *Journal of Applied Psychology, 73,* 621–629.

Kobasa, S. C. (1982). The hardy personality: Toward a social psychology of stress and health. In J. Suls & G. Sanders (Eds.), *Social psychology of health and illness* (pp. 30–32). Hillsdale, NJ: Erlbaum.

Kyriacou, C., & Sutcliffe, J. (1979). A note on teacher stress and locus of control. *Journal of Occupational Psychology, 52,* 227–228.

Lazarus, R. S. (1995). Psychological stress in the workplace. In R. Crandall & P. L. Perrewé (Eds.), *Occupational stress* (pp. 3–14). New York: Taylor & Francis.

Lee, C., Jamieson, L. F., & Earley, P. C. (1996). Beliefs and fears and type A behavior: Implications for academic performance and psychiatric health disorder symptoms. *Journal of Organizational Behavior, 17,* 151–177.

McLennan, J., Gotts, G. H., & Omodei, M. M. (1988). Personality and relationship dispositions as determinants of subjective well-being. *Human Relations, 41,* 593–602.

Meehl, P. E. (1971). High school yearbooks: A reply to Schwarz. *Journal of Abnormal Psychology, 77,* 143–148.

Meyerson, D. E. (2000). If emotions were honoured: A cultural analysis. In S. Fineman (Ed.), *Emotion in organizations* (2nd ed., pp. 167–183). Thousand Oaks, CA: Sage.

Moyle, P. (1995). The role of negative affectivity in the stress process: Tests of alternative models. *Journal of Organizational Behavior, 16,* 647–668.

Moyle, P., & Parkes, K. (1999). The effects of transition stress: A relocation study. *Journal of Organizational Behavior, 20,* 625–646.

Newton, T. J., & Keenan, A. (1990). The moderating effect of the type A behavior pattern and locus of control upon the relationship between change in job demands and change in psychological strain. *Human Relations, 43,* 1229–1255.

O'Driscoll, M. P., & Dewe, P. J. (2001). Mediators and moderators of stressor-strain linkages. In P. L. Perrewé & D. C. Ganster (Eds.), *Research in occupational stress and well-being: Vol. 1. Exploring theoretical mechanisms and perspectives* (pp. 257–287). New York: Elsevier.

O'Leary, A. (1990). Stress, emotion, and human immune function. *Psychological Bulletin, 108,* 363–382.

Parkes, K. R. (1990). Coping, negative affectivity, and the work environment: Additive and interactive predictors of mental health. *Journal of Applied Psychology, 75,* 399–409.

Plutchik, R. (1989). Measuring emotions and their derivatives. In R. Plutchik & H. Kellerman (Eds.), *Emotion: Theory, research, and experience: The measurement of emotions* (Vol. 4). Orlando, FL: Academic Press.

Rotter, J. B. (1966). Generalized expectancies for internal versus external control of reinforcement. *Psychological Monographs, 80*(1, Whole No. 609).

Sackett, P. R., & DeVore, C. J. (2001). Counterproductive behaviors at work. In N. Anderson, D. S. Ones, H. K. Sinangil, & C. Viswesvaran (Eds.), *Handbook of industrial, work, and organizational psychology* (Vol. 1, pp. 145–164). Thousand Oaks, CA: Sage.

Sauter, S. L., Lim, S. Y., & Murphy, L. R. (1996). Organizational health: A new paradigm for occupational stress research at NIOSH. *Japanese Journal of Occupational Mental Health, 4,* 248–254.

Schachter, S., & Singer, J. (1962). Cognitive, social, and physiological determinants of emotional state. *Psychological Review, 63,* 379–399.

Schaubroeck, J., Ganster, D. C., & Fox, M. L. (1992). Dispositional affect and work-related stress. *Journal of Applied Psychology, 77,* 322–335.

Schaubroeck, J., Judge, T. A., & Taylor, L. A., III. (1998). Influences of trait negative affect and situational similarity on correlation and convergence of work attitudes and job stress perceptions across two jobs. *Journal of Management, 24,* 553–576.

Shukla, A., & Upadhyaya, S. B. (1986). Similar jobs: Dissimilar evaluations. *Psychologia, 29,* 229–234.

Simon, H. (1967). Motivational and emotional controls of cognition. *Psychological Review, 74,* 29–39.

Spector, P. E. (1982). Behavior in organizations as a function of employee locus of control. *Psychological Bulletin, 91,* 482–497.

Spector, P. E. (1988). Development of the work locus of control scale. *Journal of Occupational Psychology, 61,* 335–340.

Spector, P. E. (1998). A control theory of the job stress process. In C. L. Cooper (Ed.), *Theories of organizational stress* (pp. 153–169). Oxford: Oxford University Press.

Spector, P. E., Chen, P. Y., & O'Connell, B. J. (2000). A longitudinal study of relations between job stressors and job strains while controlling for prior negative affectivity and strains. *Journal of Applied Psychology, 85,* 211–218.

Spector, P. E., Dwyer, D. J., & Jex, S. M. (1988). The relationship of job stressors to affective, health, and performance outcomes: A comparison of multiple data sources. *Journal of Applied Psychology, 73,* 11–19.

Spector, P. E., & Fox, S. (in press). An emotion-centered model of voluntary work behavior: Some parallels between counterproductive work behavior (CWB) and organizational citizenship behavior (OCB). *Human Resources Management Review.*

Spector, P. E., Fox, S., & Van Katwyk, P. T. (1999). The role of negative affectivity in employee reactions to job characteristics: Bias effect or substantive effect. *Journal of Occupational and Organizational Psychology, 72,* 205–218.

Spector, P. E., & Jex, S. M. (1998). Development of four self-report measures of job stressors and strain: Interpersonal Conflict at Work Scale, Organizational Constraints Scale, Quantitative Workload Inventory, and Physical Symptoms Inventory. *Journal of Occupational Health Psychology, 3,* 356–367.

Spector, P. E., & O'Connell, B. J. (1994). The contribution of individual dispositions to the subsequent perceptions of job stressors and job strains. *Journal of Occupational and Organizational Psychology, 67,* 1–11.

Spector, P. E., Zapf, D., Chen, P. Y., & Frese, M. (2000). Why negative affectivity should not be controlled in job stress research: Don't throw out the baby with the bath water. *Journal of Organizational Behavior, 21,* 79–95.

Spence, J. T., Helmreich, R. L., & Pred, R. S. (1987). Impatience versus achievement strivings in the type A pattern: Differential effects on students' health and academic achievement. *Journal of Applied Psychology, 72,* 522–528.

Spielberger, C. D. (1972). Anxiety as an emotional state. In C. D. Spielberger (Ed.), *Anxiety: Current trends in theory and research* (Vol. 1, pp. 23–49). Orlando, FL: Academic Press.

Spielberger, C. D. (1979). *Preliminary manual for the State-Trait Personality Inventory (STPI).* Unpublished manuscript, University of South Florida, Tampa.

Spielberger, C. D., Krasner, S. S., & Solomon, E. P. (1988). The experience, expression, and control of anger. In M. P. Janisse (Ed.), *Health psychology: Individual differences and stress* (pp. 89–108). New York: Springer.

Spielberger, C. D., Ritterband, L. M., Sydeman, S. J., Reheiser, E. C., & Unger, K. K. (1995). Assessment of emotional states and personality traits: Measuring psychological vital signs. In J. N. Butcher (Ed.), *Clinical personality assessment: Practical approaches* (pp. 41–58). New York: Oxford University Press.

Tanzer, N. K., Sim, C.Q.E., & Spielberger, C. D. (1996). Experience, expression, and control of anger in a Chinese society: The case of Singapore. In C. D. Spielberger, I. G. Sarason, J.M.T. Brebner, E. Greenglass, P. Laungani, & A. M. O'Roark (Eds.), *Stress and emotion: Anxiety, anger, and curiosity* (Vol. 16, pp. 51–65). New York: Taylor & Francis.

Wallbott, H. G., & Scherer, K. R. (1989). Assessing emotion by questionnaire. In R. Plutchik & H. Kellerman (Eds.), *Emotion: Theory, research, and experience: The measurement of emotions* (Vol. 4, pp. 55–82). Orlando, FL: Academic Press.

Watson, D., & Clark, L. A. (1984). Negative affectivity: The disposition to experience aversive emotional states. *Psychological Bulletin, 96,* 465–490.

Watson, D., & Pennebaker, J. W. (1989). Health complaints, stress, and distress: Exploring the central role of negative affectivity. *Psychological Review, 96,* 234–254.

Watson, D., Pennebaker, J. W., & Folger, R. (1987). Beyond negative affectivity: Measuring stress and satisfaction in the workplace. In J. M. Ivancevich & D. C. Ganster (Eds.), *Job stress: From theory to suggestion* (pp. 141–157). Binghamton, NY: Haworth Press.

Weiss, H. M. (in press). Deconstructing job satisfaction: Separating evaluations, beliefs, and affective experiences. *Human Resources Management Review.*

Williams, L. J., Gavin, M. B., & Williams, M. L. (1996). Measurement and non-measurement processes with negative affectivity and employee attitudes. *Journal of Applied Psychology, 81,* 88–101.

CHAPTER 3

Improving Worker Safety and Health Through Interventions

Michael J. Burke
Sue Ann Sarpy

The focus of this chapter is on how modifying individual characteristics through health and safety interventions affects the exhibition of safe work behaviors, the prevention of negative work
outcomes such as accidents and injuries, and consequently the enhancement of individual health and organizationally relevant outcomes. Interventions aimed at improving workplace health and
safety are premised on the need to change one or more of the
three primary elements in a work system—the worker, the work itself, or the work context—in order to achieve desired effects. Interventionist approaches are concerned with changing one of
these three elements through programs, procedures, or policies
designed to improve the knowledge, skill, and motivation of people (e.g., through occupational stress-reducing relaxation techniques), through the modification of the work itself (e.g., through
redesign of equipment or the content of work), or through the
modification of the work context (e.g., through a change in work-
group structures or reporting relationships).

By contrast, noninterventionist approaches deal with the three
primary elements of a work system as they exist, with the goal of
optimizing the fit among the elements. This strategy is premised

on the assumption that the better this fit, the more improved individual health and safety will be. In regard to individual differences, a primary focus of noninterventionist approaches is on the matching of people to work and work contexts through personnel recruitment and selection systems. As a result, these approaches are often concerned with the measurement and predictive utility of relatively stable individual differences such as personality characteristics, cognitive abilities, physical abilities, and psychomotor abilities.

In regard to promoting individual health and well-being, considerable bodies of literature exist within the behavioral sciences on interventions designed to affect one or more worker characteristics, work criteria, and work context variables. For instance, in relation to occupational stress-related interventions, the literature includes conceptual discussions concerning which worker characteristics (e.g., knowledge and skills for cognitively appraising stressors), work-related outcomes (e.g., accidents, blood pressure, burnout, health-care use, psychosomatic complaints), and work context variables (e.g., job demands) are expected to be affected by different types of stress management interventions (e.g., relaxation-focused biofeedback techniques) (see Ivancevich, Matteson, Freedman, & Phillips, 1990). Furthermore, considerable attention has been devoted to studying the relative effectiveness of different types of stress management interventions, and several meta-analytic investigations of the benefits of these interventions have been conducted (e.g., see van der Klink, Blonk, Schene, & van Dijk, 2001). These points are also true for a number of other areas of individual health promotion including topics not related to stress, such as nutritional counseling, smoking cessation programs, and accident prevention interventions (e.g., see Gebhardt & Crump, 1990; Wetter et al., 1998).

Yet the behavioral science literature contains very few studies on a common class of interventions intended to affect worker health and safety, namely, health and safety training programs. For the most part, studies that might contribute to furthering our understanding of relationships between worker characteristics, work criteria, and work context variables in relation to health and safety training have appeared in the public health, occupational medicine, safety, and engineering literatures. Importantly, both conceptual and empirical investigations within the latter literatures offer potentially useful insights for defining, measuring, and studying relations between worker characteristics, work criteria, and work context variables. Therefore, in this chapter we will focus on a review of the extant health and safety training literature in order to assess our current state of knowledge with respect to an integrated worker characteristic–work criteria–work context framework and identify where, with respect to estimating relationships in the cells of such a framework, future research might be directed.

Given the importance of safe work behaviors and worker health in many occupations and industries (e.g., construction, fast-food service, fire fighting, hazardous waste work, manufacturing, mining, protective services, and transportation), an appealing feature of attempting to enhance safe work behavior and worker health through health and safety training interventions is that worker knowledge, skill, and motivation are theorized to be directly related to safe work behavior (cf. Burke, Sarpy, Tesluk, & Smith-Crowe, 2002). In contrast, relatively stable individual differences such as cognitive abilities and personality traits that are assessed via noninterventionist strategies would be posited to underlie worker knowledge, skill, and motivation and thus to be indirectly related to engaging in safe work behavior and health-related outcomes. Our discussion is not intended to imply mutual exclusivity of interventionist and noninterventionist approaches to enhancing safe work behavior and worker health, which can and should, where appropriate, be used in combination. The point we are attempting to make is that interventionist strategies can, by themselves, directly and substantially contribute to enhancing safe work behavior and worker health.

Numerous types and methods of worker health and safety interventions (e.g., lecture-based fundamental knowledge programs, problem-solving skill-building simulations) have been employed in private and public sector organizations. Discussion of the effects of these interventions involves considering at least four distinguishable, although related, areas of research and practice: the description and manipulation of characteristics of workers, the description and measurement of work and work outcomes, methods for determining relationships or linkages between characteristics of workers (or manipulations of these characteristics) and work or work outcomes, and methods for translating the effects of safety and health training interventions into organizational productivity terms.

This chapter organizes these content areas in three sections. The first section provides a general framework for studying relations between worker characteristics affected by safety and health interventions, work-relevant criteria, and work context factors. The second section discusses how variables within this framework have been studied with respect to evaluations of safety and health-training interventions and presents a review of the effectiveness of these interventions. The third section is concerned with the limits of our current state of knowledge and practice and the types of research needed to expand these limits. In addition to the focus of this chapter, health and safety training, we should note that the general framework to be discussed here could be applied to the study or integration of research related to any other type of health and safety intervention such as occupational stress-related interventions or accident prevention programs.

This chapter is not intended to provide an exhaustive or meta-analytic review of each of the content areas noted. Rather, our aim is to concentrate on historical and recent developments that are, in our opinion, closest to the frontiers of the field and that have the most relevance to the issue of improving safe work behavior and individual and organizational health-related outcomes.

Classification of Workers, Work Criteria, and Work Contexts

Fundamental to the development and application of safety and health interventions is the need for appropriate taxonomic systems that classify and interrelate characteristics of the worker, work or work outcomes, and work contexts. Historically, taxonomic research concerning characteristics of the worker has resulted in significant progress toward the definition and factor structure of a number of human attribute domains (Murphy, 1996). Although research on characteristics of work (i.e., the nature of the content and processes) has produced a number of procedures for analyzing work content and processes, the development and test of work taxonomies or performance factor structures including those associated with safe work behavior is just beginning (e.g., see Burke et al., 2002; Marchand, Simard, Carpentier-Roy, & Ouellet, 1998). Likewise, taxonomic research with respect to work context factors affected by organizational and management policies, practices, and procedures is at an early stage of development (e.g., see James & McIntyre, 1996), with taxonomic research in other work context domains such as that concerned with organizational risk factors being at a more advanced level (e.g., Cohen & Colligan, 1998).

Moreover, and perhaps more critical from the perspective of this chapter, integrative research between these three approaches has been the exception. For instance, in the domain of workplace safety and health, no one has attempted to systematically study and establish linkages between characteristics of the worker, safety performance and health-related outcomes, and work context variables. Such linkages are critical in that they specify the degree to which we can expect a given worker characteristic to contribute to safe work and enhanced health and well-being (or the prevention of negative work outcomes) and under what work conditions.

One encounters at least three important issues in developing and operationalizing an integrated worker characteristic–work criteria–work context framework (Burke & Pearlman, 1988; Hattrup & Jackson, 1996). These issues are central to our understanding of and ability to accurately quantify the effects of safety and health interventions. First, what worker characteristics, work performance factors or outcomes, and work context variables should be included in the system? Addressing this question is basically a theoretical sampling issue concerning the representativeness and comprehensiveness of the system. The categories of worker characteristics should be identified such that, within any relevant type of work or work context, they are able to explain all or most of the predictable variance in work behavior. For instance, if the type of work and work context concerned fire fighting, would the categories of knowledge and skill be sufficient to account for most of the individual differences in actions that firefighters engage in, or would the inclusion of other characteristics such as values or interests enhance explanation?

The second issue is how general or specific these categories should be. This issue concerns the degree to which worker characteristic–

work criterion relationships are moderated by the level of specificity with which worker characteristics, criteria, and work context variables are defined and measured. Continuing our example, would the relationship between a knowledge variable such as knowledge of health and safety communication and different categories of safe fire-fighting behavior (such as properly using personal protective equipment, recognizing and evaluating hazards, responding to emergencies) be relatively constant or highly variable? If this relationship is relatively constant, then it implies that the taxonomy of safety performance is overly specific and hence less than optimal; that is, this pattern would imply that safety performance constructs could be considered at a more general level with little loss of information (at least with respect to their relationship with knowledge of health and safety communication).

The final issue encountered in the development of an integrated worker characteristic–work criteria–work context matrix is how the values in the matrix should be determined. Although this issue is primarily a methodological one concerning both the process by which such values are obtained and the quantitative index by which they are specified, the hypothesized nature of relationships in the matrix (i.e., linear or nonlinear) will dictate the use of particular methodologies and quantitative indices. We can broadly classify the methodological possibilities as being based on either direct judgment (e.g., expert ratings of the degree to which individual differences in given characteristics contribute to safe work behavior in given work contexts) or empirical research (e.g., experimental or field or correlational studies), with a variety of quantitative indices possible (e.g., correlation or regression coefficients, d-statistics, probabilities).

In theory, the number of cells representing potentially useful combinations of worker characteristics, work criteria, and work context variables can vary from a very few to an enormously large number. A determination of the number of useful combinations is fundamentally an issue that can be resolved only through empirical research. For the objective of organizing and presenting our critique of the literature on safety and health-training interventions, we have specified the worker characteristic–work criteria–work context matrix in Figure 3.1 at a general level.

Figure 3.1. A General Worker Characteristic-Work Criteria-Work Context Framework.

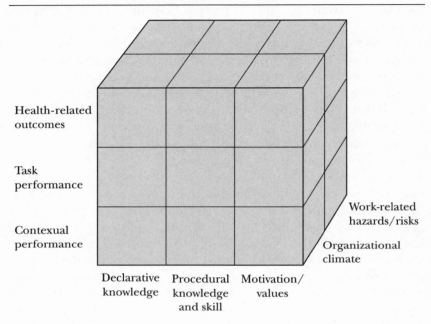

In Figure 3.1 we identify three broad worker characteristic categories: declarative knowledge, procedural knowledge and skill, and motivation/values. Following Anderson (1985) and Campbell (1990), declarative knowledge concerns the worker's understanding of safety and health-related work requirements and the labels, facts, and principles of work. Procedural knowledge and skill concerns knowing how to appropriately engage in work behavior or carry out the procedures of a defined task. Motivation is the combined effect from three choice behaviors: choice to expend effort, choice of effort level, and choice to persist at a particular level. Given that motivation is defined as a choice behavior and noting that knowledge, skill, and motivation are the only posited direct antecedents of safe work behavior in a particular work context, we would argue that health and safety knowledge and skill are the only two classes of worker characteristics manipulated by workplace health and safety interventions that have a direct impact on safe work behavior and health outcomes. Values reflect differences between individuals in preferences or what is important and are posited to underlie or serve as good indicators of worker motivation (Dawis, 1991).

Personal values are typically conceptualized as stable dispositions (Dawis, 1991). However, Burke, Borucki, and Hurley (1992) have argued that workers may desire to achieve an adaptive fit when the organization's espoused values differ from the workers' personal values. In this sense organizationally espoused values (e.g., "Safety is our highest priority"), coupled with management practices designed to reinforce such values, may engender the development or modification of schemas for making sense of the work environment as well as employee behavior. Health and safety training interventions are often explicitly and sometimes implicitly directed at modifying worker values in accordance with organizationally espoused values for health and safety (e.g., awareness training related to the importance of protecting the public and the environment or the importance of proper nutrition and exercise). Although interventions aimed at modifying worker motivation through changes in worker values would be expected to affect only the exhibition of work behaviors in an indirect manner (i.e., through motivation, choice behavior), the fact that organizationally espoused values are the subject of many health and safety-related interventions merits the inclusion of values in the preliminary framework. Recall that other more stable worker characteristics such as personality characteristics and cognitive abilities would not be included in Figure 3.1 because these factors are evaluated via noninterventionist strategies and are expected to underlie worker knowledge, skill, and motivation.

On the dependent variable or criterion side, we provisionally begin with broadly specified behavioral categories that are posited to be directly affected by workplace safety and health interventions: task performance and contextual performance. *Task performance* has been defined as behavior or actions concerned with transforming raw materials into products and services or servicing or maintaining the technical core (Borman & Motowidlo, 1993). *Contextual performance* has been defined as actions (such as helping, persisting, and following organizational rules) that support the social and psychological environment in which the technical core must function. In our view, work behaviors or actions can be of either a task or contextual nature and are likely to overlap with these types of performance along continua depending on the occupation or type of work being considered. For instance, within health-care occupations where safety concerns are an important aspect of the work, safety behaviors are likely reflected in task performance. Also, in some protective service occupations or security work, such as airport security or police work, safety behaviors of a contextual nature such as safely interacting with and helping coworkers and the public or supporting organizational objectives would be considered more routine activities.

In addition to conceptualizing work behavior in relation to both task and contextual activities, we assume that engaging in safe work behavior of a task or contextual nature can have a direct impact on individual health-related outcomes (i.e., the prevention of negative work outcomes such as accidents, illnesses, and injuries and the promotion of individual health and well-being). Although health outcomes such as illnesses and injuries are posited, for the most part, to be indirectly affected by individual-level safety-related interventions, a fair amount of health and safety training research includes these types of criteria. Therefore, we will focus on these variables as health-related outcomes in our review of the safety training literature. As we will discuss with respect to future research directions, an explicit assumption of our review is that work behavior and individual health outcomes underlie the achievement of more long-term organizational goals.

For the purposes of our review, we do not include two other sets of behaviors, which are the consequence of illness and injury, health-care use and recovery-related criteria. These criteria could be included depending on the purpose of primary or meta-analytic research on the effectiveness of certain health and safety interventions. The criteria that we will focus on in our review would be considered health maintenance behaviors and outcomes. This set of health maintenance criteria could be expanded, depending on the specific domain of interest to include wellness criteria (e.g., caloric intake, substance abuse, exercise) and stress-related criteria (e.g., blood pressure, somatic complaints, sleep disturbances).

In terms of work context variables, two broad classes of variables are expected to interact with worker characteristics to affect work performance and outcomes: organizational climate factors and organizational risk factors. Organizational climate factors refer to characteristics of the work environment and have been hypothesized to either causally affect variance in measures of performance (James, Demaree, Mulaik, & Ladd, 1992) or serve as moderators of worker characteristic–work performance/outcome relationships. Organizational risk factors refer to different types of work-related exposure risks (e.g., injury-producing forces, toxic chemicals, ergonomic stressors) recognized in various ways by the Occupational Safety and Health Administration (OSHA) standards that have been hypothesized to moderate worker characteristic–work performance relationships (Cohen & Colligan, 1998). We will summarize how work context variables, worker characteristics, and work criteria have been studied in relation to evaluating the effectiveness of health and safety training interventions.

Health and Safety Training Interventions

Organizations implement health and safety interventions, in particular safety-related training, in many cases in the United States and other industrialized countries (e.g., Canada, Germany, Sweden, and France) so as to be in compliance with government legislation and guidelines (e.g., see OSHA, 1998). The regulated nature of these interventions leads to a fair amount of conformity in terms of the content, objectives, and methods associated with workplace health and safety interventions. In particular, the highly regulated nature of safety-related work leads to a focus on health and safety training as a primary safety-related intervention in critical skills occupations (such as fire fighting, hazardous waste work, nuclear power, and protective services).

If one considers the framework that Hofmann and Tetrick proposed in Chapter One of this volume, emphasizing competing goals or objectives (i.e., intrinsic versus extrinsic and short-term versus long-term), then most of the health and safety training resides along the extrinsic, short-term dimensions. Health and safety training is largely designed to meet short-term goals (such as a change in worker knowledge) set by external regulatory agencies to ensure optimal safety-related performance in critical skills occupations. However, mandated health and safety training also focuses on more long-term goals such as refresher or certification training aimed at the maintenance of procedural skills and workers' ability to adapt to emergencies and changing scenarios. In addition, a fair amount of health and safety training is centered on the modification of worker values and attitudes to be consistent with internal, organizationally espoused safety values and the attainment of more long-term organizational outcomes.

In this section we initially discuss worker characteristics (i.e., knowledge, skills, and values) affected by health and safety training interventions and the methods for delivering these interventions. Then, we discuss research on the nature of safety performance and the role of work context variables in the transfer of health and safety training, with an emphasis on the role of safety climate. Finally, we discuss the effectiveness of health and safety training in relation to both behavioral safety performance and health-related outcomes. Together, these discussions will summarize the current state of our knowledge related to specifying and quantifying aspects of an integrated worker characteristic–work criteria–work context matrix.

Worker Characteristics and Types of Safety-Related Training

Health and safety training interventions can be delineated according to four broad and logically related areas as indicated by a program's content and behavioral objectives:

- Fundamental knowledge, skills, and values
- Recognition and awareness
- Problem-solving/analytical skills
- Decision-making skills

These categories of worker characteristics (or types of safety training) are consistent with the four categories for classifying training suggested by the Office of Technology Assessment (OTA) (1985), namely, fundamentals programs, recognition programs, problem-solving programs, and empowerment programs. Our categorization differs from the OTA groupings in that our categories are more closely aligned with the declarative knowledge, procedural knowledge and skills, and values imparted as a trainee progresses from basic to more advanced safety-related training. That is, one must be knowledgeable of the essential aspects of worker protection from hazards as addressed in fundamentals programs to progress to awareness of potential hazards in recognitions programs. Often, trainees then proceed to problem-solving/analytical skill building to address workplace hazards and finally to the development of decision-making skills related to making the workplace free from injury and illness.

By definition, fundamentals programs are the most basic type of health and safety training. Training in this area provides instruction on specified rules and procedures regarding known hazards to prevent work-related illness and injury. These programs emphasize the development of declarative knowledge such as knowledge of emergency processes and an understanding of the need for medical monitoring. These programs also focus on procedural knowledge and skills such as the proper use and maintenance of potentially hazardous tools, equipment, and materials; the use of personal hygiene measures; and the use of protective devices (e.g., masks and respirators).

The literature is replete with examples of fundamentals programs ranging from using personal protective equipment (Komaki, Heinzmann, & Lawson, 1980; Sadler & Montgomery, 1982), housekeeping (Komaki, Barwick, & Scott, 1978), operating machines (Chhokar & Wallin, 1984), to defusing potentially violent patients (Carmel & Hunter, 1990). Further, these health and safety interventions are conducted within many different types of work, including mining (Fiedler, 1987; Fiedler, Bell, Chemers, & Patrick, 1984), agriculture (Barnett et al., 1984), manufacturing (Komaki et al., 1980; Reber & Wallin, 1984), transportation (Saarela, Saari, & Alltonen, 1989; Saari & Nasanen, 1989), and public services including health care (Ewigman, Kivlahan, Hosokawa, & Horman, 1990; Foster, 1996; Linnemann, Cannon, DeRonde, & Lamphear, 1991).

Building on the technical knowledge and skills acquired in the fundamentals programs, the principal objective of recognition programs is to train workers to recognize and appropriately report these hazards. Accordingly, recognition and awareness programs emphasize being aware of workplace hazards, understanding methods available for hazard elimination or control, collecting information about workplace hazards, observing or informally inspecting the workplace for potential hazards, appropriately reporting hazards or potential hazards, and knowing employee rights and responsibilities under the law (i.e., right-to-know laws). Programs developed in response to the promulgation of the Hazardous Communication Standard (29 C.F.R. §1910.1200) are included in this category (Robins & Klitzman, 1988; Robins, Hugentobler, Kaminski, & Klitzman, 1990). For example, Caparaz, Rice, Graumlich, Radike, and Morawetz (1990) used this type of training to enhance worker knowledge of hazardous materials handling in a foundry. The intervention included instruction in safe handling procedures and spill control measures and review of the material safety data sheets to increase workers' knowledge and skill in hazard recognition and reporting.

After hazard recognition and reporting, the major objective of the problem-solving/analytical skill-building interventions is to train workers to use the appropriate mechanisms to address workplace health and safety problems. Problem-solving/analytical interventions tend to highlight worker-directed or participatory approaches in recognition and control of hazards. These programs emphasize procedural knowledge and skills necessary to participate in hazardous identification and control activities in the workplace including using union and management resources and, when necessary, eliciting support from outside agencies such as OSHA and the National Institute for Occupational Safety and Health. In an effort to promote more effective health and safety measures, Lapping and Parsons (1980) developed a problem-solving/analytical skill-building program for managers and workers in the construction industry. The program included instruction in hazard identification and reporting as well as the role of labor, management, and OSHA in effecting workplace health and safety improvements.

Health and safety decision-making interventions provide instruction in a broad range of procedural knowledge and skills related to involving workers in defending and expanding their rights to a safe and healthful workplace. The overarching goal of this type of intervention is to transfer political and economic power from employers and managers to workers. Health and safety decision-making interventions emphasize the entire array of knowledge and skills associated with fundamentals, recognition, and problem-solving interventions. Further, these programs focus on the development of skills such as observing and interpreting air monitoring and medical tests, tracking trade names of potentially hazardous chemicals, devising engineering control methods, and negotiating effective health and safety contract language. Therefore, the major objective of decision-making interventions is to have an impact on social action by empowering workers to prevent work-related illness and injury in their workplaces through collective action (Luskin, Somers, Wooding, & Levenstein, 1992; Wallerstein & Weinger, 1992; Wallerstein & Baker, 1994). Although decision-making or empowerment programs were originally associated with hazardous waste training in response to the Hazardous Waste Operations and Emergency Response Standard (29 C.F.R. §1920.120), they have been effectively developed and implemented in various work domains including agriculture (Weinger & Lyons, 1992), construction (Baker, Stock, & Szudy, 1992), and public services (Wands & Yassi, 1992).

In addition to type of training intervention, the method used to deliver the training also is relevant to the categorization of health and safety interventions and refinement of a framework such as that in Figure 3.1. Methods range from passive information-based techniques such as lecture, to computer-based programmed instruction, to learner-center performance-based techniques such as hands-on demonstrations or virtual reality training.

Lecture is the most basic and commonly used method to present factual health and safety-related information. Computer-based instruction is also designed to present factual information on a personal computer. This method offers flexibility and convenience in terms of scheduling training sessions. Computer-based instruction has been created for the entire gamut of workplace health and safety topics including occupational safety, industrial safety systems, fire protection, hazardous materials and waste, industrial hygiene, risk management and safety engineering and design (Huddock, 1994).

Methods such as hands-on demonstrations associated with behavioral simulations are learner-centered and require active participation from the trainee. Behavioral simulations are approximations to real-life situations and are under the control of the trainer. The control facilitates the ability to adjust stimuli that confront the trainee and provide timely feedback to the trainee. Burke, Bradley, and Bowers (in press) present a detailed discussion of behavioral simulations including an overview of the United States's most advanced behavioral simulation complex, the Hazardous Materials Management and Emergency Response (HAMMER) Volpentest Training and Education Center at the Department of Energy site in Hanford, Washington. When more than one method of delivery is employed to present information, this combined method is referred to as multiple techniques. An example of health and safety training with extensive use of multiple techniques would be Brown and Nguyen-Scott's use (1992) of small group discussions, sharing work life experiences, hands-on practice, brief lectures, and responding to mock incidents.

We should note that the type of knowledge or skill and the instructional method used to deliver the training are often inextricably linked. For instance, lectures may be appropriate when delivering fundamental knowledge, skills, and values training, whereas multiple methods are more appropriate for the development of health and safety decision-making skills. Therefore, not only the worker characteristic (or type of training) but also the method should be considered when examining the effectiveness of the health and safety training. In Table 3.1, we present a matrix with all possible combinations of intervention type and method. The intervention types and methods in Table 3.1 do not necessarily represent mutually exclusive categories but are developed to identify areas where primary emphasis has been placed with respect to the content and method of health and safety training.

Work Criteria and Work Context Variables Studied

As we emphasized, the modification of task and contextual behavior is a key objective of many health and safety training interventions. In terms of task-relevant behavior, employees are required to display core, basic safety behaviors to some degree across jobs in certain industries such as manufacturing, mining, chemical processing, nuclear power plant operations, municipal public services (fire, police, and emergency medical services), and others. In the literature we have cited on safety-related interventions and in the growing body of research on behavioral aspects of safety, a number

Table 3.1. An Intervention Type by Intervention Method Matrix.

	Lecture	Programmed Instruction (Including Computer-Based)	Small Group Techniques	Demonstrations (Behavioral Modeling)	Role-Playing Exercises	Simulations (Hands-On)	Virtual Reality	Multiple Techniques
Fundamental knowledge, skills, and values	X	X	X	X	X	X	X	X
Recognition and awareness knowledge and skills	X	X	X	X			X	X
Problem-solving and analytical knowledge and skills								X
Decision-making knowledge and skills								X

Note: Knowledge or skill areas (i.e., intervention types) where a method or combination of methods is typically applied are denoted by X's.

of researchers have measured general safety behaviors in different industries (e.g., Cheyne, Cox, Oliver, & Tomas, 1998; Chhokar, 1990; Griffin & Neal, 2000; Hofmann & Stetzer, 1996; Lingard & Rowlinson, 1997; McDonald, Corrigan, Daly, & Cromie, 2000; Rudmo, 2000). For the most part, these researchers have focused on the measurement of employee safety compliance, that is, the extent to which employees adhere to safety procedures and carry out work in a safe manner. Notable exceptions are the work of Hofmann, Morgeson, and Gerras (in press), which focused on an overall measure of safety citizenship behavior, and that of Marchand et al. (1998), which focused on both safety compliance and safety initiatives.

Importantly, with only a few exceptions (cf. Griffin & Neal, 2000; Marchand et al., 1998; Neal, Griffin, & Hart, 2000; Wu & Hwang, 1989), researchers conceptualize and measure safety performance with respect to a single, overall scale. Although overall measures of safety performance may be quite helpful in some contexts, an examination of the dimensionality of safety performance within an industry may assist in refining the task performance category in Figure 3.1 and tests of more refined hypotheses in safety intervention studies.

Recently, Burke et al. (2002) conducted confirmatory factor analytic tests of a four-factor model of general safety performance with data gathered from 550 hazardous waste workers. Four general safety performance factors of Burke et al.'s General Safety-Performance Scale were confirmed and labeled: using personal protective equipment, engaging in work practices to reduce risk, communicating health and safety information, and exercising employee rights and responsibilities. These findings offer progress toward a taxonomy of general safety performance by directly addressing the criterion dimensionality issue for task performance.

Furthermore, a fair amount of literature exists on the dimensionality of safety climate. However, scholars do not agree on the specific first-order factors that constitute safety climate (cf. Griffin & Neal, 2000). At the individual level of analysis, safety climate has been measured from a number of perspectives, including management concern for employee well-being (Brown & Holmes, 1986), management attitudes toward safety (Dedobbeleer & Beland, 1991; Zohar, 2000), employee perceptions of production and safety trade-offs (Hofmann & Stetzer, 1996), and employee perceptions of how various organizational characteristics adversely affect safety (Clarke, 1999). Factor analyses of employee safety climate perceptions have produced different numbers and types of factors (Flin, Mearns, O'Connor, & Bryden, 2000).

Despite the lack of consensus concerning the dimensionality of safety climate in the applied psychology and safety literatures, Burke, Borucki, and Kaufman (in press) have discussed how to use the multiple-stakeholder model of climate to conceptualize and measure generalizable climate dimensions with safety as the strategic focus. Their suggestions may prove helpful in refining the organizational climate domain of Figure 3.1. In addition, scholars generally agree on the organizational risk factors that workers encounter across occupations as defined by OSHA (e.g., injury-producing forces, toxic chemicals, ergonomic stressors) (see Cohen & Colligan, 1998), and these categories can be used to further refine the organizational risk domain in Figure 3.1.

Effectiveness of Health and Safety Training Interventions

With respect to the effectiveness of fundamental knowledge, skills, and values interventions, a considerable body of literature provides support for linkages between these types of intervention and increases in positive health and safety-related attitudes; increases in the exhibition of safe work behaviors; and reductions in accidents, illnesses, and injuries (Carmel & Hunter, 1990; Chhokar & Wallin, 1984; Ewigman et al., 1990; Foster, 1996; Komaki et al., 1978, 1980; Reber & Wallin, 1984). Notably, the effectiveness of fundamentals programs has been demonstrated across a variety of work hazards including injury, chemical, physical, and biological.

However, fundamental knowledge, skills, and values interventions has been least effective in reducing improper lifting techniques and related back injuries associated with ergonomic hazards (Scholey, 1983; St. Vincent, Tellier, & Lortie, 1989; Stubbs, Buckle, Hudson, & Rivers, 1983). Given the complex etiology of back disorders and controversies regarding safe lifting positions, these latter results are not surprising (cf. St.Vincent et al., 1989). The limited evidence supporting the effectiveness of ergonomic fundamental knowledge and skills training also highlights the important role that situational factors may play in the transfer of this type of training. For example, Carlton (1987) reported that layout of the kitchen and rapid pace of food preparation work resulted in food service workers' failure to demonstrate the proper lifting techniques learned in training on the job. Similar effects were found for warehouse workers who received training on five specific lifting principles that minimize stress in lifting and moving boxes (Chaffin, Gallay, Wooley, & Kuciemba, 1986). Results revealed that certain lifting practices were more difficult to execute on the job due to workstation layout and package size. Together, these studies underscore the notion that health and safety intervention success in increasing safe work practices depends in part on reducing or removing situational obstacles to the transfer of training.

A number of studies examining the effectiveness of recognition and awareness programs have found positive effects for this type of training with respect to enhancing knowledge and skills for hazard awareness, recognition, and reporting and reduction of exposures and injuries (Fox, Hopkins, & Anger, 1987; Maples, Jacoby, Johnson, Ter Haar, & Buckingham, 1982; Millican, Baker, & Cook, 1981). For example, using the small groups method, Caparaz et al. (1990) demonstrated an increase in posttest scores of knowledge and skill in hazard recognition and reporting (e.g., safe handling procedures and spill control measures, material safety data sheets) immediately following training for foundry workers. Due to the slight decreases in knowledge two weeks following training as well as trainees' suggestions at posttraining interviews, the authors suggest the need to consider the schedule of refresher training for the recognition and awareness programs, particularly training associated with the Hazardous Communication Standard (29 C.F.R. § 1910.1200). Although several recent studies have examined relationships between the frequency of refresher training and the exhibition of safe work behaviors (Burke et al., 2002; McClendon, Butz, & Zusman, 2001), a number of issues pertaining to the maintenance of health and safety training (e.g., optimal amount, massed versus distributed nature of refresher training), including that for recognition and awareness programs, are unresolved.

Finally, studies have provided support for the effectiveness of the problem-solving/analytical and health and safety decision-making programs with respect to enhancing the knowledge and skills necessary to actively participate in improving workplace health and safety (Lapping & Parsons, 1980; Lippin, Eckman, Calkin, & McQuiston, 2000; Luskin et al., 1992; McQuiston et al., 1994). As we noted, many of these programs emphasize participatory training and use multiple techniques in their delivery. For instance, Brown and Nguyen-Scott (1992) reported that trainee interviews conducted three months and 12 months after health and safety training indicated improvement with respect to reporting and correcting health and safety problems. However, as Cole and Brown (1996) discussed in a follow-up study, perceived management support had a substantial impact on the transfer of health and safety training. In cases where workers perceived their management to be supportive of their health and safety concerns, action was attempted for 83% of the identified problems, whereas corrective action was attempted for only 67% of the problems in workplaces where management was not perceived to be supportive. These results suggest that while the problem-solving/analytical and health and safety decision-making interventions can have positive effects on health and safety in the workplace, perceived management support plays a vital role in maximizing this effect and should be considered when developing, implementing, and evaluating these interventions.

In summary, substantial evidence supports the effectiveness of the various types and methods of health and safety training interventions (i.e., fundamental knowledge, skills, and values programs; recognition and awareness programs; problem-solving/analytical skill-building programs; and decision-making skill-building programs) to enhance safety-related task performance and thereby reduce workplace injuries and accidents. However, scholars have questioned the durability of these effects, and future research is needed on issues such as the optimal amount of refresher training for maintenance of health and safety training. Although numerous studies have demonstrated the impact of these interventions on safety-related knowledge and workplace behavior, injuries, and accidents, less emphasis has been placed on examining the effects of health and safety interventions with respect to reducing safety-related illnesses (Cohen & Colligan, 1998). In addition, a paucity of studies has included contextual performance criteria (see Hofmann, Morgeson, & Gerras, in press; Marchand et al., 1998), and greater attention should be given to these criteria in intervention effectiveness research. Finally, although the role that organizational factors play in the success of these programs and interventions has been established, further examination of the moderating role of work context variables with respect to worker characteristic–safety performance (or health-related outcome) relationships is warranted.

Extending What We Know About Health and Safety Training Interventions

The previous review has identified several gaps in our knowledge of the effectiveness of health and safety training interventions and areas of needed research. In this section we identify and discuss additional research directions for expanding our knowledge of an integrated work characteristic–work criteria–work context framework. In doing so we will highlight emerging methods of health and safety training, criterion dimensionality issues, questions concerning the implicitly assumed linear nature of the relationship between worker characteristics and safe work, the need for research on the role of organizational safety climate in the transfer of health and safety training and the effectiveness of other safety-related interventions, and research directions for enhancing our understanding of the productivity and economic gains associated with effective health and safety interventions.

Questions Concerning Emerging Health and Safety Training Interventions

One of the more promising methods for developing safety-related knowledge and skill is the use of virtual reality technology. Like the behavioral simulations we have discussed, virtual reality training is a computer-created environment used for training safe work practices in hazardous or unsafe conditions. Historically, virtual reality has been used as a training method in the military and the airline industry (e.g., flight simulators). More recently, other industries, including construction and mining, have adopted virtual reality software for health and safety training. For example, Hadipriono, Larew, and Barsoum (1996) created software named Safety in Construction Using Virtual Reality as a method to provide instruction and practice to construction laborers in preventing falls from scaffolding. Using a head-mounted display and a cyberglove or mouse, the trainee acts as a construction worker who must not only correctly erect a scaffolding platform (i.e., install each of the components using the proper sequence and position) but also appropriately inspect hazardous platforms (i.e., detect and eliminate the hazardous condition). Thus, the technology allows for on-the-job training on simulated construction platforms, one of the most hazardous conditions in all industries.

Filigenzi, Orr, and Ruff (2000) developed another innovative virtual reality program that provides instruction in hazardous recognition and evaluation routes and procedures. The program was created for surface and underground mine workers and emergency personnel. The software program creates a simulated mine emergency to which the trainee must appropriately respond using correct procedures and routes. During the hazardous recognition exercise, the trainee must identify and avoid hazards while navigating through the mine. Again, virtual reality allows the worker to experience the consequences of safe and unsafe work practices without risk of injury or fatality. Because these virtual reality software programs can be readily adapted to provide instruction in other industries, the use of virtual reality as a method for providing health and safety training is expected to increase.

Although virtual reality training and other multimedia computer-based forms of instruction offer hope for enhancing safety-related training in critical skills occupations as well as other types of safety-related work, a number of questions remain. First, assessments of the effectiveness of these forms of training relative to the criteria in Figure 3.1 are needed. Second, no one has addressed the relative effectiveness of these methods of safety training in relation to other methods of safety instruction. Finally, questions concerning the feasibility and cost-effectiveness of these methods of training especially for more routine safety-related work have yet to be addressed.

Questions Concerning the Dimensionality of Behavioral Safety Performance

As we discussed, Burke et al. (2002) have presented a model of safety performance that includes four general safety dimensions. Although the four general safety factors provide a starting point for future research on the dimensionality of general safety performance in other

types of work, the confirmation of these factors should not preclude efforts to study more specific safety performance factors in
either hazardous waste work or other types of safety-related work
or to conduct research to further evaluate the generalizability of
the four general factors. Certainly, the possibility that more specific safety performance factors could assist in explaining safety behaviors in critical skills occupations and other types of work such
as nursing or public service is very likely. For instance, recognizing
and evaluating hazards and responding to emergencies are two
possible performance factors that could be the subject of confirmatory factor-analytic efforts to explain performance variability in
other types of safety-related work such as fire fighting or emergency medical services. The consideration of more specific safety
performance factors in other types of work would occur, we hope,
along with the manipulation of more specific, isomorphic knowledge and skill variables in safety-related interventions aimed at improving performance along these dimensions.

   In addition, as we noted, the study of contextually oriented performance in the safety domain would be informative. Marchand et
al.'s confirmatory factor-analytic work (1998) on a sample of 828
workers drawn from nine manufacturing facilities suggests that a
bidimensional model of safety performance (with two correlated
factors related to safety compliance and safety initiatives dimension)
may provide greater explanatory power than a unidimensional factor
model. Clearly, exploratory factor analyses of contextually oriented
behaviors and narrative reviews within the applied psychology domain would support provisional assumptions about a general factor structure for contextual performance. That is, as Borman and
Motowidlo (1993) discuss, the factors of persisting with extra effort,
volunteering to carry out activities that are not formally part of one's
job, helping and cooperating with others, following organizational
rules and procedures, and supporting organizational objectives may
be applicable across a variety of types of safety-related work.

   In many types of safety-related work, we would expect contextually oriented behaviors such as helping others and following organizational rules to overlap highly with typical task performance. For
instance, within hazardous waste work, a high level of importance is
placed on workers' adherence to the buddy system of assisting each
other with respect to donning, doffing, and decontaminating per-

sonal protective equipment. Likewise, given that safety-related work in most industries is highly regulated, properly following rules, regulations, and procedures is an important aspect of one's work. Clearly, further research examining the factor structure of contextually oriented behavior in safety-related work and relations between Burke et al.'s (2002) four general safety performance factors and contextually oriented factors would be helpful in further refining the domain of safety performance and the framework in Figure 3.1.

Questions Concerning the Nature of Worker Characteristic–Safety Performance Relations

For the most part, research examining worker characteristic–safety performance relationships has assumed that these variables are linearly related. However, for a sample of fire fighters, Jacobs, Hofmann, and Kriska (1990), found a nonlinear relationship between months of job tenure and supervisory ratings of performance, with the plateau occurring at 60 months. More recently, Burke et al. (2002) reported in a series of exploratory analyses that general safety performance for workers in over 23 hazardous waste jobs could, in many of the cases, also be best predicted from knowledge and skill variables by a "best-fitting" line that was nonlinear. The form of nonlinear relationship was consistent with respect to the depth of knowledge–safety performance relationships. That is, a monotonically increasing prediction line that showed an initially steep slope and then tended to plateau with greater amounts of depth of knowledge and skill adequately described the overall relationships.

The possibility that safety knowledge–safety performance relationships are nonlinear over time is consistent with findings reported in the applied psychology literature with respect to relations between job experience and performance (Farrell & McDaniel, 2001; Schmidt, Hunter, & Outerbridge, 1986; Vineberg & Taylor, 1972). Overall, these findings suggest that researchers examining relationships between more proceduralized knowledge or skills and safe work behavior carefully consider theoretical rationales pertaining to possible nonlinear relationships between these types of variables and conduct appropriate nonlinear analyses to study hypothesized nonlinear relations.

Questions Concerning the Moderating
Effect of Organizational Safety Climate

Within the domain of safety performance, several studies have examined relations between climate factors and safety performance or outcomes (e.g., Griffin & Neal, 2000; Hofmann & Stetzer, 1996; Zohar, 2000). For individuals working in a chemical processing plant, Hofmann and Stetzer (1996) found that teams perceiving a stronger safety climate also reported engaging in fewer unsafe behaviors. Also, at the group level of analysis, Zohar (2000) found that safety climate predicted microaccidents for production workers in a metal-processing plant. Similarly, Griffin and Neal (2000) found that manufacturing and mining employees' perceptions of safety climate were meaningfully related to employees' safety knowledge and the extent to which employees engaged in safe work behaviors. In addition, the emphasis placed on safety during work-group meetings and the communication and coordination of work groups has been shown to influence safety through the development of norms about approaching coworkers engaged in unsafe behaviors and planning work in a way that allows for safe performance (Helmreich & Foushee, 1993; Wright, 1986). Together, the studies examining relations between safety climate and performance, at the individual and group levels of analyses, suggest that climate may moderate the effectiveness of safety-related interventions and thus relations between safety knowledge or skill and safety performance.

Recently, Smith-Crowe, Burke, and Landis (2001) examined organizational climate for the transfer of safety training as a potential moderator of safety knowledge–safety performance relationships for individuals working in two organizations in the nuclear waste industry. Their results were consistent with the hypothesis that safety knowledge–safety performance relationships would be stronger in the less restrictive (more supportive) organizational climate. More specifically, their results suggested that organizations should attend to management attitudes toward safety, the appropriateness of health and safety training, and overt attempts to discourage the transfer of health and safety training as potential moderating climate factors. Their findings concerning management attitudes, the appropriateness of health and safety training, and the impact of situational constraints are consistent with a growing body of literature suggesting the importance of these factors in the transfer of safety training in industrial settings (Chaffin et al., 1986; Cole & Brown, 1996; Ray & Bishop, 1995).

Smith-Crowe et al.'s findings (2001) indicate that the work context domain of Figure 3.1 could be refined to not only include key organizational safety climate dimensions but also incorporate a category reflective of situational factors that inhibit the exhibition of safe work behavior. This latter suggestion is consistent with Burke, Borucki, and Kaufman's commentary (in press) on climate dimensions that would be expected to generalize across settings. Future research examining the potential moderating effect of these work context factors when examining relationships between worker characteristics and safe work behavior would be informative as well as substantively address the call in the National Occupational Research Agenda (U.S. Department of Health and Human Services, Public Health Services, Centers for Disease Control and Prevention, & National Institute for Occupational Safety and Health, 1996) for research on characteristics of healthy work organizations.

Questions Concerning the Meaning of Safety Climate

Although the measurement of organizational safety climate has traditionally focused on worker perceptions of how the work environment personally affects the worker's well-being and safety performance (Flin et al., 2000), we encourage efforts to conceptualize and measure safety climate with respect to other key stakeholders as well. That is, at either the individual level or organizational level of analysis, the conceptualization and measurement of safety climate could be expanded to include not only how characteristics of the work environment affect the personal well-being of workers but also how characteristics of the work environment affect the well-being of other relevant stakeholder groups (e.g., the public, clients or customers) as well as the environment where applicable.

In particular, at the individual level of analysis, we would posit that when an organization espouses strong values related to the health and safety of different stakeholders, workers develop higher-order value-based schemas associated with the well-being of these organizational stakeholder groups (see Burke et al., 1992). For instance, if an organization has espoused values related to the health and safety of workers and the public as well as having policies and procedures in place to support workers' conformance to these organizationally espoused values, then one might expect workers' perceptions of the work environment (i.e., safety climate) to be hierarchically organized with respect to a concern for employees and a concern for the public. Furthermore, worker behavior would be posited to be focused on the exhibition of safe work behaviors directed toward these two stakeholder groups. Essentially, we are arguing that the present measurement of safety climate at either the individual level or organizational level of analysis may be deficient. Studying the role of safety climate in promoting workplace health and safety could be improved from a multiple-stakeholder perspective in a number of critical skills occupations including those involving public services.

Questions Concerning the Practical Utility and Cross-Situational Effects of Health and Safety Interventions

A number of research directions exist for enhancing our understanding of the practical implications and cross-situational effects of health and safety interventions. In terms of examining the degree to which the effects of health and safety interventions generalize across situations, there is a need for researchers to more adequately report basic descriptive statistics (i.e., means and standard deviations) on criteria by training or intervention conditions. These statistics will assist in future research efforts to meta-analytically integrate literatures, especially with respect to examining the relative effectiveness of health and safety interventions.

In regard to individual or primary studies, basic descriptive statistics can also be employed to estimate or compute common language statistics (e.g., d-values, other common language statistics) (see McGraw & Wong, 1992; Vargha & Delaney, 2000) for judging practical gains from health and safety interventions. Furthermore, the common language statistics such as d-values provide the basic information for researchers and practitioners to express the outcomes of health and safety interventions in terms of productivity gains or losses (e.g., dollars, percent improvement in performance, number of lost work days due to on-the-job injuries and illness) via decision-theoretic utility analyses (see Cascio, 2000). In this sense the effectiveness of health and safety interventions can be linked to more long-term organizational goals and functioning. Importantly, results from utility analyses can be used to guide decisions about where to more effectively allocate health and safety intervention resources.

Conclusion

Research in the domains of workplace safety and occupational health psychology has contributed substantially to our understanding of the effectiveness of health and safety interventions and the progressive development of a worker characteristic–work criteria–work context framework. We discussed numerous research prospects concerning the potential future contribution of health and safety training research for enhancing our understanding of relationships in such a framework as well as the practical improvement from such interventions. Our future success in promoting the exhibition of safe work behavior and reduction in the negative consequences of unsafe behavior will largely be a function of the extent to which we can continue to improve our conceptualizations, calculations, and communication of the effectiveness of health and safety interventions.

References

Anderson, J. R. (1985). *Cognitive psychology and its implications* (2nd ed.). New York: Freeman.

Baker, R., Stock, L., & Szudy, B. (1992). Hardware to hard hats: Training workers for action from offices to construction sites. *American Journal of Industrial Medicine, 22,* 691–701.

Barnett, P. G., Midtling, J. E., Velasco, A. R., Romero, P., O'Malley, M., Clements, C., Tobin, M. W., Wollitzer, A. O., & Barbaccia, J. C. (1984). Educational intervention to prevent pesticide-induced illness of field workers. *Journal of Family Practice, 19,* 123–125.

Borman, W. C., & Motowidlo, S. J. (1993). Expanding the criterion domain to include elements of contextual performance. In N. Schmitt & W. C. Borman (Eds.), *Personnel selection in organizations* (pp. 71–98). San Francisco: Jossey-Bass.

Brown, M. P., & Nguyen-Scott, N. (1992). Evaluating a training-for-action job health and safety program. *American Journal of Industrial Medicine, 22,* 739–749.

Brown, R. L., & Holmes, H. (1986). The use of a factor-analytic procedure for assessing the validity of an employee safety climate model. *Accident Analysis and Prevention, 18,* 455–470.

Burke, M. J., Borucki, C., & Hurley, A. (1992). Reconceptualizing psychological climate in a retail service environment: A multiple-stakeholder perspective. *Journal of Applied Psychology, 77,* 717–729.

Burke, M. J., Borucki, C., & Kaufman, J. (in press). Contemporary perspectives on the study of psychological climate: A commentary. *European Journal of Work and Organizational Psychology.*

Burke, M. J., Bradley, J., & Bowers, H. N. (in press). Health and safety training programs. In J. E. Edwards, J. Scott, & N. S. Raju (Eds.), *The human resources program-evaluation handbook.* Thousand Oaks, CA: Sage.

Burke, M. J., & Pearlman, K. (1988). Recruiting, selecting, and matching people to jobs. In J. P. Campbell & R. J. Campbell (Eds.), *Productivity in organizations* (pp. 97–142). San Francisco: Jossey-Bass.

Burke, M. J., Sarpy, S. A., Tesluk, P. A., & Smith-Crowe, K. (2002). General safety performance: A test of a grounded theoretical model. *Personnel Psychology, 55,* 429–457.

Campbell, J. P. (1990). Modeling the performance prediction problem in industrial and organizational psychology. In M. D. Dunnette & L. M. Hough (Eds.), *Handbook of industrial and organizational psychology* (pp. 687–732). Palo Alto, CA: Consulting Psychologists Press.

Caparaz, A., Rice, C., Graumlich, S., Radike, M., & Morawetz, J. (1990). Development and pilot evaluation of a health and safety training program for foundry workers. *Applied Occupational Environmental Hygiene, 5,* 595–603.

Carlton, R. S. (1987). The effects of body mechanics instruction on work performance. *American Journal of Occupational Therapy, 41,* 16–20.

Carmel, H., & Hunter, M. (1990). Compliance with training in managing assaultive behavior and injuries from inpatient violence. *Hospital Communication Psychiatry, 41,* 558–560.

Cascio, W. F. (2000). *Costing human resources: The financial impact of behavior in organizations.* Cincinnati, OH: South-Western College.

Chaffin, D. B., Gallay, L. S., Wooley, C. B., & Kuciemba, S. R. (1986). An evaluation of the effect of a training program on worker lifting postures. *International Journal of Industrial Ergonomics, 1,* 127–136.

Cheyne, A., Cox, S., Oliver, A., & Tomas, J. M. (1998). Modeling safety climate in the prediction of levels of safety activity. *Work and Stress, 12,* 255–271.

Chhokar, J. S. (1990). Behavioral safety management. *Vikalpa, 15,* 15–22.

Chhokar, J. S., & Wallin, J. A. (1984). A field study of the effect of feedback frequency on performance. *Journal of Applied Psychology, 69,* 524–530.

Clarke, S. (1999). Perceptions of organizational safety: Implications for the development of safety culture. *Journal of Organizational Behavior, 20,* 185–198.

Cohen, A., & Colligan, M. J. (1998). *Assessing occupational safety and health training* (DHHS/NIOSH Publication No. 98–145). Cincinnati, OH: National Institute for Occupational Safety and Health.

Cole, B. L., & Brown, M. P. (1996). Actions on worksite health and safety problems: A follow-up survey of workers participating in a hazardous waste worker training program. *American Journal of Industrial Medicine, 30,* 730–743.

Dawis, R. V. (1991). Vocational interests, values, and preferences. In M. D. Dunnette & L. M. Hough (Eds.), *Handbook of industrial and organizational psychology* (pp. 833–871). Palo Alto, CA: Consulting Psychologists Press.

Dedobbeleer, N., & Beland, R. (1991). A safety climate measure for construction sites. *Journal of Safety Research, 22,* 97–103.

Ewigman, B. G., Kivlahan, C. H., Hosokawa, M. C., & Horman, D. (1990). Efficacy of an intervention to promote use of hearing protective devices by firefighters. *Public Health Reports, 105,* 53–59.

Farrell, J. N., & McDaniel, M. A. (2001). The stability of validity coefficients over time: Ackerman's (1988) model and the general aptitude test battery. *Journal of Applied Psychology, 86,* 60–79.

Fiedler, F. E. (1987). *Structured management training in underground mining—five years later* (Information Circular 9145). Pittsburgh, PA: Bureau of Mines.

Fiedler, F. E., Bell, C. H., Chemers, M. M., & Patrick D. (1984). Increasing mine productivity and safety through management training and organization development: A comparative study. *Basic and Applied Social Psychology, 5,* 1–18.

Filigenzi, M. T., Orr, T. J., & Ruff, T. M. (2000). Virtual reality for mine safety training. *Applied Occupational and Environmental Hygiene, 15,* 465–469.

Flin, R., Mearns, K., O'Connor, P., & Bryden, R. (2000). Measuring safety climate: Identifying the common features. *Safety Science, 34,* 177–192.

Foster, L. (1996). Manual handling training and changes in work practices. *Occupational Health, 11,* 402–406.

Fox, D. K., Hopkins, B. L., & Anger, W. K. (1987). The long-term effects of a token economy on safety performance in open-pit mining. *Journal of Applied Behavioral Analysis, 20,* 215–224.

Gebhardt, D. L., & Crump, C. E. (1990). Employee fitness and wellness programs in the workplace. *American Psychologist, 45,* 262–272.

Griffin, M. A., & Neal, A. (2000). Perceptions of safety at work: A framework for linking safety climate to safety performance. *Journal of Occupational Health Psychology, 5,* 347–358.

Hadipriono, F. C., Larew, R. E., & Barsoum, A. S. (1996). *Safety in construction using Virtual Reality (SAVR): A model for labor safety* (Working Paper Series No. WP-022). Columbus, OH: Ohio State University, Center for Labor Research. (ERIC Document Reproduction Service No. ED 397 267)

Hattrup, K., & Jackson, S. E. (1996). Learning about individual differences by taking situations seriously. In K. Murphy (Ed.), *Individual differences and behavior in organizations* (pp. 507–547). San Francisco: Jossey-Bass.

Hazardous Communication Standard (Revised as of July 1, 2002). (29 C.F.R. §1910.1200) (pp. 461–483). Washington, DC: U.S. Government Printing Office. Available on-line at http://frwebgate.access. gpo.gov/cgi-bin/get-cfr.cgi.

Hazardous Waste Operations and Emergency Response Standard. (Revised as of July 1, 2002). (29 C.F.R. §1920.120) (pp. 370–412). Washington, DC: U.S. Government Printing Office. Available on-line at http://frwebgate.access.gpo.gov/cgi-bin/get-cfr.cgi.

Helmreich, R. L., & Foushee, H. C. (1993). Why crew resource management? Empirical and theoretical bases of human factors training in aviation. In E. L. Weiner, B. G. Kanki, & R. L. Hiemreich (Eds.), *Cockpit resource management* (pp. 3–45). Orlando, FL: Academic Press.

Hofmann, D. A., Morgeson, F. P., & Gerras, S. J. (in press). Climate as a moderator of the relationship between LMX and content specific citizenship behavior: Safety climate as an exemplar. *Journal of Applied Psychology.*

Hofmann, D. A., & Stetzer, A. (1996). A cross-level investigation of factors influencing unsafe behaviors and accidents. *Personnel Psychology, 49,* 307–339.

Huddock, S. D. (1994). The application of educational technology to occupational safety and health training. *Occupational Medicine: State of the Art Reviews, 9,* 201–210.

Ivancevich, J. M., Matteson, M. T., Freedman, S. M., & Phillips, J. S. (1990). Worksite stress management interventions. *American Psychologist, 45,* 252–261.

Jacobs, R., Hofmann, D. A., & Kriska, S. D. (1990). Performance and seniority. *Human Performance, 3,* 107–121.

James, L. R., Demaree, R. G., Mulaik, S. A., & Ladd, R. T. (1992). Validity generalization in the context of situational models. *Journal of Applied Psychology, 77,* 3–14.

James, L. R., & McIntyre, M. D. (1996). Perceptions of organizational climate. In K. Murphy (Ed.), *Individual differences and behavior in organizations* (pp. 416–450). San Francisco: Jossey-Bass.

Komaki, J., Barwick, K. D., & Scott, L. R. (1978). A behavioral approach to occupational safety pinpointing and reinforcing safe performance in a food manufacturing plant. *Journal of Applied Psychology, 63,* 434–445.

Komaki, J., Heinzmann, A. T., & Lawson, L. (1980). Effect of training and a component analysis of a behavioral safety program. *Journal of Applied Psychology, 65,* 261–270.

Lapping, J. E., & Parsons, M. A. (1980). The impact of training in the construction industry. *Professional Safety, 25,* 13–18.

Lingard, H., & Rowlinson, S. (1997). Behavior-based safety management in Hong Kong's construction industry. *Journal of Safety Research, 28,* 243–256.

Linnemann, C. C., Cannon, C., DeRonde, M., & Lamphear, B. (1991). Effect of educational programs, rigid sharps containers and universal precautions on reported needle stick injuries in health care workers. *Infection Control and Hospital Epidemiology, 12,* 214–219.

Lippin, T. M., Eckman, A., Calkin, K. R., & McQuiston, T. H. (2000). Empowerment-based health and safety training: Evidence of workplace change from four industrial sectors. *American Journal of Industrial Medicine, 38,* 697–706.

Luskin, J., Somers, C., Wooding, J., & Levenstein, C. (1992). Teaching health and safety: Problems and possibilities for learner-centered training. *American Journal of Industrial Medicine, 22,* 665–676.

Maples, T. W., Jacoby, J. A., Johnson, D. E., Ter Haar, G. L., & Buckingham, F. M. (1982). Effectiveness of employee training and motivation programs in reducing exposure to inorganic lead and lead alkyls. *American Industrial Hygiene Association Journal, 43,* 692–694.

Marchand, A., Simard, M., Carpentier-Roy, M. C., Ouellet, F. (1998). From a unidimensional to a bidimensional concept and measurement of workers' safety behavior. *Scandinavian Journal of Work, Environment, and Health, 24,* 293–299.

McClendon, C., Butz, R. M., Zusman, R. (2001, April). *Effects of safety training spacing on safety-related performance.* Paper presented at the annual conference of the Society for Industrial and Organizational Psychology, San Diego, CA.

McDonald, N., Corrigan, S., Daly, C., & Cromie, S. (2000). Safety management systems and safety culture in aircraft maintenance organizations. *Safety Science, 34,* 151–176.

McGraw, K. O., & Wong, S. P. (1992). A common language effect size statistic. *Psychological Bulletin, 111,* 361–365.

McQuiston, T. H., Coleman, P., Wallerstein, N. B., Marcus, A. G., Morawetz, J. S., & Ortlieb, D. W. (1994). Hazardous waste worker education: Long-term effects. *Journal of Occupational Medicine, 36,* 1310–1323.

Millican, R., Baker, R. C., & Cook, G. T. (1981). Controlling heat stress: Administrative versus physical control. *American Industrial Hygiene Association Journal, 42,* 411–416.

Murphy, K. R. (1996). Individual differences and behavior in organizations: Much more than *g*. In K. R. Murphy (Ed.), *Individual differences and behavior in organizations* (pp. 3–30). San Francisco: Jossey-Bass.

Neal, A., Griffin, M. A., & Hart, P. M. (2000). The impact of organizational climate on safety climate and individual behavior. *Safety Science, 34,* 99–109.

Occupational Safety and Health Administration (OSHA). (1998). *Training requirements in OSHA standards and training guidelines* (OSHA 2254). Washington, DC: U.S. Department of Labor.

Office of Technology Assessment (OTA). (1985). *Preventing illness and injury in the workplace* (OTA-H-256). Washington, DC: U.S. Congress.

Ray, P. S., & Bishop, P. A. (1995). Can training alone ensure a safe workplace? *Professional Safety, 40,* 56–59.

Reber, R. A., & Wallin, J. A. (1984). The effects of training, goal setting, and knowledge of results on safe behavior: A component analysis. *Academy of Management Journal, 27,* 544–560.

Robins, T. G., Hugentobler, M. K., Kaminski, M., & Klitzman, S. (1990). Implementation of the federal hazard communication standard: Does training work? *Journal of Occupational Medicine, 32,* 1133–1140.

Robins, T. G., & Klitzman, S. (1988). Hazard communication in a large U.S. manufacturing firm: The ecology of health education in the workplace. *Health Education Quarterly, 15,* 451–472.

Rudmo, T. (2000). Safety climate, attitudes, and risk perception in Norsk Hydro. *Safety Science, 34,* 47–59.

Saarela, K. L., Saari, J., & Alltonen, M. (1989). The effects of an informational safety campaign in the shipbuilding industry. *Journal of Occupational Accidents, 10,* 255–266.

Saari, J., & Nasanen, M. (1989). The effect of positive feedback on industrial housekeeping and accidents: A long-term study at a shipyard. *International Journal of Industrial Ergonomics, 4,* 201–211.

Sadler, O. W., & Montgomery, G. M. (1982). The application of positive practice overcorrection to the use of hearing protection. *American Industrial Hygiene Association Journal, 43,* 451–454.

St. Vincent, M., Tellier, C., & Lortie, M. (1989). Training in handling: An evaluative study. *Ergonomics, 32,* 191–210.

Schmidt, F. L., Hunter, J. E., & Outerbridge, A. N. (1986). Impact of job experience and ability on job knowledge, work sample performance, and supervisory ratings of job performance. *Journal of Applied Psychology, 71,* 432–439.

Scholey, M. (1983). Back stress: The effects of training nurses to lift patients in a clinical situation. *International Journal of Nursing Studies, 20,* 1–13.

Smith-Crowe, K., Burke, M. J., & Landis, R. S. (2001). *Organizational climate as a moderator of safety knowledge: Safety performance relationships.* Paper presented at the 61st annual meeting of the Academy of Management, Washington, DC.

Stubbs, D. A., Buckle, P. W., Hudson, M. P., & Rivers, P. M. (1983). Back pain in the nursing profession II: The effectiveness of training. *Ergonomics, 26,* 767–779.

U.S. Department of Health and Human Services, Public Health Services, Centers for Disease Control and Prevention, & National Institute for Occupational Safety and Health. (1996). *National Occupational Research Agenda* (DHHS/NIOSH Publication No. 96-191-952). Cincinnati, OH: Author.

van der Klink, J. J., Blonk, R. W., Schene, A. H., & van Dijk, F. J. (2001). The benefits of interventions for work-related stress. *American Journal of Public Health, 91,* 270–276.

Vargha, A., & Delaney, H. D. (2000). A critique and improvement of the CL common language effect size statistics of McGraw and Wong. *Journal of Educational and Behavioral Statistics, 25,* 101–132.

Vineberg, R., & Taylor, E. N. (1972). *Performance in four military jobs by men of different aptitude (AFQT) levels: The relationship of AFQT and job experience to job performance.* Alexandria, VA: Human Resources Research Organization.

Wallerstein, N., & Baker, R. (1994). Labor education programs in health and safety. *Occupational Medicine: State of the Art Reviews, 9,* 305–320.

Wallerstein, N., & Weinger, M. (1992). Health and safety education for empowerment. *American Journal of Industrial Medicine, 22,* 619–635.

Wands, S. E., & Yassi, A. (1992). "Let's talk back": A program to empower laundry workers. *American Journal of Industrial Medicine, 22,* 703–709.

Weinger, M., & Lyons, M. (1992). Problem-solving in the fields: An action-oriented approach to farmworker education about pesticides. *American Journal of Industrial Medicine, 22,* 677–690.

Wetter, D. W., Fiore, M. C., Gritz, E. R., Lando, H. A., Stitzer, M. L., Hasselblad, V., & Baker, T. B. (1998). The agency for health care policy and research smoking cessation clinical practice guidelines: Findings and implications for psychologists. *American Psychologist, 53,* 657–669.

Wright, C. (1986). Routine deaths: Fatal accidents in the oil industry. *Sociological Review, 4,* 265–289.

Wu, T., & Hwang, S. (1989). Maintenance error reduction strategies in nuclear power plant, using root cause analysis. *Applied Ergonomics, 20,* 115–121.

Zohar, D. (2000). A group-level model of safety climate: Testing the effect of group climate on microaccidents in manufacturing jobs. *Journal of Applied Psychology, 85,* 587–596.

CHAPTER 4

Designing Healthy Work

Sharon K. Parker
Nick Turner
Mark A. Griffin

I leave my brain at [the] gate when I arrive, and pick it up at end of shift, if I'm alive.
—A PETROCHEMICAL TECHNICIAN WHEN ASKED TO
DESCRIBE HIS WORK

This provocative quotation from our field notes highlights the two key variables of interest here. Poor work design is the first. Work design is concerned with the content and organization of employees' mental, interpersonal, and physical tasks and with how these aspects affect employees (i.e., their motivation, behavior, cognitions, well-being, and performance) and organizations (e.g., productivity). Two key aspects of work design include (1) the degree of autonomy that employees have over how and when they perform their tasks and (2) their level of job demands. The second variable of interest here is employee health. The employee jokes that he will collect his

Acknowledgments: We thank the United Kingdom Health and Safety Executive and the Social Sciences and Humanities Research Council of Canada for financial support that helped make this collaboration possible. We thank Peter Warr for his helpful comments concerning mental health; Gavin Clarke for library assistance; and Julian Barling, Toby Wall, and Helen Williams for feedback on earlier drafts of this chapter.

brain at the end of the day if he's still alive—not such a joke when one considers that in 1998 alone, U.S. workplaces had 3.8 million disabling injuries and 5,100 fatal injuries (National Safety Council, 1999). Work design and health are related, as this chapter will show.

The link between work design and employee health is not a new topic. Indeed, reporting on work factors causing neurotic illness, Fraser (1947, p. 363) concluded, "by far the most influential attribute [of jobs] is the opportunity work offers—or fails to offer—for the use of workers' abilities and for associated feelings of interest, sense of accomplishment, personal growth, and self-respect." Today, the topic is just as important. Countries such as Sweden have explicitly incorporated the importance of work design in occupational health statutes, and other countries acknowledge the potential role of work design for health. For example, the National Institute for Occupational Safety and Health (NIOSH) in the United States recently identified the "organization of work" as a priority area in its national research agenda.

In this chapter we draw on reviews of work design and employee health to provide a synopsis of the main findings, with particular focus on job autonomy (control) and job demands. In line with Hofmann and Tetrick's framework (Chapter One of this volume), we examine not only whether particular work designs can cause or prevent ill health for individuals (i.e., a short-term, intrinsic focus) but also how work characteristics might promote positive health (i.e., a long-term, intrinsic focus). We then consider the various processes by which work design can affect the broader issue of organizational health, which essentially moves our focus from intrinsic to extrinsic health, as identified in Hofmann and Tetrick's model.

Designing Work for Individual Health: The Theory

To set the platform for our discussion about work design and health, we start by conceptualizing individual health at work and briefly recap key work design theories.

What Is a Healthy Individual?

Influenced by a medical model of health, many work design studies focus on how work characteristics affect psychological ill health or cardiovascular disease (CVD) (Warr, 1987). However, good health is not merely the absence of ill health, injury, or disease. For example, good mental health by Western standards is more than not being depressed; it is also aspiring to learn, being reasonably independent, and possessing confidence (Karasek & Theorell, 1990). Similarly, safe working is not just avoiding injury but includes active participation in initiatives aimed at improving safety. We draw on Warr's framework (1987, 1994) to identify types of mental health that differ in their relative emphasis on ill health and good health.

Affective well-being is the most commonly investigated mental health outcome. Warr (1987) identified two orthogonal dimensions of affective well-being: pleasure (from feeling bad to feeling good) and level of arousal (from low to high). Any affective state can be viewed in terms of its location on these dimensions. Warr (1990) identified three assessable aspects: the horizontal axis of pleasure or displeasure, which is often measured in terms of satisfaction or happiness; an axis from anxiety (high arousal, low pleasure) to comfort (low arousal, high pleasure); and an axis from depression (low arousal, low pleasure) to enthusiasm (high arousal, high pleasure). Anxiety-comfort and depression-enthusiasm are distinct aspects of well-being, although measures of distress and psychological ill health often combine these two forms. Burnout is a specific form of affective well-being that involves feelings of emotional exhaustion, depersonalization, and diminished personal accomplishment (Maslach & Leiter, 1997).

Indicators of affective well-being that assess anxiety, depression, burnout, psychological distress, and physiological or psychosomatic symptoms emphasize the detection of ill health rather than the presence of good health. Indicators of well-being that assess high arousal–high pleasure states, such as enthusiasm, emphasize positive mental health. Measures of job satisfaction can be considered as either an indicator of ill health (e.g., when job dissatisfaction is the focus) or as a positive indicator of health (e.g., when job satisfaction and its link to behaviors such as organizational citizenship behavior are the focus). Either way, job satisfaction remains a relatively passive form of mental health because most measures assess the degree of pleasure or displeasure about a job, but they do not assess arousal (Warr, 1997). In Bruggeman, Groskurth, and Ulich's terms (1975), employees can experience a sense of resigned job satisfaction, in which they have lowered their level of aspiration and have become resigned to a job.

In addition to affective well-being, Warr (1987) identified the following types of mental health: positive self-regard (e.g., high self-esteem), competence (e.g., effective coping), aspiration (e.g., goal directedness), autonomy/independence (e.g., proactivity), and integrated functioning (i.e., states involving balance, harmony, and inner-relatedness). These components of mental health can affect one's well-being (e.g., effective coping can reduce distress), but they are also important in their own right. First, they are potentially more enduring aspects than affective well-being. Second, in contrast to a "passive contentment" (Warr, 1994, p. 86) view of mental health, outcomes such as competence, aspiration, and autonomy/independence represent more active states and behaviors than most measures of well-being (such as job satisfaction). We return to these outcomes later when we review how work design might affect positive mental health.

Note that these additional types of mental health are not necessarily interchangeable with affective well-being. As Warr (1994, p. 86) noted: "proactive, risk-taking people may be considered healthy in terms of competence, aspiration, and autonomy; but their difficult interactions with the environment may also make them anxious for a considerable proportion of time." High anxiety would be considered a sign of poor mental health only if it were excessive or sustained over a long period. This point highlights the value of including multiple indicators of health, from those focused on ill health to those assessing positive health.

Work Design Theory

Campion and Thayer (1987) identified several ways of approaching the topic of work design, such as using ergonomic principles to design workstations (a biological approach) or regulating demands to ensure reasonable concentration requirements (a perceptual/motor approach). This chapter takes a work characteristics (or what Campion and Thayer refer to as a motivational approach) that emphasizes key job characteristics, such as autonomy, for motivation and other outcomes (see Parker & Wall, 1998, for more detail on this approach). The work characteristics approach emerged in response to job simplification, which involved enhancing efficiency by breaking jobs into narrow tasks and removing decision-making responsibilities.

One response to the dominance of job simplification, which had negative effects for employee well-being (e.g., Fraser, 1947), was to look for ways to redesign jobs, such as via job enlargement (i.e., expanding the job to include additional tasks) or job enrichment (i.e., increasing job autonomy). Another response was to develop theories to explain how work design affects employees. One of the most well-known theories in this respect is the job characteristics model (Hackman & Oldham, 1976), which proposes five core job characteristics (skill variety, task identity, task significance, autonomy, and job feedback) that enhance employee performance and health-related outcomes (i.e., higher work satisfaction, internal work motivation, lower absenteeism and turnover). The sociotechnical systems approach (e.g., Trist & Bamforth, 1951) is an equally influential work design theory, based on the proposition that there should be simultaneous design and joint optimization of the social and technical subsystems in organizations. This proposition gives rise to principles such as that methods of working should be minimally specified and that variances in the work processes (e.g., breakdowns) should be handled at the source by the employees rather than controlled by supervisors or specialists.

Parker, Wall, and Cordery (2001) proposed an elaborated work characteristics model. This model proposes a broader range of individual and group-level work characteristics that can have psychological and behavioral consequences, such as the level of emotional demands in one's job. The model also contains additional outcomes of work redesign, including safety and active mental health (e.g., proactivity, personal initiative, learning), both outcomes pertinent to this chapter.

A further important theory relevant to the design of healthy work is the demand-control model of strain (Karasek, 1979). This interaction model proposes that high job demands (e.g., high workload), when accompanied by low decision latitude (i.e., low job autonomy and skill discretion), will cause strain and, in the long term, stress-related illnesses such as heart disease. In contrast, if the high demands occur in the presence of high decision latitude, a so-called active job, then strain will not accrue, and other benefits will occur. An elaborated model proposes that work design can affect mental health outcomes via two spirals of learning and behavior. In the first an active job and its learning opportunities lead to an increased feeling of mastery and confidence that in turn helps the person to cope with further job demands and challenges (Karasek & Theorell, 1990). The second, a negative behavioral spiral, results from a combination of high job demands and low decision latitude, which leads to reduced mastery and poorer coping, followed by higher residual strain levels and so on.

In summary, models of work design propose various individual and group work characteristics that are important for promoting well-being outcomes such as job satisfaction and, more recently, active health outcomes such as mastery and learning. In the next two sections, we review evidence regarding the link between work design and health outcomes, focusing first on ill health outcomes and second on positive health outcomes such as active mental health.

Designing Work to Prevent Ill Health: The Evidence

Our focus here is on the link between work characteristics and aspects of mental health that emphasize ill health or that are relatively passive, including psychological health and well-being (e.g., job satisfaction), CVD, musculoskeletal problems, and injuries.

Psychological Health and Well-Being

In a useful overview, Warr (1999) documented the relationship between ten work characteristics and employee well-being. The work characteristics included opportunity for personal control, opportunity for skill use, externally generated goals (e.g., workload, role conflict), variety, environmental clarity (e.g., role clarity, job security), availability of money, physical security, supportive supervision, opportunity for interpersonal contact, and valued social position. For each of these work characteristics, considerable evidence shows that "more" is associated with higher affective well-being, at least up to moderate levels. For example, many studies have shown links between low perceived job control and negative health outcomes such as anxiety, depression, psychosomatic complaints, burnout, and excess alcohol consumption (see also Terry & Jimmieson, 1999). Excess levels of externally generated goals such as job demands and time pressure can lower well-being. For example, Lee and Ashforth (1996) found in a meta-analysis that workload and work pressure were associated with depersonalization and emotional exhaustion. Some evidence (e.g., Xie & Johns, 1995) suggests curvilinear relationships between work characteristics and well-being. Moreover, some job characteristics relate particularly to some types of well-being. In particular, high job demand is often more strongly linked to anxiety, whereas low job control is more strongly related to depression or dissatisfaction (Warr, 1999).

The sheer number of studies investigating the link between work characteristics and well-being, with largely consistent results, is encouraging. Nevertheless, two important issues arise. First, many of the studies have cross-sectional designs that cannot address direction of causality or confounding variables. Second, the work design literature suggests more consistent results for job satisfaction than for other types of well-being. Thus, we report first the findings for job satisfaction and then for other forms of well-being. In both cases we focus on methodologically rigorous studies.

Job Satisfaction

Job satisfaction, which Locke (1976, p. 1304) defines as a "pleasurable or positive emotional state resulting from the appraisal of one's job or job experiences," is the most commonly used measure of well-being in work design research. Scholars have distinguished two types of job satisfaction: intrinsic job satisfaction (satisfaction with aspects intrinsic to the job, such as job control) and extrinsic satisfaction (satisfaction with aspects external to the job content, e.g., pay). Not surprisingly, work design has stronger relationships with intrinsic job satisfaction, or overall job satisfaction, than with extrinsic job satisfaction (e.g., Wong, Hui, & Law, 1998).

Results from both cross-sectional and longitudinal studies of work redesign converge to show that enriched work designs, or the presence of core job characteristics such as job autonomy and skill use, promote job satisfaction. For example, in an analysis of 32 field experiments that manipulated core job characteristics, Kopelman (1985) showed that job satisfaction increased in 80% of the studies. Many reviews and meta-analyses support this conclusion

(Fried & Ferris, 1987; Kelly, 1992; Kinicki, McKee-Ryan, Schries-heim, & Carson, 2002; Loher, Noe, Moeller, & Fitzgerald, 1985; Spector, 1986; Stone, 1986). Illustrative intervention studies include those by Campion and McClelland (1993) and Pearson (1992).

A positive development in this research has been to use objective job characteristics rather than perceived job characteristics to deal with criticisms of common method variance. Perceptions of work characteristics correlate more strongly with job satisfaction than do objective reports, but the latter nevertheless show consistently positive correlations with job satisfaction (Glick, Jenkins, & Gupta, 1986). In the same vein, studies have tested the direction of causality between job satisfaction and perceived job characteristics. Evidence suggests a reciprocal relationship: satisfied individuals rate their jobs more positively, and positive work characteristics enhance job satisfaction (James & Tetrick, 1986; Steel & Rentsch, 1997; Wong et al., 1998).

One area in which more development is needed is the timing of job satisfaction changes. In a longitudinal analysis of job enrichment (Griffin, 1991), job satisfaction improved in the first six months but declined to initial levels over time (in contrast, performance improved after 24 and 48 months). One explanation of this decline in job satisfaction is that employees' expectations rise over time. Employees therefore might report levels of satisfaction that are the same as prior to the work redesign, but they expect more from their work. Assessing active mental health alongside job satisfaction might help to capture these changes in aspiration.

A further issue is that, compared to job enrichment variables, the link between job demands and job satisfaction is more mixed. For example, Fox, Dwyer, and Ganster (1993) found no significant link between workload and job satisfaction, whereas other longitudinal studies have shown that excess demands reduce job satisfaction (De Jonge et al., 2001; Parkes, 1995).

Psychological Distress and Ill Health

Fewer studies examine the link between work design and outcomes like psychological distress or psychosomatic symptoms. Spector (1986), for example, located only four studies looking at job autonomy and emotional distress, although, as expected, these aspects were negatively correlated.

When one focuses on rigorous studies, such as those with a longitudinal design, the studies are even fewer and the results rather mixed. For example, Sargent and Terry (1998) found contemporaneous effects of job control on depressive symptoms but no longitudinal association after controlling for initial levels of depression. Parkes (1991) and Parkes, Mendham, and Von Rabenau (1994) reported similar findings of no longitudinal association with job control. In contrast, Dwyer and Fox (2000) found that job control predicted reduced health problems five years later; Parker, Chmiel, and Wall (1997) showed that increased role clarity and participation in decision making were associated with improved psychological health during a period of downsizing; and Karasek (1979) showed that low decision latitude led to exhaustion and depression; and Parker (in press) showed that the deskilling and reduced participative decision making associated with lean production led to increased job depression.

A further type of longitudinal study evaluates the effects of work redesign, or interventions deliberately intended to change job characteristics. Here again, the evidence is mixed (see Sonnentag, 1996, for a similar conclusion). For example, in one study, Wall and Clegg (1981) found that autonomous work groups had positive effects on internal work motivation and performance in the first six months and later positive effects on job satisfaction and psychological health. However, in an evaluation that included control groups, Wall, Kemp, Jackson, and Clegg (1986) showed lasting effects of autonomous work groups on intrinsic job satisfaction but no effects on work motivation, commitment, psychological health, or performance.

Focusing on demands rather than control, results of longitudinal studies appear more consistent. Martin and Wall (1989) showed that as machine operators moved between different roles with different levels of demand and responsibility, the predicted changes in well-being occurred. Similarly, Parkes (1995) showed that reducing objective workload levels for a group of United Kingdom driving test examiners led to reduced perceptions of demands, reduced anxiety, increased job satisfaction, and enhanced cognitive performance. Frese (1985) conducted a series of studies, including one longitudinal study, to rule out alternative explanations of a link between work demands (e.g., speed of work, role conflict) and psychosomatic complaints. Using assessments of both perceived demands and observer-rated demands, Frese found strong evidence that demands caused psychosomatic complaints such as headaches. An analysis of U.K. civil servants showed that high social support was positively linked with psychological health, whereas high job demands predicted poorer psychological health at follow-up (Stansfeld, Fuhrer, Head, Ferrie, & Shipley, 1997). Diary studies in which employees report on their work and well-being several times per day indicate longitudinal links between time pressure, workload, and well-being outcomes (Sonnentag, 2001; Teuchmann, Totterddell, & Parker, 1999). Finally, in a study with six measurement waves, Garst, Frese, and Molennar (2000) showed that time pressure increased worrying and health complaints. Thus, on the whole, longitudinal studies present quite a convincing picture that work pressures and demands can impair psychological health.

In summary, the evidence is clear: designing enriched jobs with work characteristics such as job autonomy will enhance intrinsic or overall job satisfaction. However, whether this will improve psychological health is less clear-cut; evidence is somewhat more convincing for the negative effects of demands than for the positive effects of autonomy. Consistent with these conclusions, a multilevel analysis by Elovainio, Kivimaeki, Steen, and Kalliomaeki-Levanto (2000) showed that psychological distress varied mostly at the individual level and was explained primarily by personality, whereas job satisfaction varied at the individual and group level, with job control accounting for variation at both levels.

Demand-Control Interaction

Before looking at additional indicators of ill health, we report on the vast set of studies investigating the interactive effects of job demand and job control on well-being. Karasek (1979) proposed that job control (or what Karasek referred to as job decision latitude) moderates the effects of demands on strain, such that high demands do not cause strain when control is high. This interaction or buffer hypothesis can be contrasted with the strain hypothesis that posits additive main effects of demands and control. There has been some confusion in the literature in regard to these hypotheses. Although Karasek proposed an interaction hypothesis in his 1979 publication on the model, Karasek argued in 1989 that it is the additive effects of demands and control that are key. As Van Der Doef and Maes (1999) recognized, tests of both hypotheses have been proclaimed as tests of the demand-control model, so it is important to differentiate these studies in any review. Here, we focus on tests of the buffering hypothesis. However, it is relevant to note that, in their 1999 review, Van Der Doef and Maes found strong support for the additive hypothesis in cross-sectional studies but less support in longitudinal studies (which accords with the summary we presented).

In an initial review of the buffering hypothesis, Ganster and Fusilier (1989) evaluated studies conducted within homogeneous occupations as well as multioccupation studies (for additional reviews, see Jones & Fletcher, 1996, and De Jonge & Kompier, 1997). Both types of studies have suffered various methodological problems, and therefore Ganster and Fusilier (1989) concluded that support for an interaction effect was not at all convincing. In two follow-up reviews almost ten years later, researchers reached the same conclusion of only sporadic support for an interaction (Terry & Jimmieson, 1999; Van Der Doef & Maes, 1999).

These conclusions might have been sufficient to stop interest in the issue once and for all, but recent study developments have yielded more promising findings. For example, using objective measures to avoid problems of common-method variance, researchers have shown that the demand-control interaction predicts health outcomes such as sick leave, absenteeism, and systolic blood pressure (e.g., Dwyer & Ganster, 1991; Fox et al., 1993). Wall, Jackson, Mullarkey, and Parker (1996) showed evidence of an interaction effect with a measure of job control that was not confounded by items assessing skill discretion or job complexity. More recently, Van der Doef and Maes (1999) concluded that how demands and control are conceptualized is a key factor in discriminating studies that support the interaction hypothesis from those that do not.

A further important development is the consideration of multiple levels of analysis. For example, work characteristics can be assessed at the group level by aggregating the perceived work characteristics of job incumbents working in the same job and in similar workplaces (Frese, 1985). These shared perceptions of work characteristics are more representative than individuals' job perceptions because individual idiosyncrasies are reduced and some of the attributional problems of third-party ratings of work characteristics are avoided. Two recent multilevel studies of the demand-control model show the promise of this approach. De Jonge, van Breukelen, Landeweerd, and Nijhuis (1999) found a significant and strong interaction at the aggregate level for job satisfaction but no interaction effects (group or individual) for emotional exhaustion or job-related anxiety. The aggregated data explained some additional variance for work motivation and job satisfaction, which suggests that contextual influences play a role in affecting these outcomes. However, aggregated effects did not explain any additional variance for emotional exhaustion or anxiety, suggesting these types of processes might be more person-centered (see also de Jonge et al., 2001). Adopting a slightly different approach, Van Yperen and Snijders (2000) found support for a demand-control interaction when using scores that compared the individual's score with the group mean. An illustrative finding is that psychological health symptoms occurred when the perceived demands of the job—as compared with the mean of the employee's working group—exceeded the perceived control possibilities available to the worker.

Another interesting development has been to examine whether organizational or dispositional characteristics moderate the effect of a demand-control interaction on employee well-being. Most attention has been given to social support as a moderator, no doubt in part a result of its inclusion as a third variable in the demand-control model (Karasek & Theorell, 1990). Some evidence indicates that the negative effects of high demands and low job control are greatest when social support is low (see Terry & Jimmieson, 1999), but the number of studies is limited and the overall findings mixed (Van Der Doef & Maes, 1999). Some studies have also shown that dispositional factors, such as self-efficacy (e.g., Schaubroeck, Lam, & Xie, 2000) and proactive personality, can moderate the interaction. For instance, Parker and Sprigg (1999) found that proactive employees in high demand–high control jobs experienced low strain but that those with a more passive personality experienced higher strain in these jobs. Proactive employees were suggested to be more likely to take advantage of higher job control, using it to better manage job demands, whereas more passive employees do not seize the opportunity that job control offers.

At the end of the day, therefore, past failures to demonstrate interactions between demands and control may result from a lack of methodological and theoretical sophistication.

Cardiovascular Disease (CVD)

Several scholars have reviewed the links between work character-istics and CVD (e.g., Kristensen, 1995; Schnall, Landsbergis, & Baker, 1994). In the latest of these (Theorell & Karasek, 1996), 16 out of 22 studies showed significant associations between high demand–low control jobs and CVD or CVD symptoms. Results are more consistent in studies focusing on CVD than in those focus-ing on risk factors for heart disease, such as smoking or blood pres-sure, and are more consistent in studies of blue-collar employees than of white-collar employees. An example of a supportive finding comes from a prospective cohort study from the Whitehall program, which showed that those with low job control at baseline had 1.5 to 1.8 higher risks of new heart disease during the 5.3 year follow-up (Bosma, Stansfeld, & Marmot, 1998).

Despite fairly consistent evidence in this area, many of the stud-ies cited in the reviews have been criticized because occupational comparisons confound social class (cf. Karasek & Theorell, 1990). It is also notable that few intervention studies exist. An exception is a Swedish study (Orth-Gomer, Eriksson, Moser, Theorell, & Fred-lund, 1994) that showed that employees in sites with individual counseling and work reorganization had improved physiological functioning relative to those who were in nonintervention sites, al-though note that the effects of work reorganization cannot be sep-arated from the individual counseling. A further problem with research linking work design and heart disease is its imprecision. Current research does not allow a detailed understanding of what specific aspects of decision latitude or demands are most impor-tant and for what categories of workers (Theorell & Karasek, 1996).

The latter point is somewhat addressed by studies investigating the moderating role of individual characteristics. For example, there is some suggestion that job enrichment might be harmful to type A individuals (i.e., those behaving with impatience and com-petitiveness) because these individuals become overstimulated and hyperactive. Schaubroeck, Ganster, and Kemmerer (1994) found that for a sample of police and fire department employees, job complexity was negatively associated with CVD morbidity for those low on type A but was positively associated with it for those high on type A. Dwyer and Fox (2000) found a similar pattern regarding the moderating effect of hostility on the link between job control and health-care costs but different results for job control when it came to CVD health. Increased control resulted in reduced CVD and respiratory health complaints, even for nurses with high hostility. The authors postulated that the discrepant results for job control when predicting different outcomes occurred because nurses have control over different aspects of their work. Control over work schedules, running stations, and the like might have more positive effects than increased decision-making control and accountability over patient care. Like previous studies, this suggests that close attention to the type of control is important.

Musculoskeletal Symptoms

Most research investigating musculoskeletal problems has focused on physical aspects of work, such as posture and vibration. However, in the past few decades, researchers have given attention to psychosocial aspects, including work characteristics. Studies (e.g., Bongers, de Winter, Kompier, & Hildebrandt, 1993) have shown positive relationships between work characteristics such as job control and lack of musculoskeletal problems, and negative associations for demands.

Elovainio and Sinervo (1997) observed two mechanisms through which psychosocial factors might affect musculoskeletal problems. First, demands (time pressure and patient-stressors) were associated with psychological ill health, which in turn were linked with musculoskeletal problems. Second, demands were associated with greater physical workload, which then predicted musculoskeletal problems. Hollmann, Heuer, and Schmidt (2001) found that control had positive effects on musculoskeletal problems but only when the physical workload was relatively light. It seems that any effects of control are overwhelmed by the high levels of workload, which is a plausible idea. Demands affected musculoskeletal symptoms for both high and low workload groups.

This brief analysis shows that it might be possible to design jobs to reduce the incidence of work-related musculoskeletal problems, albeit within constraints such as relatively light workloads. Research in this area is in its early days and suffers from some of the methodological problems that have plagued other work design research.

Injuries

Work design could affect the occurrence of injury through a number of mechanisms. For example, job variety might alleviate boredom and increase attentiveness, thereby reducing risks, whereas excess job demands might cause workers to take shortcuts, thereby increasing the chance of injuries. However, few studies so far have investigated how work design affects injuries, and the limited studies that exist are very diverse. For example, some studies (e.g., Frone, 1998) use distinct measures of work characteristics, but others (e.g., Neal, Griffin, & Hart, 2000) use broader constructs such as organizational climate to capture perceptions of the work environment. Different types of safety outcomes are also used, ranging from self-reported injury measures (e.g., Frone, 1998) to injury data from company safety records (e.g., Iverson & Erwin, 1997).

Overall, the inconsistent measures used and the relative youth of this area of inquiry make it a challenge to establish how work characteristics affect injury occurrence. Although most of the studies suggest that work characteristics do affect injuries, there is little consistency regarding which work characteristics are key. For example, Frone (1998) found that workload, job boredom, and physical hazards were three important predictors of injury occurrence among working adolescents. However, in a similar analysis of an adult sample, Iverson and Erwin (1997) did not find the same relationship for physical hazards and role demands; instead, they found a significant effect for routinization (boredom). Neither study found that role ambiguity predicted injuries, whereas Hemingway and Smith (1999) found this variable affected injuries among nurses.

In sum, although recognition of work design factors in the prevention of injury is growing, the evidence is far from clear. As we discuss later, it is important to consider how work characteristics affect positive indices of safe work rather than focusing solely on injuries.

Designing Work to Promote Positive Health: The Evidence

We now turn to the relationship between work design and positive health. Although we recognize that work design could promote positive health in several ways (e.g., by enhancing physical fitness or promoting positive social relationships outside of work), we restrict our attention to two areas for which research provides a reasonable degree of supportive evidence: active mental health and safety behaviors. We also acknowledge that work characteristics such as job autonomy have been shown to promote outcomes like organizational commitment, intrinsic motivation, and involvement (e.g., Spector, 1986), which could be considered as indicators of positive mental health. Here, however, we focus on active mental health and safety behaviors, which are somewhat more neglected potential outcomes of work design.

Active Mental Health

Researchers have long argued that simplified, low autonomy jobs engender outcomes such as passivity, apathy, reduced aspiration, and learned helplessness. For example, Baldamus (cited in Emery, 1959, p. 180) described many workers as being in a state of "dull contentment," that is, "a sort of borderline satisfaction apparently quite distinct from the elation experienced in pleasurable activities or the quieter satisfaction of an engaging task." Likewise, Argyris (1957) suggested that if individuals in simplified jobs do not adapt by fighting the organization or leaving it physically (e.g., through absence), then they leave it "psychologically" by becoming uninvolved, indifferent, and focused on extrinsic elements of work. These types of changes in individuals are distinct from changes in affective well-being. They involve a relatively enduring reduction in active states and behaviors, or a decline in what we refer to as active mental health.

The idea that work design affects employees' active mental health was supported in early cross-sectional research. Kornhauser (1965) found that blue-collar employees in low-skilled jobs, compared to those in higher-skilled jobs, reported low self-esteem, high anxiety, few friendships, low life satisfaction, and an absence of an active or goal-directed orientation to life. Kohn and Schooler (1973) also showed a positive link between two work characteristics (the requirement to make complex decisions and freedom from close supervision) and strong values for self-direction, favorable self-image, and intellectual flexibility. These findings were subsequently supported by two longitudinal studies. First, Brousseau (1978) found that task significance and feedback assessed at time 2 (on average, 5.9 years after time 1) were positively associated with change in active orientation (e.g., taking initiative) and freedom from depression. Second, Kohn and Schooler (1978) found that substantive job complexity had a strong effect on an employee's intellectual flexibility over a 10-year period. They found an even stronger but gradual reciprocal effect by which intellectual flexibility affected job complexity.

A later analysis (Kohn & Schooler, 1982) expanded the outcomes to include self-directedness (e.g., not being conformist) and distress (e.g., anxiety, lack of self-confidence). Once again, autonomous jobs increased intellectual flexibility and promoted a self-directed orientation, and once again, reciprocal effects were demonstrated. As Kohn and Schooler (1982, p. 1282) summarized: "the findings highlight the centrality for job and personality of a mutually reinforcing triumvirate—ideational flexibility, a self-directed orientation to self and society, and occupational self-direction." Although confidence in this conclusion is limited by some methodological issues (e.g., the use of self-reported work characteristics, the use of only two time points with a very long lag time), the study itself is highly rigorous (Zapf, Dormann, & Frese, 1996).

Despite the promise of these early studies, not much subsequent research has linked work design with active outcomes, especially relative to the number of studies examining outcomes like job satisfaction (Parker, 2000). We describe next exceptions to this trend, mapping the studies onto Warr's categories (1997) of mental health that can be seen as active outcomes.

Competence

Can work design affect one's ability to handle problems and act on the environment with at least a moderate degree of success (Warr, 1997), such as might be indicated by effective coping, mastery, and self-efficacy? This proposition is very plausible. For example, if jobs are redesigned to enhance autonomy, then perceived mastery (or a belief that one can control the demands that occur) is an obvious potential outcome. A cross-sectional study provided initial support. Parker and Sprigg (1999) found that higher job autonomy (as well as lower job demands and proactive personality) positively predicted employees' sense of mastery over their work. In turn, mastery was associated with lower levels of job strain.

Job autonomy might also enable and motivate a problem-focused coping strategy, which research suggests will be better for employee well-being than an emotion-focused strategy (Latack, 1986). Job autonomy not only results in employees having greater opportunity for solving and preventing problems (Jackson, 1989), but it also potentially enhances employees' motivation to take ownership for a broad range of problems (Parker, Wall, & Jackson, 1997), promotes the development of knowledge needed to prevent faults (Wall, Jackson, & Davids, 1992), and has been associated with higher levels of "integrated understanding" (i.e., breadth of knowledge about the organization, such as understanding the bigger picture and knowing what other departments do) (Parker & Axtell, 2001, p. 1089). All of these learning-related changes are likely to enhance one's ability to cope with demands. To date, work design studies have tended to treat coping style as a moderating variable, such as showing that control buffers the effects of demands for those high on active coping but not those low on coping (De Rijk, Le Blance, Schaufeli, & de Jonge, 1998). If job autonomy were a determinant of active problem-focused coping, this would support Karasek and Theorell's idea (1990) of a positive spiral in which autonomous, challenging work results in employees being able to cope with even higher job demands.

Autonomy, Independence, and Proactivity

Mentally healthy people are usually considered to be reasonably autonomous, that is, able to determine their own opinions and actions rather than totally influenced by the environment (Warr, 1997). Of course, extreme levels of autonomy and independence might be undesirable, and Warr proposes that a balance between dependence and independence shows good mental health of this kind.

As we described, some of the early studies suggested that enriched work design might promote values for self-direction and achievement orientation. Consistent with these studies, work characteristics appear to affect employees' proactive motivation, or their motivation to act on the environment in a self-directed way to bring about change, such as by showing initiative, preventing problems, and scanning for opportunities (Parker, 2000). Parker proposed that, before individuals engage in more proactive tasks in their job, they need to see this as part of their role (i.e., having a "flexible role orientation," p. 448) and they need to feel confident that they can execute such behaviors (i.e., possessing "role breadth self-efficacy," p. 448). Work design has been linked to both of these aspects. First, Parker et al. (1997) showed that employees whose jobs were enriched via self-managing teams developed more flexible role orientations. Second, studies have shown that enhanced job autonomy and improved communication quality both predicted higher role breadth self-efficacy (Parker, 1998); whereas the deskilling and loss of decision-making influence associated with lean production reduced role breadth self-efficacy (Parker, in press). Others have similarly shown that enhanced job complexity leads to the display of personal initiative (Frese, Kring, Soose, and Zempel, 1996) and that this is partly attributable to the development of greater work-related self-efficacy (Speier & Frese, 1997). Studies have also shown a link between job autonomy and employee creativity (e.g., Oldham & Cummings, 1996).

Aspiration and Positive Self-Regard

A high level of aspiration is indicated by establishing and trying to attain realistic goals, engaging in motivated activity, and seeking to grow and develop (Warr, 1997). Positive self-regard is about having positive views about oneself, such as indicated by high self-esteem, self-worth, and self-acceptance. With the exception of some of the earlier studies (e.g., Kornhauser, 1965), scholars have given little attention to the idea that work characteristics might affect variables such as aspiration and self-esteem, which are typically considered relatively stable characteristics of individuals. Researchers have tended to focus on how these variables affect work design. For example, research shows that individuals with positive core self-evaluations (e.g., high self-esteem) perceived more intrinsic value in their work (Judge, Locke, Durham, & Kluger, 1998) and attain more complex jobs (Judge, Bono, & Locke, in press), suggesting that those with high self-esteem seek out or are allocated challenging jobs. The reverse path—that enriched jobs promote self-esteem—has not been tested.

In summary, longitudinal studies in the 1970s provided a solid base for proposing a link between work design and active mental health, yet there was surprisingly little follow-up research until quite recently. Evidence of a link between work design and outcomes such as self-efficacy and proactivity is now accumulating, although less attention has been given to outcomes like problem-focused coping, aspiration, and self-esteem. These outcomes might be important in their own right, as indicators of active mental health, and might also be mechanisms that enhance affective well-being.

Safety-Related Work Behaviors

Although occupational health research has traditionally used injury rates as an indicator of safety, there is growing interest in using more proximal and positive outcomes such as safety-related work behaviors. Consistent with the distinction between task and contextual performance (e.g., Borman & Motowidlo, 1993), researchers have identified two types of related safety behavior dimensions. Safety compliance (Griffin & Neal, 2000; also referred to as compliance-with-safety-rules, Marchand, Simard, Carpentier-Roy, & Ouellet, 1998) is analogous to task performance and refers to behaviors such as adhering to safety regulations, wearing protective equipment, and reporting incidents. Safety proactivity, also referred to as safety participation (Griffin & Neal, 2000), is analogous to contextual performance and focuses on discretionary, self-directed behaviors that make the workplace safer beyond the prescribed safety precautions, such as by volunteering to carry out safety audits.

To date, research about what predicts safety compliance has tended to focus on safety-specific antecedents rather than work design factors. For example, DeJoy, Searcy, Murphy, and Gershon (2000) found that availability of personal protective equipment predicted higher compliance among hospital personnel in terms of using the equipment. The same study also found that job-related hindrances were related to higher levels of personal protective equipment compliance as well as general compliance. This finding points to the importance of role demands, a factor indirectly implicated in Neal, Griffin, and Hart's study (2000) of Australian hospital staff. Their study found that safety knowledge predicted safety compliance, suggesting that employees who have a clear understanding of safety aspects will take appropriate safety precautions.

Although role conflict has not been specifically investigated as an antecedent of safety behaviors, the link is plausible. If different people such as peers or line managers have inconsistent expectations of an employee with regard to that person's tasks, this tension could negatively affect safety-related behaviors, in much the same way that role demands affect job performance more generally (Jackson & Schuler, 1985). Similarly, when employees face undue work pressure, such as heightened production goals, they may ignore or forget to follow safety procedures as they work to get the job done. Evidence for the negative effect of workload on safety comes from both qualitative (Collinson, 1999) and quantitative studies (Hofmann & Stetzer, 1996).

Although excessive role demands may hinder safety compliance and safety proactivity, work design can provide employees ways to prevent or manage these demands, such as in the form of enhanced autonomy. Again, the results are not definitive, at least for safety compliance. Simard and Marchand (1997) found that autonomy was not a significant predictor of workgroups' compliance with safety procedures. Conversely, Parker, Axtell, and Turner (2001) found in a longitudinal study that job autonomy was indirectly related to safety compliance via organizational commitment. Evidence of job autonomy as a predictor of proactive safety behaviors is more coherent. Simard and Marchand (1995) found that autonomy positively influenced workgroups' propensity to take safety initiatives; and Geller, Roberts, and Gilmore (1996) similarly found that individuals' task control positively linked with a construct that tapped proactive safety behaviors. It appears that employees with high job control have the opportunity or the motivation to get involved in safety tasks that fall outside of their job descriptions.

Designing Work for Organizational Health

Having reviewed the link between work design and individual health, we now address the broader construct of organizational health. To do this, we introduce a model of organizational health that conceptualizes multiple levels of effectiveness.

A Model of Organizational Health

NIOSH (1998) defines a healthy organization as one that has low rates of injury, illness, and disability in its workplace but also is competitive in the marketplace. Therefore, it is necessary to include the success of an organization, or what we refer to as its effectiveness, in the concept of organizational health. Our approach to this integration is depicted in Figure 4.1. Individual health outcomes are distinguished from effectiveness outcomes, and both aspects are related to individual behavior in the workplace. Together, these elements constitute the health of an organization.

Prior to discussing the ways in which work design might lead to organizational effectiveness, it is important to understand what we mean by this latter concept. Most definitions of effectiveness emphasize outcomes that achieve organizational goals (Pritchard, 1992). It is useful to differentiate outcomes at three levels of analysis, each of which can contribute to organizational goals: individual, team, and organizational (Klein & Kozlowski, 2000). For example, at the individual level, doing core tasks well is important for individual effectiveness. At the team level, effective teams are often defined in terms of their productivity and their sustainability. At the

**Figure 4.1. Model of Work Design
and Organizational Health.**

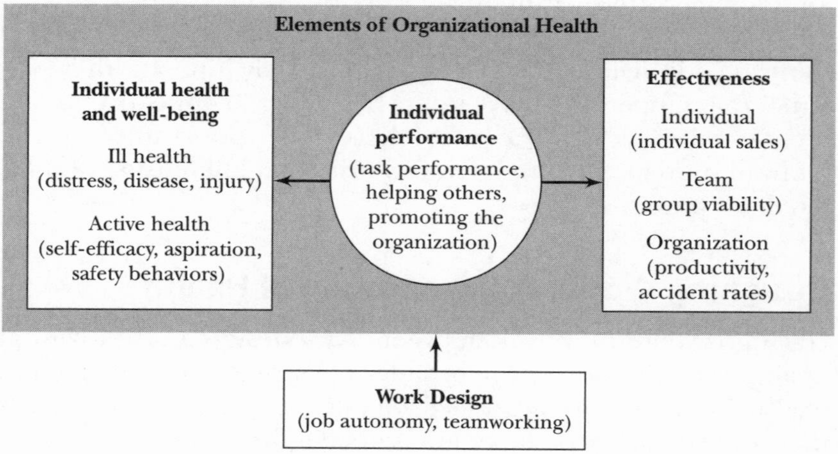

organizational level, discussions of effectiveness focus on productivity, profitability, and success in the face of external competition (Pritchard, 1992). Effectiveness indicators at lower levels are often aggregated to indicate effectiveness at higher levels of analysis (e.g., individual sales can be aggregated to provide a measure of organizational performance).

It is also important to differentiate effectiveness from individual performance, even though the terms are often used interchangeably. Individual performance describes the behaviors and activities that employees carry out at work (Campbell, McCloy, Oppler, & Sager, 1993). These behaviors might or might not be effective depending on circumstances. For example, an employee might come up with an idea for working more safely, but the department might refuse to buy the equipment needed to implement this idea. In this case the employee would not have achieved an effective outcome for the organization despite performing important work behaviors. An organizational feature constrained the link between individual performance and individual effectiveness (Griffin, Neal, & Neale, 2000).

Work Design and Effectiveness

Reviews suggest that work design can influence effectiveness, but the evidence is inconsistent (Parker & Turner, 2002). For example, Kopelman (1985) found performance was increased in 63% of the field studies that were reviewed, but Kelly (1992), in a review of 31 methodologically rigorous studies, found little evidence of a link between job perceptions and performance. One reason that work design might be inconsistently related to effectiveness is because important contingencies are neglected (Parker & Turner, 2002). For example, job enrichment has been shown to enhance productivity to a greater extent when there is high uncertainty than when there is low uncertainty (Wall, Corbett, Martin, Clegg, & Jackson, 1990). In addition, confusion about the link between work design and effectiveness outcomes might be resolved if we investigate more closely how work characteristics lead to effectiveness, including the multilevel processes underpinning this link. The latter issue has been particularly neglected, and we focus on it here.

In relation to organizational health, we propose three ways that work design can link to effectiveness. First, work design characteristics can have a direct impact on effectiveness outcomes. For example, one study showed lower costs for supervision following the introduction of autonomous work groups (Wall et al., 1986), and another showed faster employee response times to problems because employees were allowed to address problems themselves rather than waiting for supervisors or specialists (Jackson & Wall, 1991). In these examples, work design influenced effectiveness without reference to individual health and well-being.

Second, work design can have an indirect relationship with effectiveness through its impact on individual health. Work redesigns that enhance health might then translate into enhancing the effectiveness of organizations. This link is particularly important for understanding the bottom-up processes through which individuals contribute to organizational outcomes. For example, teamwork designs that enhance the job satisfaction and proactivity of individuals might improve the quality of team outputs (e.g., Podsakoff, Ahearne, & MacKenzie, 1997). Similarly, meta-analytic evidence (e.g., Spector, 1986) suggests that work characteristics such as autonomy are weakly associated with reduced withdrawal behaviors (e.g., absence and intention to leave), and although not necessarily tested, it is usually assumed this effect is mediated via well-being and mental health. Not all indirect processes linking work design and effectiveness will involve health. For example, work redesigns that enhance individual knowledge about the task can increase individual effectiveness (Leach, Jackson, & Wall, 2001), and work characteristics might affect outcomes like absence and turnover for reasons other than health.

We might consider a third process, which involves an indirect link between work design and health via effectiveness. This link occurs when effectiveness outcomes have a positive impact on individual well-being. For example, an increase in company productivity may improve job satisfaction and decrease distress associated with job insecurity among employees. This process incorporates top-down mechanisms in which features of the organization affect individual processes (Klein & Kozlowski, 2000).

Linking Work Design, Well-Being, and Effectiveness

All three processes we have described can occur simultaneously and involve complex causal directions. Clearly, the outcomes that constitute effectiveness cover a wide territory, and incorporating them all is beyond the scope of this chapter. Our focus is on the way that individual attitudes and behaviors contribute to effectiveness. Through this process a work design intervention that enhances well-being, active mental health, or safety might also lead to a more effective organization. Despite the intrinsic appeal of this process, it has proved difficult to articulate and demonstrate, particularly across levels of analysis (DeNisi, 2000).

The difficulties and possibilities of linking health and effectiveness are exemplified by the link between effectiveness and job satisfaction, which is the well-being outcome that has been most convincingly linked to work design. Early reviews of the link between job satisfaction and performance outcomes demonstrated, at best, modest relationships (e.g., Iaffaldano & Muchinsky, 1985). Successful job performance is also likely to affect individual job satisfaction rather than just the reverse relationship (Naylor, Pritchard, & Ilgen, 1980). Nevertheless, a recent meta-analysis shows larger associations between job satisfaction and performance than demonstrated previously (Judge, Thoresen, Bono, & Patton, 2001). Among other things, researchers are increasingly aware that they are more likely to observe the impact of job satisfaction on effectiveness when they consider outcomes more broadly than they do individual task performance (Borman & Motowidlo, 1993) and at more aggregate levels of analysis including over longer periods of time (Ostroff, 1992).

Turning to the first of these, we find that broader approaches to work performance encompass not only core task behaviors but also a wide range of activities under headings such as contextual performance (Borman & Motowidlo, 1993), citizenship behavior (Organ, 1988), and proactive behavior (Parker, 2000). Safety-related behaviors could also be considered as part of a broader definition of work performance (Griffin & Neal, 2000). Job satisfaction is an important predictor of many of these behaviors because they are largely under the motivational control of individuals and are influenced by affective states. Therefore, satisfaction can play a stronger role in effectiveness than that typically associated with job satisfaction and core task performance. In addition to job satisfaction, active mental health (e.g., self-efficacy) is likely to lead to these broader behaviors, especially those that involve being proactive (Parker, 2000).

A broader definition of work performance is particularly important for understanding the impact of health on team effectiveness. Many behaviors that constitute contextual performance, citizenship behavior, and prosocial behavior are directed toward the efficient functioning of groups. These behaviors contribute to team effectiveness by increasing the overall performance of a team (e.g., helping coworkers can enhance team performance) and by enhancing the long-term viability of the team (e.g., helping coworkers can promote a more positive working environment that attracts new members and helps retain current members). These types of prosocial behaviors are enhanced by enriched jobs (e.g., Parker & Axtell, 2001).

Broader work behaviors may also contribute directly to some aspects of organizational effectiveness. Behaviors such as promoting and defending the organization when it is criticized (Van Dyne, Graham, & Dienesch, 1994) do not necessarily contribute to individual or team effectiveness but can enhance the overall success of an organization.

Turning to the second way that we are more likely to observe an impact of job satisfaction on effectiveness, we note the importance of aggregating individual behaviors to the group and organizational level and of aggregating behaviors over time. Substantial evidence indicates a moderate link between job satisfaction and individual behaviors associated with ill health, absenteeism, and exit from the organization (see Warr, 1999). These types of behaviors and outcomes can be aggregated to the team and organizational level and retain their meaning as indicators of effectiveness. For example, injury rates aggregated at the departmental level can provide a measure of functioning where team-level interventions are important. Zohar (2002) found that supervisory practices could be modified to emphasize safety goals with a resulting decrease in group-level safety incidents. In this example the intervention was targeted at the level of analysis at which the effectiveness measure was aggregated. Considering aggregation over time is also important. For example, one consequence of individuals supporting each other is the increased viability of the team. However, one can observe this viability only by taking a long-term view.

In summary, considering a broader range of work behaviors suggests that work design can enhance effectiveness at each level of the organization via its impact on job satisfaction and the more active components of health. In addition, aggregating individual outcomes across levels of analysis and over time can provide more appropriate indicators of the impact of work design and the role of individual health.

Conclusions

In this final section, we summarize what we know about work design and health and identify areas for further investigation.

What Do We Know?

Most research investigating the link between work design and health outcomes has focused on indicators assessing the absence of health (e.g., psychological distress, CVD) or on relatively passive indicators of health (e.g., job satisfaction). We know from this research that work enrichment variables such as high job autonomy and skill use promote job satisfaction, particularly intrinsic job satisfaction. This conclusion is based on longitudinal studies investigating reverse causality, intervention studies, and studies using objective assessments of work characteristics. However, we can be less definitive about whether enhanced job autonomy affects organizational health via the reduction of psychological ill health (e.g., anxiety, depression, and psychosomatic symptoms), injuries, CVD, and musculoskeletal problems, although the evidence is somewhat more convincing in relation to the negative psychological health consequences of excess job demands. In some cases the ambiguity is attributable to the research area being quite new (e.g., only a small number of work design studies list injuries and musculoskeletal problems as outcomes); in other cases the ambiguity arises out of there being insufficient rigorous studies to draw causal inferences (e.g., the area of psychological health and CVD). Nevertheless, methodological advancements might well lead to more rather than less support for theoretical predictions, if research on the demand-control model is anything to go by. Results from increasingly rigorous and conceptually sophisticated studies in this field have given fresh impetus to the intuitively appealing but often unsupported idea that job control buffers the effects of job demands on mental health.

A particular contribution of this chapter has been to look at how work design might promote positive health, such as active mental health and safety-related behaviors. More recent studies linking work design to active mental health outcomes adds to the evidence from investigations conducted in the early 1980s on this topic. First, they suggest that work redesigned to prevent ill health might have important spin-offs for employee outcomes like active coping, aspiration, and proactivity. If individuals can develop and grow as a result of their job content, then work redesign is a more important and far-reaching intervention than hitherto assumed. Second, these findings suggest another route, and perhaps a more direct route, by which work design can increase effectiveness. For example, enhancing employee proactivity and mastery is likely to contribute directly to effectiveness outcomes such as customer satisfaction. Demonstrating links between work design, active mental health, and effectiveness will help to convince management to consider work redesign interventions that also can improve health.

Finally, we know that work characteristics affect safety through changing individuals' behavior, although we do not know exactly which characteristics are important or why. Initial data suggests that role demands such as workload and role ambiguity hinder safety compliance and safety proactivity, whereas job autonomy is particularly key for promoting safety proactivity.

Future Directions

In part some of the ambiguity about work design and aspects of health arises because of the lack of rigorous studies. A starting point is context-appropriate and conceptually nonconfounded measures of work characteristics (as far as possible, using existing measures so as to enhance consistency). Self-reported measures of work characteristics can be complemented with third-party ratings, or, if there is a group structure, the researcher can use multilevel analysis procedures that involve aggregating individual perceptions. Multiwave longitudinal studies are still required to rule out alternative causal explanations (see Garst, Frese, & Molennar, 2000, for an excellent example study). Of course, simply using longitudinal designs does not solve all problems. Zapf et al. (1996) identified common deficiencies in their critique of longitudinal stress studies, such as using weak designs (e.g., measuring only work characteristics at time 1 and well-being at time 2), not testing for reverse causality, and not considering time lags.

Intervention studies are also important: demonstrating a causal relationship is one thing, and showing that work redesign can be successfully designed and implemented with positive outcomes is quite another. One issue in intervention studies, however, is that researchers sometimes assume that work redesigns affect job characteristics (e.g., that autonomous work groups enhance job autonomy) without testing this assumption. We recommend testing a mediational model in which the intervention has its effects on outcomes via work characteristics. This approach has been adopted by researchers aiming to understand the effects of various practices, such as lean production (Jackson & Mullarkey, 2000) and temporary employment contracts (Parker, Griffin, Sprigg, & Wall, in press). We also recommend identifying which work characteristics, or combination of them, are driving the change in outcomes and disentangling change in work characteristics from co-occurring change in other systems (e.g., payment schemes).

In theoretical terms little attention has been given to how long work design is expected to have its effects or how long any effects are expected to last. Such time lags will depend on the outcome. Change in cardiovascular functioning, for example, might be observed only over several years, whereas change in affective well-being might occur relatively quickly and be more transient. Any discussion of time lags requires a closer understanding of the mechanisms that underpin the link between the particular work characteristic and the particular work outcome. For example, work redesign might affect proactivity via different time-dependent processes: quite quickly, by removing restrictive practices that inhibit an individual's natural propensity to be proactive; over the medium term, by motivating individuals to act in this way; and over the long term, by raising an individual's self-efficacy and promoting a more active coping style. The theoretical consideration of time lags concurs with a general move to understand how work characteristics affect performance, such as by considering cognitive processes rather than solely motivational ones (Parker et al., 2001). In relation to health outcomes, researchers have proposed many interesting and plausible propositions (e.g., Frese, 1989; Jackson, 1989; Karasek & Theorell, 1990; Terry & Jimmieson, 1999; Theorell & Karasek, 1996) but rarely tested them.

The value of considering mechanisms can be illustrated by considering how work design might affect active mental health. Earlier writings suggested that work can provide individuals with, or deprive them of, experiences that satisfy their basic needs and help them to develop into adults who feel capable of shaping their world (Kohn & Schooler, 1973). Brousseau (1978), in contrast, proposed that enriched jobs enhance cognitive complexity, such as the capacity for abstract thought, which in turn increases the extent to which an individual can generate elaborate plans for dealing with the environment. Other theoretically based mechanisms could be hypothesized. There is not likely to be a single process, and different mechanisms will operate according to the particular work characteristic, outcome, and time lag in question.

As well as considering processes over time, the construct of organizational health implicates processes across levels of analysis. For example, as we discussed earlier, work design might influence effectiveness within an organization via various bottom-up processes. Multilevel processes are not well articulated or investigated in the work design literature (or indeed in the wider stress domain, Bliese & Jex, 1999). This review shows the value of considering multiple levels to understand perceived work characteristics and the extent to which these characteristics are individually based perceptions or shared within units. Jackson (1989) proposed an uncertainty framework for studies of job control that identifies origins of and responses to uncertainty at different levels of analysis. This model implicates top-down processes by which individuals might form their perceptions of work characteristics. Thus, the individual's feelings of control reflect his or her own job control as well as perceptions of control that derive from the larger systems the individual is embedded within, such as the department's power within the organization or the organization's power within the industry. We have also primarily focused on individual health, but group-level health can be considered (Sonnentag, 1996).

Finally, theoretical developments for work design proposed elsewhere also apply here. For example, to manage the scope of our review, we focused mostly on job demands and job control. Of course, many additional important work characteristics exist (e.g. Parker et al., 2001; Theorell & Karasek, 1996); some have been investigated a great deal (e.g., role conflict), but others that have not been examined much yet are important in the modern context (e.g., performance monitoring, emotional demands). We have also seen in this review an increasing attention to the interplay between work and individual factors, such as how individual characteristics (e.g., type A/proactive personality) might moderate responses to work design. As well as the need to continue to investigate these types of moderating influences, organizational contingencies deserve attention. For example, job autonomy has been shown to have greater performance benefits in uncertain environments where flexible responses are required (e.g., Wall et al., 1990). Jackson (1989) made a similar argument in relation to health, proposing that autonomy is important because, when faced with the stressor of uncertainty, people who feel in control will be more likely to use proactive problem solving rather than emotion-focused coping. One could therefore propose that where organizational uncertainty is high, job autonomy might be particularly important for well-being. Other organizational factors such as the level of downsizing (Dwyer & Fox, 2000) or the degree of work interdependence (Sprigg, Jackson, & Parker, 2000) might also affect the receptiveness and appropriateness of work redesign for employee health.

References

Argyris, C. (1957). *Personality and organization.* New York: HarperCollins.

Bliese, P. D., & Jex, S. M. (1999). Incorporating multiple levels of analysis into occupational stress research. *Work and Stress, 13,* 1–6.

Bongers, P. M., de Winter, C. R., Kompier, M. A., Hildebrandt, V. H. (1993). Psychosocial factors at work and musculoskeletal disease. *Scandinavian Journal of Work, Environment, and Health, 19,* 297–312.

Borman, W. C., & Motowidlo, S. J. (1993). Expanding the criterion domain to include elements of contextual performance. In N. Schmitt & W. C. Borman and Associates (Eds.), *Personnel selection in organizations* (pp. 71–98). San Francisco: Jossey-Bass.

Bosma, H., Stansfeld, S. A., & Marmot, M. G. (1998). Job control, personal characteristics, and heart disease. *Journal of Occupational Health Psychology, 3,* 402–409.

Brousseau, K. R. (1978). Personality and job experience. *Organizational Behavior and Human Decision Processes, 22,* 235–252.

Bruggeman, A., Groskurth, P., & Ulich, E. (1975). *Arbeitszufriedenheit.* Bern: Huber.

Campbell, J. P., McCloy, R. A., Oppler, S. H., & Sager, C. E. (1993). A theory of performance. In N. Schmitt, W. C. Borman and Associates (Eds.), *Personnel selection in organizations*. San Francisco: Jossey-Bass.

Campion, M. A., & McClelland, C. L. (1993). Follow-up and extension of the inter-disciplinary costs and benefits of enlarged jobs, *Journal of Applied Psychology, 78*, 339–351.

Campion, M. A., & Thayer, P. W. (1987). Job design: Approaches, outcomes, and trade-offs. *Organizational Dynamics, 15*, 66–79.

Collinson, D. L. (1999). "Surviving the rigs": Safety and surveillance on North Sea oil installations. *Organization Studies, 20*, 579–600.

De Jonge, J., Dormann, C., Janssen, P.P.M., Dollard, M. F., Landeweerd, J. A., & Nijhuis, F.J.N. (2001). Testing reciprocal relationships between job characteristics and psychological well-being: A cross-lagged structural equation model. *Journal of Occupational and Organizational Psychology, 74*, 29–46.

De Jonge, J., & Kompier, M.A.J. (1997). A critical examination of the demand-control-support model from a work psychological perspective. *International Journal of Stress Management, 4*, 235–258.

De Jonge, J., van Breukelen, G.J.P., Landeweerd, J. A., & Nijhuis, F.J.N. (1999). Comparing group and individual level assessments of job characteristics in testing the job demand-control model: A multilevel approach. *Human Relations, 52*, 95–122.

DeJoy, D. M., Searcy, C. A., Murphy, L. R., & Gershon, R.R.M. (2000). Behavioral-diagnostic analysis of compliance with universal precautions among nurses. *Journal of Occupational Health Psychology, 5*, 127–141.

DeNisi, A. S. (2000). Performance appraisal and performance management: A multilevel analysis. In K. J. Klein & S.J.W. Kozlowski (Eds.), *Multilevel theory, research, and methods in organizations: Foundations, extensions, and new directions* (pp. 121–156). San Francisco: Jossey-Bass.

De Rijk, A. E., Le Blance, P. M., Schaufeli, W. B., & de Jonge, J. (1998). Active coping and need for control as moderators of the job demand-control model: Effects on burnout. *Journal of Occupational and Organizational Psychology, 71*, 1–18.

Dwyer, D., & Fox, M. L. (2000). The moderating role of hostility in the relationship between enriched jobs and health. *Academy of Management Journal, 43*, 1086–1096.

Dwyer, D., & Ganster, D. C. (1991). The effects of job demand and control on employee attendance and satisfaction. *Journal of Organizational Behavior, 12*, 595–608.

Elovainio, M., Kivimaeki, M., Steen, N., & Kalliomaeki-Levanto, T. (2000). Organizational and individual factors affecting mental health and job satisfaction: A multilevel analysis of job control and personality. *Journal of Occupational Health Psychology, 5,* 269–277.

Elovainio, M., & Sinervo, T. (1997). Psychosocial stressors at work, psychological stress and musculoskeletal symptoms in the care for the elderly. *Work and Stress, 11,* 351–361.

Emery, F. E. (1959). *Characteristics of socio-technical systems* (Document No. 527). London: Tavistock Institute of Human Relations.

Fox, M. L., Dwyer, D. J., & Ganster, D. C. (1993). Effects of stressful job demands and control on physiological and attitudinal outcomes in a hospital setting. *Academy of Management Journal, 36,* 289–318.

Fraser, R. (1947). *The incidence of neurosis among factory workers* (Industrial Health Research Board, Report No. 90). London: HMSO.

Frese, M. (1985). Stress at work and psychosomatic complaints: A causal interpretation. *Journal of Applied Psychology, 70,* 314–328.

Frese, M. (1989). Theoretical models of control and health. In S. L. Sauter, J. J. Hurrell Jr., & C. L. Cooper (Eds.), *Job control and worker health* (pp. 108–128). New York: Wiley.

Frese, M., Kring, W., Soose, A., & Zempel, J. (1996). Personal initiative at work: Differences between East and West Germany. *Academy of Management Journal, 39,* 37–63.

Fried, Y., & Ferris, G. R. (1987). The validity of the job characteristics model: A review and meta-analysis. *Personnel Psychology, 40,* 287–322.

Frone, M. R. (1998). Predictors of work injuries among employed adolescents. *Journal of Applied Psychology, 83,* 565–576.

Ganster, D. C., & Fusilier, M. R. (1989). Control in the workplace. In C. L. Cooper & I. T. Robertson (Eds.), *International review of industrial and organizational psychology* (Vol. 14, pp. 235–280). New York: Wiley.

Garst, H., Frese, M., & Molennar, P.C.M. (2000). The temporal factor of change in stressor-strain relationships: A growth curve model on a longitudinal study in East Germany. *Journal of Applied Psychology, 85,* 417–438.

Geller, E. S., Roberts, D. S., & Gilmore, M. R. (1996). Predicting propensity to actively care for occupational safety. *Journal of Safety Research, 27,* 1–8.

Glick, W. H., Jenkins, G. D., & Gupta, N. (1986). Method versus substance: How strong are underlying relationships between job characteristics and attitudinal outcomes? *Academy of Management Journal, 29,* 441–464.

Griffin, M. A., & Neal, A. (2000). Perceptions of safety at work: A framework for linking safety climate to safety performance, knowledge, and motivation. *Journal of Occupational Health Psychology, 5,* 347–358.

Griffin, M. A., Neal, A., & Neale, M. (2000). The contribution of task performance and contextual performance to effectiveness: Investigating the role of situational constraints. *Applied Psychology: An International Review, 49,* 516–532.

Griffin, R. W. (1991). Effects of work redesign on employee perceptions, attitudes, and behaviors: A long-term investigation. *Academy of Management Journal, 34,* 425–435.

Hackman, J. R., & Oldham, G. R. (1976). Motivation through the design of work: Test of a theory. *Organizational Behaviour and Human Performance, 16,* 250–279.

Hemingway, M. A., & Smith, C. S. (1999). Organizational climate and occupational stressors as predictors of withdrawal behaviours and injuries in nurses. *Journal of Occupational and Organizational Psychology, 72,* 285–299.

Hofmann, D. A., & Stetzer, A. (1996). A cross-level investigation of factors influencing unsafe behaviors and accidents. *Personnel Psychology, 49,* 307–339.

Hollmann, S., Heuer, H., & Schmidt, K. (2001). Control at work: A generalized resource factor for the prevention of musculoskeletal symptoms? *Work and Stress, 15,* 29–39.

Iaffaldano, M. R., & Muchinsky, P. M. (1985). Job satisfaction and job performance: A meta-analysis. *Psychological Bulletin, 97,* 251–273.

Iverson, R. D., & Erwin, P. J. (1997). Predicting occupational injury: The role of affectivity. *Journal of Occupational and Organizational Psychology, 70,* 113–128.

Jackson, P. R., & Mullarkey, S. (2000). Lean production teams and health in garment manufacture. *Journal of Occupational Health Psychology, 5,* 231–245.

Jackson, P. R., & Wall, T. D. (1991). How does operator control enhance performance of advanced manufacturing technology? *Ergonomics, 34,* 1301–1311.

Jackson, S. E. (1989). Does job control control job stress? In S. L. Sauter, J. J. Hurrell Jr., & C. L. Cooper (Eds.), *Job control and worker health* (pp. 25–51). New York: Wiley.

Jackson, S. E., & Schuler, R. S. (1985). A meta-analysis and conceptual critique of research on role ambiguity and role conflict in work settings. *Organizational Behavior and Human Decision Processes, 36,* 16–78.

James, J. R., & Tetrick, L. E. (1986). Confirmatory analytic tests of three causal models relating job perceptions to job satisfaction. *Journal of Applied Psychology, 71,* 77–82.

Jones, F., & Fletcher, B. C. (1996). Job control and health. In M. Schabracq, J.A.M. Winnubst, & C. L. Cooper, (Eds.), *Handbook of work and health psychology* (pp. 33–50). New York: Wiley.

Judge, T. A., Bono, J. E., & Locke, E. A. (in press). Personality and job satisfaction: The mediating role of job characteristics. *Journal of Applied
Psychology*.

Judge, T. A., Locke, E. A., Durham, C. C., & Kluger, A. N. (1998). Dispositional effects on job and life satisfaction: The role of core evaluations. *Journal of Applied Psychology, 83,* 17–34.

Judge, T. A., Thoresen, C. J., Bono, J. E., & Patton, G. K. (2001). The job
satisfaction–job performance relationship: A qualitative and quantitative review. *Psychological Bulletin, 127,* 376–407.

Karasek, R. (1979). Job demands, job decision latitude, and mental strain:
Implications for job redesign. *Administrative Science Quarterly, 24,*
285–308.

Karasek, R. (1989). Control in the workplace and its health-related aspects. In S. L. Sauter, J. J. Hurrell Jr., & C. L. Cooper (Eds.), *Job control and worker health* (pp. 129–160). New York: Wiley.

Karasek, R. A., & Theorell, T. (1990). *Healthy work: Stress, productivity, and
the reconstruction of working life.* New York: Basic Books.

Kelly, J. E. (1992). Does job re-design theory explain job re-design outcomes? *Human Relations, 45,* 753–774.

Kinicki, A. J., McKee-Ryan, F. M., Schriesheim, C. A., & Carson, K. P.
(2002). Assessing the construct validity of the Job Descriptive Index:
A review and meta-analysis. *Journal of Applied Psychology, 87,* 14–32.

Klein, K. J., & Kozlowski, S.J.W. (2000). A multilevel approach to theory
and research in organizations: Contextual, temporal, and emergent
processes. In K. J. Klein & S.J.W. Kozlowski (Eds.), *Multilevel theory,
research, and methods in organizations: Foundations, extensions, and new
directions* (pp. 3–90). San Francisco: Jossey-Bass.

Kohn, M. L., & Schooler, C. (1973). Occupational experience and psychological functioning: An assessment of reciprocal effects. *American Sociological Review, 38,* 97–118.

Kohn, M. L., & Schooler, C. (1978). The reciprocal effects of the substantive complexity of work on intellectual complexity: A longitudinal assessment. *American Journal of Sociology, 84,* 24–52.

Kohn, M. L., & Schooler, C. (1982). Job conditions and personality: A longitudinal assessment of their reciprocal effects. *American Journal of
Sociology, 87,* 1257–1286.

Kopelman, R. E. (1985). Job redesign and productivity: A review of the
evidence. *National Productivity Review, 4,* 237–255.

Kornhauser, A. (1965). *Mental health of the industrial worker.* New York:
Wiley.

Kristensen, T. S. (1995). The demand-control-support model: Methodological challenges for future research. *Stress Medicine, 11,* 17–26.

Latack, J. C. (1986). Coping with job stress: Measures and future directions for scale development. *Journal of Applied Psychology, 71,* 377–385.

Leach, D. J., Jackson, P. R., & Wall, T. D. (2001). Realizing the potential of empowerment: The impact of a feedback intervention on the performance of complex technology. *Ergonomics, 44,* 870–886.

Lee, R. T., & Ashforth, B. E. (1996). A meta-analytic examination of the correlates of the three dimensions of job burnout. *Journal of Applied Psychology, 81,* 123–133.

Locke, E. A. (1976). The nature and causes of job satisfaction. In M. D. Dunnette (Ed.), *Handbook of industrial and organizational psychology* (pp. 1297–1343). Skokie, IL: Rand McNally.

Loher, B. T., Noe, R. A., Moeller, N. L., & Fitzgerald, M. P. (1985). A meta-analysis of the relation of job characteristics to job satisfaction. *Journal of Applied Psychology, 70,* 280–289.

Marchand, A., Simard, M., Carpentier-Roy, M. C., & Ouellet, F. (1998). From a unidimensional to a bidimensional concept and measurement of workers' safety behavior. *Scandinavian Journal of Work, Environment, and Health, 24,* 293–299.

Martin, R., & Wall, T. D. (1989). Double machine operation and psychological strain. *Work and Stress, 3,* 323–326.

Maslach, C., & Leiter, M. P. (1997). *The truth about burnout: What organizations do to cause personal stress and what to do about it.* San Francisco: Jossey-Bass.

National Institute for Occupational Safety and Health (NIOSH). (1998). http://www2.cdc.gov/NORA/default.html

National Safety Council. (1999). *Injury facts.* Itasca, IL: National Safety Council.

Naylor, J. D., Pritchard, R. D., & Ilgen, D. R. (1980). *A theory of behavior in organizations.* Orlando, FL: Academic Press.

Neal, A., Griffin, M. A., & Hart, P. M. (2000). The impact of organizational climate on safety climate and individual behavior. *Safety Science, 34,* 99–109.

Oldham, G. R., & Cummings, A. (1996). Employee creativity: Personal and contextual factors at work. *Academy of Management Journal, 39,* 607–634.

Organ, D. W. (1988). *Organizational citizenship behavior: The good soldier syndrome.* San Francisco: New Lexington Press.

Orth-Gomer, K., Eriksson, I., Moser, V., Theorell, T., & Fredlund, P. (1994). Lipid lowering through work stress reduction. *International Journal of Behavioural Medicine, 1,* 204–214.

Ostroff, C. (1992). The relationship between satisfaction, attitudes, and performance: An organizational level analysis. *Journal of Applied Psychology, 77,* 963–974.

Parker, S. K. (1998). Role breadth self-efficacy: Relationship with work enrichment and other organizational practices. *Journal of Applied Psychology, 83,* 835–852.

Parker, S. K. (2000). From passive to proactive motivation: The importance of flexible role orientations and role breadth self-efficacy. *Applied Psychology: An International Review, 49,* 447–469.

Parker, S. K. (in press). Longitudinal effects of lean production on employee outcomes and the mediating role of work characteristics. *Journal of Applied Psychology.*

Parker, S. K., & Axtell, C. M. (2001). Seeing another point of view: Antecedents and outcomes of employee perspective taking. *Academy of Management Journal, 44,* 1085–1100.

Parker, S. K., Axtell, C. M., & Turner, N. (2001). Designing a safer workplace: Importance of job autonomy, communication quality, and supportive supervisors. *Journal of Occupational Health Psychology, 6,* 211–228.

Parker, S. K., Chmiel, N., & Wall, T. D. (1997). Work characteristics and employee well-being with a context of strategic downsizing. *Journal of Occupational Health Psychology, 2,* 289–303.

Parker, S. K., Griffin, M. A., Sprigg, C., and Wall, T. D. (in press). Effect of temporary contracts on perceived work characteristics and job strain: A longitudinal study. *Personnel Psychology.*

Parker, S. K., & Sprigg, C. A. (1999). Minimizing strain and maximizing learning: The role of job demands, job control, and proactive personality. *Journal of Applied Psychology, 84,* 925–939.

Parker, S. K., & Turner, N. (2002). Work design and individual job performance: Research findings and an agenda for future inquiry. In S. Sonnentag (Ed.), *The psychological management of individual performance: A handbook in the psychology of management in organizations* (pp. 69–94). New York: Wiley.

Parker, S. K., & Wall, T. D. (1998). *Job and work design: Organizing work to promote well-being and effectiveness.* Thousand Oaks, CA: Sage.

Parker, S. K., Wall, T. D., & Cordery, J. (2001). Future work design research and practice: Towards an elaborated model of work design. *Journal of Occupational and Organizational Psychology.*

Parker, S. K., Wall, T. D., & Jackson, P. R. (1997). "That's not my job": Developing flexible employee work orientations. *Academy of Management Journal, 40,* 899–929.

Parkes, K. R. (1991). Locus of control as moderator: An explanation for additive versus interactive findings in the demand-discretion model of work stress? *British Journal of Psychology, 82,* 291–312.

Parkes, K. R. (1995). The effects of objective workload on cognitive performance in a field setting: A two-period cross-over trial. *Applied Cognitive Psychology, 9,* S153–S157.

Parkes, K. R., Mendham, C. A., & Von Rabenau, C. (1994). Social support and the demand-discretion model of job stress: Tests of additive and interactive effects in two samples. *Journal of Vocational Behavior, 44,* 91–113.

Pearson, C.A.L. (1992). Autonomous work groups: An evaluation at an industrial site. *Human Relations, 45,* 905–936.

Podsakoff, P., Ahearne, M., & MacKenzie, S. B. (1997). Organizational citizenship behavior and the quantity and quality of work group performance. *Journal of Applied Psychology, 82,* 262–270.

Pritchard, R. D. (1992). Organizational productivity. In M. D. Dunnette & L. M. Hough (Eds.), *Handbook of industrial and organizational psychology* (2nd ed., Vol. 3, pp. 443–471). Palo Alto, CA: Consulting Psychologists Press.

Sargent, L. D., & Terry, D. J. (1998). The effects of work control and job demands on employee adjustment and work performance. *Journal of Occupational and Organizational Psychology, 71,* 219–236.

Schaubroeck, J., Ganster, D. C., & Kemmerer, B. E. (1994). Job complexity, "type A" behavior, and cardiovascular disorder: A prospective study. *Academy of Management Journal, 37,* 426–439.

Schaubroeck, J., Lam, S.S.K., & Xie, J. L. (2000). Collective efficacy versus self-efficacy in coping responses to stressors and control: A cross-cultural study. *Journal of Applied Psychology, 85,* 512–525.

Schnall, P. L., Landsbergis, P. A., & Baker, D. (1994). Job strain and cardiovascular disease. *Annual Review of Public Health, 15,* 381–411.

Simard, M., & Marchand, A. (1995). A multilevel analysis of organizational factors related to the taking of safety initiatives by work groups. *Safety Science, 21,* 113–129.

Simard, M., & Marchand, A. (1997). Workgroups' propensity to comply with safety rules: The influence of micro-macro organizational factors. *Ergonomics, 40,* 172–188.

Sonnentag, S. (1996). Work group factors and individual well-being. In M. A. West (Ed.), *Handbook of work group psychology* (pp. 345–370). New York: Wiley.

Sonnentag, S. (2001). Work, recovery activities, and individual well-being: A diary study. *Journal of Occupational Health Psychology, 6,* 196–210.

Spector, P. E. (1986). Perceived control by employees: A meta-analysis of studies concerning autonomy and participation at work. *Human Relations, 39,* 1005–1016.

Speier, C., & Frese, M. (1997). Generalized self-efficacy as a mediator and moderator between control and complexity at work and personal initiative: A longitudinal study in East Germany. *Human Performance,* *10,* 171–192.

Sprigg, C. A., Jackson, P. R., & Parker, S. K. (2000). Production team-working: The importance of interdependence for employee strain and satisfaction. *Human Relations, 53,* 1519–1543.

Stansfeld, S. A., Fuhrer, R., Head, J., Ferrie, J., & Shipley, M. (1997). Work and psychiatric disorder in the Whitehall II study. *Journal of Psychosomatic Research, 43,* 73–81.

Steel, R. P., & Rentsch, J. R. (1997). The dispositional model of job attitudes revisited: Findings of a ten-year study. *Journal of Applied Psychology, 82,* 873–879.

Stone, E. F. (1986). Job scope–job satisfaction and job scope–job performance relationships. In E. A. Locke (Ed.), *Generalizing from laboratory to field settings.* San Francisco: New Lexington Press.

Terry, D. J., & Jimmieson, N. L. (1999). Work control and employee well-being: A decade review. In C. L. Cooper & I. T. Robertson (Eds.), *International review of industrial and organizational psychology* (Vol. 14, pp. 95–148). New York: Wiley.

Teuchmann, K., Totterddell, P., & Parker, S. K. (1999). Rushed, unhappy, and drained: An experience sample study of relations between time pressure, perceived control, mood, and emotional exhaustion in a group of accountants. *Journal of Occupational Health Psychology, 4,* 37–54.

Theorell, T., & Karasek, R. A. (1996). Current issues relating to psychosocial job strain and cardiovascular disease research. *Journal of Occupational Health Psychology, 1,* 9–26.

Trist, E. L., & Bamforth, K. W. (1951). Some social and psychological consequences of the long-wall method of coal-getting. *Human Relations, 4,* 3–38.

Van Der Doef, M., & Maes, S. (1999). The job demand-control (-support) model and psychological well-being: A review of twenty years of empirical research. *Work and Stress, 13,* 87–114.

Van Dyne, L., Graham, J. W., & Dienesch, R. M. (1994). Organizational citizenship behavior: Construct redefinition, measurement, and validation. *Academy of Management Journal, 37,* 765–802.

Van Yperen, N. W., & Snijders, T.A.B. (2000). A multi-level analysis of the demands-control model: Is stress at work determined by factors at the group level or the individual level? *Journal of Occupational Health Psychology, 5,* 182–190.

Wall, T. D., & Clegg, C. W. (1981). A longitudinal study of group work redesign. *Journal of Occupational Behaviour, 2,* 31–49.

Wall, T. D., Corbett, M. J., Martin, R., Clegg, C. W., & Jackson, P. R. (1990). Advanced manufacturing technology, work design, and performance: A change study. *Journal of Applied Psychology, 75,* 691–697.

Wall, T. D., Jackson, P. R., & Davids, K. (1992). Operator work design and robotics system performance: A serendipitous field study. *Journal of Applied Psychology, 77,* 353–362.

Wall, T. D., Jackson, P. R., Mullarkey, S., & Parker, S. K. (1996). The demands-control model of job strain: A more specific test. *Journal of Occupational and Organizational Psychology, 69,* 153–166.

Wall, T. D., Kemp, N. J., Jackson, P. R., & Clegg, C. W. (1986). An outcome evaluation of autonomous work groups: A long-term field experiment. *Academy of Management Journal, 29,* 280–304.

Warr, P. B. (1987). *Work, unemployment, and mental health.* Oxford: Clarendon Press.

Warr, P. B. (1990). The measurement of well-being and other aspects of mental health. *Journal of Occupational Psychology, 52,* 129–148.

Warr, P. B. (1994). A conceptual framework for the study of work and mental health. *Work and Stress, 8,* 84–97.

Warr, P. B. (1997). Age, work, and mental health. In K. W. Schaie & C. Schooler (Eds.), *The impact of work on older adults* (pp. 252–296). New York: Springer.

Warr, P. B. (1999). Well-being and the workplace. In D. Kahneman, E. Diener, & N. Schwarz (Eds.), *Well-being: The foundations of hedonic psychology* (pp. 392–412). New York: Russell Sage Foundation.

Wong, C., Hui, C., & Law, K. S. (1998). A longitudinal study of the job perception–job satisfaction relationship: A test of the three alternative specifications. *Journal of Occupational and Organizational Psychology, 71,* 127–146.

Xie, J. L., & Johns, G. (1995). Job scope and stress: Can job scope be too high? *Academy of Management Journal, 38,* 1288–1309.

Zapf, D., Dormann, C., & Frese, M. (1996). Longitudinal studies in organizational stress research: A review of the literature with reference to methodological issues. *Journal of Occupational Health Psychology, 1,* 145–169.

Zohar, D. (2002). Modifying supervisory practices to improve sub-unit safety: A leadership-based intervention model. *Journal of Applied Psychology, 87,* 156–163.

CHAPTER 5

Group and Normative Influences on Health and Safety

Perspectives from Taking a Broad View on Team Effectiveness

Paul Tesluk
Narda R. Quigley

It is not surprising that industrial/organizational (I/O) psychologists focus a great deal of their attention on understanding groups and teams in organizations. A fundamental shift has been occurring as groups emerge as the critical unit of analysis in relation to task accomplishment and performance. The structure of work has been changing such that work groups and teams, rather than individuals, now form the basic building blocks of the modern organization (Guzzo & Shea, 1992; Ilgen, 1999; Ilgen & Pulakos, 1999). This is reflected in the increasing rates at which organizations have implemented production, service, cross-functional, self-managing, and other types of teams during the past 20 years (e.g., Lawler, Mohrman, & Ledford, 1995). There are many reasons for the rapid adoption of different forms of work groups and teams, and we do not need to repeat them all here. But included among them are that teams can provide a greater combination of different types of skills, perspectives, and resources; enable more flexibility in using talents and resources; and spark potential synergies that result in the whole being more than the sum of the individual contributors' parts (Hackman, 1998). Given the potential benefits that work groups and teams offer organizations, it is therefore not surprising that a tremendous amount of research in recent years has been devoted to understanding the factors that facilitate their effectiveness (for reviews, see, e.g., Cohen & Bailey, 1997; Guzzo & Dickson, 1996).

Besides the performance benefits they bring to the organization, work groups are important to us as I/O psychologists because of the influence they have on their members' well-being. Since the time of the classic Hawthorne studies (Roethlisberger & Dickson, 1939), I/O psychologists have been well aware of the important role of groups on individuals' attitudes and behaviors at work (see Guzzo & Shea, 1992; Hackman, 1992). Individuals typically feel more committed to their work groups than to their organizations, supervisors, or senior managers (Becker, 1992; Lawler, 1992). In addition, groups often have a greater influence on socializing new employees than do organizations (Moreland & Levine, 2001), and they serve as a principal source for satisfying needs of belonging, support, and encouragement (Nelson, Quick, & Eakin, 1988; Riordan & Griffith, 1995; Sonnentag, 1996; Zaccaro & Dobbins, 1989).

Work groups and teams play an even more important role in terms of employees' attitudes, behavior, and adjustment as organizations and work are becoming more individualized (Lawler & Finegold, 2000). For instance, many organizations are rapidly moving to establish virtual workplaces: environments that provide employees with the flexibility to work at home and adjust their work schedules, thereby providing individuals with more self-control and self-direction (i.e., tailoring work and organizational experiences to the needs of individual workers). Yet even when people work outside of the office, work groups, often in the form of virtual teams whose members are connected via various forms of information and communications technology, serve as a primary means by which employees stay connected and communicate with their colleagues and organizations (Cascio, 2000; Kirkman, Rosen, Gibson, Tesluk, & McPherson, in press). In short, groups form a critical role in supporting our well-being and shaping our social attachments (Sonnentag, 1996; Zaccaro & Riley, 1987), and all indications are that they will become more (not less) important in the evolving modern workplace.

Finally, consistent with the theme of teams as a critical unit of inquiry is work attending to how they both serve as vehicles for and provide contexts that support important long-term criteria including learning and adaptability. Organizations have strived to increase their flexibility, innovation, and learning by adopting flat, decentralized structures that use teams as a means for coping with uncertainty (Mohrman, Cohen, & Mohrman, 1995). Teams are also seen as the most efficient vehicles for creating knowledge in modern organizations (Nonaka & Takeuchi, 1995), for example, by serving to bring together diverse perspectives to bear on problems, as in the case of cross-functional teams (Denison, Hart, & Khan, 1996). In addition, teams serve as important contexts that can influence the willingness and ability of team members to engage in learning by seeking feedback, sharing information, asking for help, talking openly about errors and mistakes, and experimenting (Edmondson, 1999). For instance, teams in climates in which team members routinely respond to errors with punitive actions and the assigning of blame have been found to interfere with members' willingness to admit and then discuss mistakes and problems in ways that facilitate learning (Edmondson, 1996; Hofmann & Stetzer, 1998).

Work groups and teams, then, in and of themselves, are an important unit of analysis for understanding health as this book broadly defines it. That is, work groups and teams relate to health at different levels of analysis and do so by (1) helping to avoid injuries and illnesses (e.g., by providing a climate that supports attention to safety concerns, resulting in fewer accidents and injuries) (Hofmann & Stetzer, 1996) and (2) serving to promote effectiveness and effective functioning (e.g., by facilitating an organization's competitive advantage through the development of new innovations) (Nonaka & Takeuchi, 1995).

Our chapter connects to this volume's themes by relating group and normative influences to the notion of health as both the absence of injury or illness and the presence of well-being or the ability to flourish and be self-sustaining (see Chapter One). Because other chapters address organizational-level issues, we are primarily concerned with the safety, health, and well-being of work groups and teams themselves as well as how they serve as contexts that influence the health and well-being of their members. Also, consistent with one of the themes in this volume, that health involves balancing competing goals, we consider group and normative influences from the perspective of short- versus long-term goals and as serving extrinsic versus intrinsic needs.

Importance of Group and Normative Influences: The Case of Safety

To help set the stage for our discussion of the role of group and normative influences, we begin by considering the specific domain of safety. We do so for two reasons. First, relative to other criteria, the role of group and team factors in safety has been significantly understudied (Turner & Parker, in press). Second, this analysis reveals a set of what we might call pressure points or levers whereby group and normative variables shape the health and safety of teams and their members. By elucidating these mechanisms, we can more clearly identify how group and team phenomena influence health and safety, develop a set of prescriptive guidelines for promoting team effectiveness, and make suggestions for needed research in these areas.

I/O psychologists have increasingly examined the role of group and normative influences on many of the phenomena that have traditionally been viewed primarily at the individual level of analysis. Examples include absenteeism (Mathieu & Kohler, 1990; Nicholson & Johns, 1985), turnover (Martocchio, 1994), and work attitudes such as commitment and satisfaction (Mathieu, 1991). The shift to focusing on group and normative influences on safety, however, has been slow to develop. Although improving safety was traditionally considered an engineering issue, I/O psychology's initial attention to it has been on developing interventions focusing on individual behavior. Commonly identified as behavioral safety performance management techniques, they typically include a combination of strategies directed toward employees, such as safety behavior training, promotion of safety awareness, goal setting, feedback, and rewards or incentives (e.g., Komaki, Barwik, & Scott, 1978; Komaki, Heinzmann, & Lawson, 1980; Reber & Wallin, 1984, 1994; Reber, Wallin, & Chhokar, 1990).

Reviews of these types of intervention efforts have made two important observations. The first is that the vast majority of behaviorally based safety interventions (e.g., safety training, feedback) have focused on the individual worker (DeJoy, 1996; Hofmann, Jacobs, & Landy, 1995; McAfee & Winn, 1989; Sulzer-Azaroff, Harris, & McCann, 1994). This reflects an implicit assumption that the underlying causes for unsafe acts and accidents reside primarily with the worker who committed the unsafe act. However, strong evidence suggests that factors at the work-group level of analysis often function as contributing, if not primary, factors in influencing accidents (Reason, 1990). For example, in reviewing the causes for fatal accidents of British offshore oil workers in the North Sea, Wright (1986) found several common themes that directly pointed to the importance of group and normative influences. For instance, most accidents occurred when workers were engaged in what they considered to be routine work procedures. In short, they were doing their work the same way as everyone else in their work group; they were following well-accepted expectations (i.e., norms) within their work groups for how to go about performing their job. However, as the accident reviews demonstrated, when examined after the fact, these were clearly unsafe procedures.

A second observation that can be made regarding worker-focused safety interventions is that even though these interventions are often effective, improvements could be made by directing more attention to group and normative factors (DeJoy, 1996; Hofmann et al., 1995; Hurst, Bellamy, Geyer, & Astley, 1991; Kletz, 1985; Sulzer-Azaroff et al., 1994). Interventions focusing on individual behavior can improve workplace safety, but most do so only up to a certain point, due to the influence of broader social factors that reside at the work-group and organizational levels. When work environments are more supportive of safety, interventions targeted at promoting workers' use of safe work practices (e.g., safety awareness and skills training, goal setting, and feedback) are more effective (DeJoy, 1996; Hofmann & Stetzer, 1998). Furthermore, even though these interventions have been shown to reduce accidents and injuries (e.g., Komaki et al., 1978), these improvements do not last once the specific intervention is discontinued unless the broader environment is strongly supportive of safety (Cohen & Jensen, 1984).

Using an Input-Process-Outcome Lens

When I/O psychologists study the effectiveness of work groups, they do so most often through a lens that uses an input-process-outcome (I-P-O) perspective (Guzzo & Shea, 1992; McGrath, 1991). Because it is the dominant perspective for studying work groups, the I-P-O framework can be used to examine different types of health from a common vantage point. This will likely help, for instance, to understand trade-offs involved in how groups influence members adopting either safety-focused versus production-focused role behaviors (Zohar, 2000). As I/O psychologists studying safety and well-being have lamented (e.g., Parker, Wall, & Cordery, in press; Sonnentag, 1996; Turner & Parker, in press), aside from isolated team settings where safety is a clear and overriding priority such as with nuclear power plant control room teams (e.g., Gaddy & Wachtel, 1992) and flight crews (Helmreich & Foushee, 1993), outcomes such as safety and health are often not included in models of effectiveness and are virtually ignored by empirical research. Using the I-P-O framework may help bring health and safety more squarely within more mainstream groups or teams research by connecting these criteria to well-researched and understood group processes and normative factors. Finally, the I-P-O paradigm is inherently temporal and can be used to understand developmental team dynamics (Marks, Mathieu, & Zaccaro, 2001). Because the notion of health at all levels of analysis involves consideration of criteria that can be viewed from both short-term (e.g., task performance) and long-term (e.g., team sustainability) perspectives (see Chapter One of this volume), consideration of group and normative influences must take into account this time-based perspective.

Within the I-P-O lens, inputs refer to factors such as group composition, design, leadership, and organizational context conditions (rewards, training, information systems) that influence the processes group members engage in with each other and their environment as they work toward their objectives. Later in this chapter, we consider specific input factors that offer promising opportunities to researchers for studying health and safety in team contexts. As for outcomes, many researchers rely on Hackman's multidimensional definition (1987) of team effectiveness being (1) the productive output of the team as defined by those who evaluate it, (2) the satisfaction of team members, and (3) the capabilities of the members of the team and their willingness to continue working together over time. Note that these outcomes span the continua of the two dimensions of the integrative multilevel framework posed by Hofmann and Tetrick (Chapter One). Specifically, team outcomes may be defined as more extrinsic (e.g., productive output) or intrinsic (e.g., team-member satisfaction) based on whether they achieve goals pursued to satisfy internally held values versus external stakeholder demands. Second, team outcomes can also be considered from a short-term (e.g., monthly team productivity) to a long-term (e.g., development of team capabilities) focus, involving the achievement of short-term versus long-term goals and values. Although researchers have implicitly considered this framework by following Hackman's multidimensional conceptualization of team outcomes (1987), virtually no research has explored how teams can balance the competing demands that underlie each of these dimensions. For example, teams that strive to produce high quality solutions to problems presented by senior management with maximum efficacy (short-term, externally focused) may push themselves so hard that they do so at the cost of sacrificing member satisfaction, the viability of the team to work together on future projects (long-term, internally focused); to put it another way, teams may work to the point that they "burn themselves up" (Hackman, 1987, p. 323). These types of dysfunctional dynamics likely occur much more in real teams than is reflected in our literature based on what managers report as the most critical problems they face leading teams (e.g., Robbins & Finley, 2000; Thompson, 2000). From this perspective a key characteristic of healthy teams involves the extent to which they can balance striving for competing goals and outcomes at the different stages of their development. Teamwork processes and emergent states, the third component in the I-P-O framework, serve a critical role in understanding how this might be accomplished in different team contexts.

Processes have been defined in a number of different ways but typically are considered to be interactions among members of a team or with other groups or individuals outside the team that serve to transform inputs (e.g., members' skills) and resources (e.g., materials, information) into meaningful outcomes (Cohen & Bailey, 1997; Gladstein, 1984; Marks et al., 2001). They are therefore typically considered as mediating the relationship between input factors and group outcomes (McGrath, 1991), and in most models of team effectiveness, processes are depicted as serving a central role (e.g., Gist, Locke, & Taylor, 1987; Gladstein, 1984; Guzzo & Shea, 1992; Hackman, 1987). Indeed, an accumulating amount of research in different team settings has shown that processes such as coordination and communication facilitate team performance in terms of outcomes such as productivity and manager ratings of group effectiveness (e.g., Campion, Medsker, & Higgs, 1993; Hackman & Morris, 1975; Hyatt & Ruddy, 1997). Yet although there is clear evidence of the connection between group process and effectiveness, group process remains very much an enigma when it comes to looking past short-term or immediate outcomes to understanding safety and health in the team context. One of our objectives is to suggest new directions for research to better understand how team processes affect safety and health in team settings. Before we can do that, however, it is helpful to first focus attention on ways in which group and normative influences have the most meaningful influences on safety and health, which we refer to as pressure points.

Pressure Points for Group and Normative Influences

When researchers mention workplace safety, they usually mean the occurrence of negative events, which include combinations of precipitating causes and subsequent resulting effects. For example, falling on the floor is an accident (precipitating cause) that could lead to an injured lower back (resulting effect). To trace the precipitating causes back further, the accident may have been caused by the worker ignoring the sign warning passers-by to keep off an area of the floor that was recently mopped, thereby committing a safety violation. An injured lower back may require the worker to take time off due to the injury and result in filing a worker compensation claim. In other words, the cause-effect relationships involving negative events may involve a longer series of causal connections than one generally expects.

Increasing evidence suggests that group and normative factors play an important role in first influencing whether negative events and occurrences that affect safety take place, second in shaping how they transpire, and third in affecting whether learning occurs that will minimize the possibility of future accidents or injuries. We consider each of these roles in turn, beginning with how group and normative factors directly affect negative events and occurrences such as unsafe behavior and accidents. Figure 5.1 helps frame this discussion.

Proactive orientations toward safety represent a broad set of characteristics that serve to maintain safety, prevent accidents, and proactively manage safety-related issues (Turner & Parker, in press). They may be behavioral (e.g., engaging in safety initiatives and safety-related citizenship behaviors), cognitive (e.g., awareness of safety issues), or attitudinal (e.g., safety commitment) and are often interconnected and mutually reinforcing. As an example, when team members are cognizant of safety concerns and committed to following safe working procedures, they are more likely to demonstrate safety behaviors such as raising safety issues with a supervisor or approaching a coworker who fails to use safe work methods. At the same time, when workers routinely engage in safety-focused behaviors, they begin to consider these as part of their formal set of roles and responsibilities and become more committed to safety objectives. Groups whose members demonstrate these different characteristics that support a positive orientation toward safety through, for instance, high levels of safety commitment, engagement in safety initiatives and activities, and consistent compliance with safety procedures have fewer incidents of unsafe behaviors and fewer accidents and resulting injuries (Hofmann & Morgeson, 1999; Hofmann & Stetzer, 1996; Simard & Marchard, 1995). As such, group and normative influences on proactive orientations toward safety constitute one important pressure point to consider.

Group and normative factors are also important for influencing how teams react and adapt to errors, problems, or stressors in an effort either to avoid negative events (e.g., responding to an emergency to avoid injury) or to minimize the extent to which an accident or mistake results in injury. Turner and Parker (in press) refer to these as indicators of "resilience towards working safely." At the group level of analysis, they capture what Hackman (1993) has referred to as teams' ability to self-correct. A good example from the literature concerns the cockpit crew. Foushee and colleagues

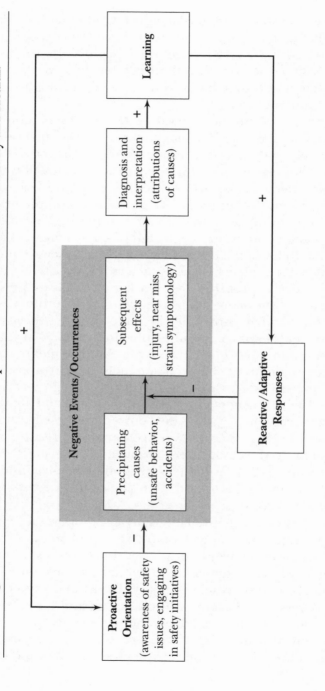

Figure 5.1. Pressure Points for Group and Normative Influences on Safety and Health.

(1986) found that when a flight crew had significant experience working together as an intact team, crew members were better able to compensate for errors due to fatigue. By having experience working as an intact team, members of these crews had a greater level of familiarity with their crewmates' behavior and actions that helped them to catch each other's mistakes and better coordinate their actions. These examples of team monitoring and backup behavior constitute an important set of processes that occur during action sequences when teams are working toward goal accomplishment (Marks, Mathieu, & Zaccaro, 2001). Healthy teams, whether defined in terms of productivity, quality, member adjustment and satisfaction, or safety, are those that are aware of the demands in their environment and are able to adapt to pressures and changing circumstances (Kozlowski, Gully, Brown, Salas, Smith, & Nason, 1999). Consequently, how group and normative influences enhance versus impede team members' ability to react and adapt as situational demands require is another important pressure point to consider.

Finally, we can take the notion of adaptability one step further by considering how groups benefit from mistakes and errors. Indeed, long-term adaptability can be defined as a team's ability to improve its internal processes and configurations through feedback to meet specified performance criteria (Kozlowski, Gully, Nason, & Smith, 1999). Referring back to the organizing framework for organizational health (Chapter One), we find that the orientation is decidedly more long-term. Specifically, we are concerned here with whether groups learn from mistakes or accidents in ways that minimize the likelihood that safety will be compromised in the future. Learning in the group context is a complex process. But in order to derive the lessons that workers can apply to avoid future mistakes or errors from occurring again, the process needs to originate with an effective diagnosis of the root causes. As Hofmann and Stetzer (1998) found in studying attributions for the causes of accidents, open, free-flowing communication over safety issues within the work group helps establish a positive climate toward safety that minimizes the likelihood of making defensive attributions that interfere with learning following accidents. Learning from mistakes, accidents, errors, and near misses at both the individual and group levels is therefore likely to be a function of the shared implicit beliefs in the group about how to respond to those types of problems and issues when they occur (Cannon & Edmondson, 2001; Edmondson, 1996, 1999). Learning capabilities, then, serve as a third pressure point where group and normative factors may function in a critical capacity to contribute to the long-term health of a work group and its members.

In summary, using safety as an exemplar, we have identified three primary ways whereby group and normative factors may influence safety outcomes in a team context. We have identified these as pressure points or areas of leverage that we see as important avenues for research, particularly when considered in combination. For instance, although researchers have considered group influences on safety behavior and outcomes from either a proactive, reactive/adaptive, or learning perspective, no studies have considered all three in concert, even though each likely addresses a different functional and complementary mechanism affecting safety. Although our focus in this section has been on safety, these pressure points apply when considering other effectiveness criteria as well (e.g., performance as judged by external stakeholders).

Not only are the leverage points themselves of interest, but so too are the group and normative factors that influence or shape these pressure points. To examine these group and normative influences in greater depth, we return to the I-P-O model of group effectiveness reviewed earlier. We do so first by discussing the mutually supporting role of groups' teamwork processes and emergent states as they relate to and influence health and safety. We then discuss some potentially critical input conditions that facilitate group emergent states and processes that support worker safety and health following the broad definition adopted in this book.

Using an I-P-O Perspective to Understand Influences on Safety and Health

The discussion that follows focuses on new developments in the areas of team processes and emergent states that appear as promising areas for additional research on group and normative influences on safety and health through the three pressure points we described earlier.

Processes

In the next two sections, first we focus on the relationship between team process and taskwork. Then, we discuss team processes using a temporal perspective.

Processes and Taskwork

One promising avenue for researchers to explore is how team processes facilitate and shape the taskwork that occurs in teams. Taskwork refers to team members' interactions and interfaces with tasks, tools, machines, and systems (Bowers, Braun, & Morgan, 1997). One can think of taskwork as capturing what members of a team do individually with respect to their tools, materials, and technology, whereas teamwork describes how they do their work in relation to each other (Marks et al., 2001). Team processes such as monitoring teammates' performance and providing feedback to others who are not working according to proper procedures are important for ensuring that team members are performing their tasks in a proper and efficient way that best facilitates goal accomplishment (McIntyre & Salas, 1995). Other processes that help teams manage their taskwork functions in ways that facilitate safety include communicating and exchanging information, helping teammates when needed, assisting with production-related taskwork by ensuring that team members have what they need to know in order to complete their tasks effectively, and providing necessary assistance to complete work in a timely manner. We next discuss how teamwork processes facilitate taskwork accomplishment in teams to enhance safety.

In the safety context, taskwork includes behaviors such as properly using personal protective equipment and engaging in work practices that reduce risk (Burke, Sarpy, Tesluk, & Smith-Crowe, 2002). They can be thought of as the types of safety behaviors, whether proactive or reactive, that serve as the immediate precursors of avoiding accidents and subsequent injuries. Consistent with the focus of much of the research on safety at the individual level of analysis, these examples of taskwork are dependent in part on individuals' knowledge and skills (e.g., developed through training). However, they are also influenced by processes occurring at the team level of analysis, which serve to direct, align, and monitor safety-related taskwork within the team (Helmreich & Foushee, 1993). For instance, Hofmann and Stetzer (1996) found that in work groups with effective processes, team members were more likely to approach coworkers engaged in unsafe acts, thereby helping to ensure that teammates completed their taskwork in safe manner. By approaching coworkers engaging in unsafe acts, teams with strong processes in place function as self-correcting units that catch errors before they result in serious consequences (Hackman, 1990). Team processes can also facilitate team members' safety-related taskwork more indirectly as well. Processes such as noticing when teammates need assistance and then providing help can reduce workloads that contribute to burnout and engaging in unsafe behavior to save time (Turner & Parker, in press).

Future research on safety in team settings would benefit by a closer examination of the ways in which different forms of team processes facilitate safety-related taskwork. As team members become more interdependent in carrying out their taskwork, the importance of team processes to facilitate the ability of team members to engage in proactive and responsive forms of safety-related taskwork will likely increase as well (Tesluk, Mathieu, Zaccaro, & Marks, 1997). Teams in settings where safety is an important consideration and with high levels of member interdependence (e.g., emergency response teams) are prime candidates for this type of research. Even though the discussion of the teamwork-taskwork association has been focused on safety outcomes, research is needed on how processes influence other criteria that consider other aspects of team health, such as team adaptation to changing performance demands and challenges. How teams manage their external environment, whether in the form of external stakeholders or situational constraints that impede performance, is important for long-term success and viability (Ancona, 1990). And evidence suggests that team processes (e.g., backup behavior) serve an important role for supporting the types of taskwork in teams required to effectively manage performance barriers such as problems with equipment or materials (Tesluk & Mathieu, 1999).

Processes from a Temporal Perspective

A clear reality and challenge for many teams is how to manage multiple tasks and objectives simultaneously and adjust to shifting priorities. To understand the role of processes in enabling teams to do this requires incorporating a temporal perspective. Although it demonstrates the importance of team processes, existing research tells us little about how teams balance the interests of competing outcomes from intrinsic/extrinsic and short-term/long-term perspectives. Because this requires understanding how teams manage multiple goals and sets of activities over time (McGrath, 1991), recent theoretical work taking a temporal perspective to team processes is particularly promising for understanding the functioning of healthy teams. Kozlowski et al. (1999) offer one perspective, identifying the critical phases in team development; the ways they build upon and make the transition from one to another; and the primary processes that take place at individual, dyadic, and team levels as teams progress in their development.

According to Kozlowski et al.'s framework (1999), after the team's initial formation, team members focus on establishing an interpersonal foundation and shared understanding of the team's purpose and goals through socialization and orientation. Then, they shift their attention to members' development of task competencies and self-regulation through individual skill acquisition. This in turn segues into team members developing an understanding of their own and their teammates' roles and responsibilities. Finally, once team members have a good understanding of each other, their own individual task requirements, and their teammates' roles and responsibilities, they focus on how best to manage interdependencies both in routine and novel situations by attending to the network of relationships that connect team members to each other.

What the Kozlowski et al. (1999) perspective makes clear is that healthy teams are those that are able to successfully manage the different challenges that each phase of their development presents. This model also suggests that, because the phases are progressive and yield team capabilities that build on each other and are necessary for meeting the requirements for the next set of developmental challenges, teams that fail to effectively engage in critical sets of processes at earlier stages of development and achieve requisite cognitive, affective, and behavioral outcomes will be at a distinct disadvantage at later developmental stages. For instance, even mature teams with a substantial shared history will be limited in their adaptability and learning capabilities if team members do not develop a high level of familiarity with their team members and a strong team orientation through team socialization.

Marks and colleagues (2001), who also use a temporal approach to describe team processes, offer another complementary theoretical perspective. Although they do not take a developmental view of how teams build progressively complex capabilities, they do focus on team performance as a recurring set of I-P-O action and transition episodes where outcomes from initial episodes serve as inputs for the next cycle. Thus, consistent with Kozlowski et al. (1999), team outcomes at earlier stages influence team functioning in subsequent phases. Further, Marks et al. (2001) present a taxonomy that organizes different forms of team processes depending on the phase of task accomplishment. Sometimes groups are actively engaged in working toward goal accomplishment, so action processes are dominant (e.g., coordination, communication, team and systems monitoring). At other times, groups are either planning for upcoming activities or reflecting on past performances, so members are involved with transition processes (e.g., planning, performance analysis, goal specification). Finally, other interactions involving interpersonal processes such as managing conflict, affect, and teammates' motivation occur during both transition and action phases of teamwork. Together, these different types of processes capture unique forms of member interaction but merge into a sequenced series of team member activities and interactions as a team performs its work.

This distinction between different types of processes based on when they occur during a team's set of performance episodes is helpful both in terms of theory and practice, depending on the aspect of team effectiveness of particular interest. For instance, we mentioned that one critical pressure point for establishing healthy well-functioning teams is the support of learning. Interactions supporting a climate that encourages openly discussing problems, mistakes, and errors—a necessary condition for learning to occur in teams (Edmondson, 1996; Hofmann & Stetzer, 1998)—depend on interpersonal processes such as affect and conflict management in order to help raise alternative perspectives and dissenting viewpoints and keep disagreements focused on the task rather than becoming personally directed at team members (Jehn, 1995). These types of interpersonal processes are necessary both when a team is actively working toward task accomplishment as well as during transition periods such as during routine meetings before and after a change of shift. In the case of action teams such as surgical teams, they occur between intense performance events such as operations. Effective interpersonal processes are important not only for contributing to a team's ability to recognize and interpret errors and use its experiences to derive learning that enhances capabilities to meet future demands (upper-right quadrant in Figure 1.1) but also for maintaining the team's long-term viability by supporting cohesion and members' commitment to the team by constructively managing disagreements and conflict and contributing to individual team members' skill development and team longevity (outcomes that would be situated in the lower-right quadrant in Figure 1.1).

In contrast, if we are interested in studying how teams prioritize and value safety initiatives, another pressure point whereby group factors can affect the well-being of teams and their members, we should focus attention on transition processes such as whether and how teams specifically identify, discuss, and emphasize safety as an explicit goal during transition phases in the team life cycle. From an intervention standpoint, this might mean training teams to more effectively use opportunities such as preshift meetings or downtime to work on prioritizing safety goals and reviewing their safety performance and conditions influencing safety. To help teams more effectively weigh potentially competing goals such as achieving high levels of productivity and working safety, interventions could also be designed to help teams explicitly identify, prioritize, and try to balance different and perhaps competing team objectives.

In summary, more recent approaches to team processes are moving beyond simply showing that they are important for team effectiveness. Although this work is still largely theoretical, it suggests that team processes influence safety and health through facilitating effective taskwork in teams and enabling teams to effectively manage conflicting interests when multitasking and working toward several objectives simultaneously. Research is now needed to apply these dynamic perspectives of team processes to understand how the different pressure points highlighted earlier are influenced by temporally dependent team processes.

Emergent States

Certain team outcomes are more distal end states, such as team longevity, or more externally focused, such as productivity or achievement of customer service. Others, such as collective efficacy, group safety climate, or shared cognition, are more proximal in nature and internally focused in that they describe the motivational, affective, or cognitive properties of a team. These emergent states, as defined by Marks et al. (2001), are also dynamic and serve as inputs in shaping future interactions between team members (i.e., team processes). For instance, members of a production team may have a high level of confidence in their team's ability to successfully maintain safety (i.e., collective efficacy regarding the ability to work safely, a motivational emergent state) in the face of increasing production demands. This confidence motivates team members to carefully monitor each other's work (a teamwork process) to make sure that safety procedures are followed and to intervene when necessary if coworkers' safety is in jeopardy. Engaging in teamwork to work safely in turn further reinforces the confidence that group members have in their team's ability to successfully manage production demands without sacrificing safety, resulting in a positive self-reinforcing spiral (Lindsley, Brass, & Thomas, 1995). In this way the team's collective efficacy over time as an emergent state functions both as a proximal outcome and dynamic input supporting and being reinforced by team member interactions that enhance the ability of team members to work safely. Emergent states and processes, then, although conceptually distinct, are temporally interconnected and coevolve over the course of a team's history (Marks et al., 2001). As we will explain, three types of emergent states appear to be important for healthy teams. They include those that are normative (e.g., shared beliefs), those that are primarily motivational (e.g., collective efficacy), and those that involve shared cognition (e.g., team mental models).

Normatively Based Emergent States

Norms are informal rules that groups adopt to regulate and regularize behaviors within the group. Group members are more likely to consider norms important and to enforce them if the norms (1) facilitate group survival, (2) enable group members to predict each other's actions and anticipate responses, (3) help the group avoid embarrassing or uncomfortable interpersonal problems, or (4) communicate the central values of the group and clarify what is distinctive about the group's identity (Feldman, 1984; Hackman, 1992). In addition, norms can exert a powerful influence on the behavior of group members. Just how powerful depends in part on the norm's intensity, its overall strength of approval or disapproval associated with the behavior as captured through the level of affect attached to the norm, and crystallization, the degree of consensus within the group about the level of approval or disapproval associated with the different levels of behavior (Hackman, 1992).

These concepts of norm intensity and crystallization have application to studying safety in team settings. For instance, in a typical high reliability industrial setting (e.g., an oil refinery), work groups are likely to have norms regarding the valance of certain safety behaviors (e.g., assisting teammates to make sure they perform their work safely, raising safety concerns during planning meetings, taking actions to stop safety violations in order to protect others on the team) (Hofmann, Morgeson, & Gerras, in press). Ideally, these norms are both intense, for example, when members show strong approval toward coworkers who engage in safety citizenship (e.g., publicly thanking them and doing so consistently), and well crystallized, as shown through a high degree of consensus within the team as to what the appropriate forms of these behaviors are (e.g., all team members agree that they value raising safety concerns during meetings). In these cases the intense positive and well-crystallized norms regarding safety citizenship are likely to result in high levels of member compliance and quick action taken against norm violators. Even in cases of highly crystallized norms on safety (team members have very similar opinions as to what is acceptable), if those norms are weak in intensity, team members will do little to safety violators because the group members do not strongly hold those values. The implication, then, and one that translates into a provoking research proposition, is that consistent engagement in safety behaviors will most likely be seen in teams with both relatively intense and well-crystallized norms on safety citizenship. Finally, it should also be the case that a proactive orientation toward safety becomes strengthened when teams with strong norms toward safety are also highly cohesive. Highly cohesive groups, where members like one another and want to remain in the group, put pressure on members to conform to team norms (Hackman, 1992). Although some anecdotal evidence suggests that safety norms foster greater compliance with safety rules and regulations when groups are highly cohesive (Simard & Marchand, 1997), this question too remains to be formally studied.

At the same time, however, strong norms and high levels of cohesiveness can threaten other forms of healthy team functioning. High levels of cohesiveness create strong pressures toward uniformity and conformity that can severely limit dissent (Hackman, 1992; Festinger, 1950). Dissenting points of view are important for expanding the range of perspectives being considered by stimulating constructive conflict around the task. Cognitive or task-based conflict has consistently been found to be important for group decision making because it stimulates group members to consider multiple alternatives and rethink basic assumptions (Jehn, 1995).

Another concern to healthy team functioning is that strong norms can serve to reinforce developing patterns of group behavior to the point that they can become entrenched and highly refined as to become automatized habitual routines (Gersick & Hackman, 1990). Although habitual routines contribute to efficient performance and can minimize overt competition and disagreement in ways that benefit smooth team functioning and short-term performance, they can also blind team members to recognizing when situations require novel action and can discourage the type of dissent and disagreement that help generate alternative perspectives facilitating innovation. Both of these concerns—pressures to minimize dissenting viewpoints and task conflict as well as reinforce habitual routines—threaten the long-term ability of teams to grow in competence, skills, or perspectives. In situations where group effectiveness is judged according to criteria such as the ability to develop novel solutions, adapt to changing performance demands, and accomplish long-term learning, the power of strong norms and high cohesiveness may have serious detrimental effects. In other words, strong norms may support one pressure point, effective group functioning in routine environments, at the expense of another critical pressure point, team adaptability.

But this tension might be managed in a few ways. One is by taking steps to actively build cohesion around the task rather than around interpersonal attraction (Hackman, 1992; Zaccaro & Lowe, 1988). When social cohesion is dominant, group members question their actions in terms of how they might affect their standing and interpersonal relationships within the group. In contrast, task-based cohesion leads individuals to question how their actions might affect the performance of the group. Because of the obvious implications for team development, more research is needed to pursue this distinction on the different types of group cohesion and implications for the influence of group norms on outcomes related to team health, such as adaptability and learning.

In a related vein, establishing metaroutines that encourage group members to periodically review their work processes is a way to continuously examine activities and make incremental adjustments as necessary (Gersick & Hackman, 1990). Dealing with such issues as what the group's norms are, why group members occasionally deviate from these norms, and whether this is in fact beneficial or counterproductive for the group is difficult. But Hackman (1992, p. 250) claims, "If groups can be helped to learn how to deal skillfully with issues of conformity and deviance, then both team effectiveness and the personal well-being of group members should be greatly enhanced." Transition phases, such as during midpoint meetings and other natural breaks in groups' activities (Gersick, 1988), offer the best opportunity for teams to engage in these types of processes.

Complementing this is building a climate of mutual trust and respect where team members feel comfortable raising concerns and admitting mistakes. Edmondson (1999) has called this shared belief within the team "psychological safety." In groups with high levels of psychological safety, members are less likely to feel excessive concern over bringing up mistakes, deviations, or problems that they could identify, diagnose, correct, and use to make changes. They are also more at ease suggesting alternative methods to the group for how it can go about doing its work. Both of these activities—reflecting on mistakes and incorporating new knowledge from them—contribute to team and individual learning, an important pressure point for team health and safety identified earlier. A climate of psychological safety is also likely to be important for team members to feel interpersonally comfortable approaching coworkers engaging in, for example, unsafe acts (cf., Hofmann & Stetzer, 1996). Finally, evidence suggests that those working in team settings characterized by more supportive interpersonal climates experience higher levels of well-being (Sonnentag, Brodbeck, Heinbokel, & Stolte, 1994), suggesting that psychological safety and other related normative emergent states may be important in helping individuals cope with the pressures associated with working in turbulent and stressful environments.

Motivationally Based Emergent States

A second general category of emergent states concerns the collective motivation of team members arising through their repeated interactions and exchanges concerning the tasks, context, processes, and prior performance of the group. Examples include team empowerment (Kirkman & Rosen, 1999), group potency (Guzzo, Yost, Campbell, & Shea, 1993), and collective efficacy (Lindsley et al., 1995). Despite an increasing amount of research applying motivational concepts such as empowerment and efficacy to the group level of analysis, nearly all of this work to date has considered how these motivationally based emergent states affect team performance (e.g., Gibson, 1999; Kirkman & Rosen, 1999) and has virtually ignored their potential role with respect to safety and health-related outcomes. This is surprising given that work groups form an important source of social support and simply working in a group can serve as an important resource in coping with stress (Sonnentag, 1996).

Building on this observation, one notable exception is recent work by Jex and Bliese (1999) on collective efficacy. Although initial evidence has suggested that collective efficacy may be less important to supporting the types of behaviors that contribute to team learning (Edmondson, 1999), collective efficacy appears to be important in helping to buffer group members against the harmful effects of workplace stressors. Drawing on work at the individual level of analysis finding that self-efficacy helps to negate the harmful effects of stressors such as work overload and role ambiguity, Jex and Bliese studied soldiers in army units and found that the shared belief in the unit's ability to successfully perform its mission moderated two individual-level stressor-strain relationships: (1) those between work overload and job satisfaction and (2) those between task significance and organizational commitment. Specifically, soldiers in units with higher collective efficacy reacted less negatively to having a heavy workload and working on highly significant tasks. Presumably, this was due to collective efficacy facilitating a positive interpersonal climate where members of the unit cooperate with each other, provide helpful emotional support during stressful situations, and feel confident in their ability to perform despite the challenging demands being placed upon them. In other words a strong esprit de corps helped to support soldiers' adaptive responses to workplace stressors and in so doing helped to reinforce that critical pressure point (as identified in Figure 5.1).

This example of cross-level moderation, where a group-level variable such as collective efficacy moderates an individual-level relationship, is rarely studied, but it demonstrates that understanding the effects of stress and other health and safety concerns often requires combining individual and team-level factors. From a practical standpoint, it suggests that taking actions to stimulate a strong sense of efficacy in groups (e.g., through goal setting and rewards) (Guzzo & Shea, 1992; Jex & Bliese, 1999) may benefit not only team performance but also team member well-being, particularly when the team is working under stressful and trying conditions. More research is needed to better understand why collective efficacy mitigates team members' stress (e.g., through social support) and to consider other motivationally based emergent states that may be important with respect to safety and health issues. For instance, it may be that teams with a strong sense of empowerment may feel greater ownership over workplace conditions that relate to safety, and this may have important implications for safety outcomes at both the team and individual levels.

Cognitively Based Emergent States

A third form of emergent states includes various forms of shared cognition, such as team (or shared) mental models (Klimoski & Mohammed, 1994) and transactive memory systems (Liang, Moreland, & Argote, 1995). Although several different forms of shared cognition exist, all refer in some way to certain types of knowledge structures or mechanisms that members of a team hold in common. For instance, based on individual mental models as organized knowledge structures of tasks, concepts (e.g., team interaction), or technology or equipment that enable individuals to interact with their environment by understanding and forming expectations, shared mental models refer to the overlapping or similar cognitive representations of individuals in a team (Cannon-Bowers, Salas, & Converse, 1993). Because they enable individuals to anticipate what their teammates are doing and what they need in order to do it, shared mental models enable team members to better coordinate with each other. Recent empirical work has shown that shared mental models facilitate team processes, such as communication, which in turn affect performance (Mathieu, Heffner, Goodwin, Salas, & Cannon-Bowers, 2000). Also, shared mental models appear to be more beneficial for teams facing novel environments as opposed to those working in routine settings (Marks, Zaccaro, & Mathieu, 2000).

Although research on shared mental models has focused almost exclusively on short-term performance criteria, they are likely to be important for other outcomes such as safety and long-term viability in settings that require a high degree of real-time synchronization in dynamic and demanding environments. Examples include a wide range of action teams such as emergency response teams, military combat teams, and fire-fighting crews, which must often adjust quickly and spontaneously to changing performance conditions. For instance, Goodman and his colleagues (e.g., Goodman & Shah, 1991) have found that mining crews with greater levels of familiarity, defined as the work experience that crews have with unique configurations of machinery, materials, physical environment, and crew members, both have better safety records and are more productive. Although we have not seen Goodman's findings discussed in association with shared mental models, familiarity with specific combinations of mining conditions, crew members, and equipment, as developed through experience, likely creates a set of mental models on critical aspects of their work such as their machinery, environmental conditions in the mine, and crew roles and responsibilities. To the extent that members of crews work together as intact teams over time, they likely develop a shared mental model that helps experienced crews adapt to unanticipated demands (e.g., equipment breakdowns) and demonstrate more effective team processes (e.g., backing up crew members when necessary).

Flight crews are another type of action and performing team whose shared mental models appear to be important for multiple aspects of effectiveness including safety. Research on aviation cockpit crews has well established that team functioning, rather than mechanical problems or the technical proficiency of individual pilots, is responsible for most airline accidents (Helmreich & Foushee, 1993). A troubling set of statistics reported by the National Transportation Safety Board (NTSB) from a review of major airline accidents over more than 10 years (NTSB, 1994) indicates that familiarity appears to be particularly important to crew functioning and safety. The NTSB found that 73% of accidents happened on the crew's first day of flying together, with 44% of accidents occurring after the crew's first flight. A study by Foushee et al. (1986) suggests why this might be the case. They found, as one would expect, that crew member fatigue was positively associated with individual crew members making errors. However, members of experienced crews compensated for these mistakes by catching their team members' errors and coordinating their actions with their teammates'. In fact, the researchers found that fatigued crews that had worked together as intact units performed significantly fewer errors than did well-rested crews who had not worked together previously. As with the mining crews studied by Goodman and colleagues (Goodman & Leyden, 1991; Goodman & Shah, 1991), it appears that by working together over time, intact flight crews develop shared mental models of teamwork that may help them respond to breakdowns in procedures, problems, or emergencies quickly and in a coordinated fashion, thereby avoiding potentially serious mistakes. The development of shared cognitive structures like team mental models appears to be important for the short-term success and long-term viability of action teams functioning in challenging and turbulent environments.

Another form of shared cognition that is beginning to be more widely researched is transactive memory (Wegner, 1986). Whereas shared mental models are shared representations of tasks, working relationships, equipment, and situations, transactive memory systems refer to how groups learn, remember, and communicate the information and knowledge required for their team tasks. Transactive memory is especially important for teams working on complex tasks, facing high information-processing demands, and relying on specialized member expertise because it enables team members to use a cooperative division of labor to store, access, and use team member knowledge (Liang et al., 1995). For instance, members of a commercial flight crew occupy different specialized positions (e.g., pilot, navigator) and areas of expertise, and crew members cannot (and should not) be highly proficient in each of the different positions' task requirements or be concerned with all of the information-processing demands associated with each role in the crew. At the same time, crew members are highly dependent on the other members adequately performing their tasks and sharing vital information. Transactive memory systems facilitate these interdependencies and allow for specialization through a collective understanding of which crew members have specific knowledge and expertise in what areas, how best to channel incoming information to the appropriate experts, and how to retrieve and share that specialized knowledge in ways that best manage team interdependencies. The laboratory research completed to date has borne out this prediction: teams that have stronger transactive memory systems collectively retrieve and apply greater amounts of task-relevant information, and this is associated with superior team performance (Liang et al., 1995; Moreland, Argote, & Krishnan, 1996). Although no research on the topic has been done to date, transactive memory may serve a useful purpose for enhancing team safety as well. An operations specialist in a nuclear power plant control team, for instance, does not need to know the specific technical reasons for why a particular gauge is reading outside its normal range, as long as he or she is aware that the radiological control technician has that knowledge and the operations specialist knows how to access the knowledge of the control technician to take corrective action.

Most of the research to date on shared mental models, transactive memory, and other forms of shared cognition has been done in controlled laboratory settings. Although studying shared knowledge structures in field settings poses a formidable challenge, examining the role of shared cognition with respect to safety might be promising because safety procedures and practices can often be clearly defined. This can enable researchers to define an expert mental model for working safely that they can use to determine mental-model accuracy as well as study mental-model convergence across team members. Transactive memory may serve a complementary role by helping to identify how knowledge of safety practices and procedures is distributed across team members and how teams can best manage that collective knowledge to work more safely.

In summary, three distinct types of emergent states—normative, motivational, and cognitive—appear promising for future research on group influences on safety and health in team settings. As more fluid qualities that reflect team experiences and interactions and that influence future team processes and subsequent outcomes in a dynamic and reciprocating fashion, emergent states need to be considered as distinct from (but studied in combination with, using a temporal perspective) team processes. Although researchers considering safety and health outcomes (e.g., safety climate, team norms) have perhaps most studied normative emergent states, virtually no significant research has been done to date on how motivational and cognitive qualities of teams are related to effectiveness criteria concerning safety and health. And although all three types of emergent states are important for effective team functioning from a broad perspective, specific types of emergent team states appear particularly promising for understanding key pressure points. Initial research on collective efficacy, for example, suggests that it helps to buffer team members against different workplace stressors, whereas psychological safety is an important condition facilitating team learning. Finally, we highlight that certain emergent states may be particularly critical for health and safety concerns in specific team settings. Shared team mental models, transactive memory systems, and other forms of shared cognition, for instance, appear to be most useful in dynamic and complex settings, where teams have high levels of interdependence; team members often have specialized areas of expertise; and team performance and safety is dependent on adaptability and sharing knowledge.

Enhancing Positive Group and Normative Influences on Health and Safety

We have now considered how group and normative factors influence the health and safety of groups and their members both in terms of avoiding a lack of health (e.g., through minimizing accidents and subsequent injuries) and by enhancing the capabilities of individuals and teams to function effectively (e.g., through learning from failures). In this final section, we discuss ways by which group and normative factors that positively influence health and safety may be enhanced through task and team design and team leadership. Returning to the fundamental themes outlined at the start of this chapter, we do so from the perspective of how these strategies can support the health of individuals and groups through balancing competing demands using the framework provided in Chapter One and suggest directions for future research.

Task and Team Design

The design of the team's task and corresponding team design or structure are important ways to facilitate the types of team processes and emergent states that appear as important for promoting safety and health in teams. For instance, in order for members of a work group to be actively engaged and committed to their tasks in ways that promote effective team norms for productivity, flexibility, and openness to change, the group needs to be responsible for working on a whole and meaningful piece of work (Hackman, 1992). A proactive orientation toward valued outcomes such as safety may be enhanced through encouraging feelings of team empowerment by ensuring that teams work on important and meaningful tasks (Kirkman & Rosen, 1999). Tasks must also offer opportunities for direct feedback to facilitate knowledge of results in order to support collective feelings of team efficacy (Lindsley et al., 1995).

Because Chapter Four of this volume reviews the topic of work and task design in full, we do not detail task design elements further except to note the importance of and need for research on considering how task design and team structure can facilitate team safety and health. This is reflected in the sociotechnical systems theory principle of matching technical (task) and social (group design) systems and is well demonstrated in some of the earliest studies by Emory, Trist, Bamforth, and their colleagues on team design and coal mining crews' productivity and miner safety and well-being (e.g., Emory & Trist, 1960; Trist, Higgins, Murray, & Pollack, 1963; Trist & Bamforth, 1951). They found that, in addition to demonstrating higher productivity and efficiency in terms of coal extracted per labor hour, miners in teams operating under the composite system (teams that rotated members across different tasks and functions) experienced fewer lost days due to injury as well as lower absenteeism as compared to miners in teams using highly structured work roles and tasks (conventional system). This was likely due to miners experiencing higher levels of stimulation and reduced fatigue from greater task and skill variety, increased feelings of responsibility for work outcomes and commitment to the team by working on a set of related tasks that contributed to team objectives, flexibility to adjust to the constantly changing mining conditions, and greater ability to coordinate work within the team due to team members having greater knowledge of different tasks and responsibilities through rotation.

As "groups of individuals who can self-regulate work on their interdependent tasks" (Goodman, Devedas, & Hughson, 1988, p. 296), self-managing work-group designs provide teams with a high degree of autonomy and control over day-to-day work processes, which are necessary for working on meaningful and challenging group tasks. The degree of autonomy that teams have over managing their own production and service responsibilities has been associated with the degree to which teams experience a sense of empowerment (Kirkman & Rosen, 1999) and has been found to predict team performance (e.g., Campion et al., 1993). In addition to teams being able to adopt more appropriate strategies in response to challenging task demands and changing performance conditions, self-management as a team design principle appears important because it facilitates motivationally and cognitively based emergent states (Goodman et al., 1988; Hackman, 1987). For instance, greater control and involvement from self-management facilitates feelings of autonomy, efficacy, intrinsic motivation, and sense of responsibility that foster effort and persistence. In addition, by having more broadly defined roles and responsibilities and being cross-trained on other team members' jobs, members of self-managing teams may develop greater knowledge and a more detailed and shared understanding of their teammates' tasks and demands, and they are therefore better able to coordinate their work more effectively. In other words, self-managing team designs might be an important design tool for promoting the types of cognitively based emergent team states reviewed earlier that help improve team processes, particularly in unpredictable and challenging settings. This possibility awaits future research and, if identified, would demonstrate an important mechanism linking team design, shared cognition, and team outcomes.

In summary, these mechanisms (through motivational and cognitively based emergent states) are ways that self-managing team designs may enhance the ability of teams to perform more safely as well as to perform their work more proficiently and productively (Turner & Parker, in press). However, in a recent review of subsequent studies involving self-managing team designs and safety outcomes,

Turner and Parker found that although some research has found a positive relationship with self-management and safety (e.g., Pearson, 1992), other studies have been either inconclusive (e.g., Goodman, 1979) or have found no such relationship (e.g., Cohen & Ledford, 1994). Turner and Parker (in press) suggest that one important consideration that may help explain these inconsistent findings is that whether self-managing team designs influence safety depends on whether safety is clearly identified as a goal and responsibility of the teams. For instance, in a rigorously designed longitudinal study comparing self-managing to traditional leader-managed teams, Pearson (1992) found that the lower injury rate for self-managing teams appeared to be due to team members assuming responsibilities and adopting a strong safety orientation as demonstrated by discussing safety issues in team meetings and assigning specific safety responsibilities to members of the team. In contrast, in another study that compared self-managing to traditionally designed teams (Cohen & Ledford, 1994), that was not the case. Although self-managing teams were associated with higher quality of work life outcomes, because safety was not emphasized as a specific goal and objective, team members did not incorporate a safety orientation into their assumed set of responsibilities, and therefore these teams had no better safety records than did traditionally managed teams.

A self-managing team design appears to be an important means by which teams can function both productively and safely if safety and health objectives are designated as explicit goals and teams are encouraged to assume responsibility for those safety-related objectives. Particularly in work contexts where teams face complex and ambiguous tasks with changing performance conditions, the flexibility afforded by self-managing designs along with the greater breadth of knowledge of interdependent tasks, responsibilities, and motivational advantages are important considerations that can help teams function both more productively and safely. Team members appear to internalize safety as well as productivity and other performance-related goals as team responsibilities as a result of their team leaders' messages regarding the importance of safety. As the use of self-managing teams continues to grow, this will become an increasingly important area for research.

Team Leadership

Despite the fact that scholars well recognize that team leaders play a pivotal function in the success of teams performing different functions in various settings and contexts (see, e.g., Hackman & Walton, 1986; Zaccaro & Marks, 1999), substantive research on the topic of the role of leadership in teams is lacking. In two areas of particular relevance to this chapter's theme, safety and team learning, recent research is making important contributions suggesting fruitful directions for additional work contributing to our understanding of group and normative influences on health and safety in teams. Building on our earlier discussion on self-managing designs and safety outcomes, we begin with the role of team leaders in supporting safety. Then, we consider the role of leaders as we shift to examine more long-term criteria of team health through building self-sustaining capabilities for learning and balancing competing demands.

Team leaders create climates within the team that communicate to members what types of goals and objectives are important and what types of behaviors are expected. For example, Zohar (2000, 2002) demonstrated that the way in which team leaders translate management policies regarding safety through work-group practices creates distinct and differentiable team climates for safety (see also Chapter Seven of this volume for a more complete discussion of the relationship between leadership and work-group climate). Given that employee perceptions of the work group's climate are heavily based on what the supervisor rewards, supports, and expects (Kozlowski & Doherty, 1989), when supervisors demonstrate strong support for safety and are open to discussing safety issues and concerns, they are likely sending a strong message about the importance of safety (Hofmann & Stetzer, 1996, 1998). Thus, one important function of team leaders in supporting safety outcomes is in helping establish intense and crystallized norms regarding the importance of following safe work procedure behaviors and demonstrating other safety-related behaviors (e.g., bringing safety issues to the group's attention).

However, as a recent study by Hofmann, Morgeson, and Gerras (in press) demonstrated, creating a climate that emphasizes safety is only part of the equation of effective leadership in supporting team member promotion of safety (and in all likelihood, other outcomes as well). The other component involves developing high quality social exchanges with group members in order to encourage team members to define their roles and responsibilities to include fulfillment of safety obligations. Hofmann and colleagues proposed that when team leaders work to develop high quality exchanges with subordinates, team members are more willing to accept as in role the types of behaviors that leaders communicate are expected. If the climate emphasizes safety, then team members will see safety citizenship as a means by which to reciprocate a positive social exchange created through a high quality working relationship with the team leader. As predicted, they found that team leaders' high quality exchanges with subordinates was associated with team members assuming broader role definitions that included safety citizenship behaviors when team leaders also created a climate for their team that emphasized safety. In turn, the researchers found that team members who viewed safety citizenship as being more in role engaged in more safety-related behavior.

Based on our earlier discussion of self-managing team designs, we might speculate that a powerful set of conditions is created to support safety when certain leadership and team design conditions exist. First, work groups need to have a leader who clearly communicates and reinforces the message that safety is a key priority and establishes a climate of mutual trust, respect, and obligation within the group that encourages group members to assume responsibility for safety behaviors. Second, a self-managing team design should be provided so that team members have the necessary autonomy to assume responsibility for the goals that the workgroup leader is communicating. Research exploring the combination of team design and the complementary roles of leadership in creating strong norms and high quality exchange may be particularly well suited for ongoing teams with relatively well-defined tasks.

In addition to facilitating group and normative influences on safety in teams, there is a need to better understand how leaders can facilitate team learning. Teams working in complex, dynamic settings face situations where team goals change over time and often compete with each other. For instance, hospital surgical teams face competing pressures of performing their work as flawlessly as possible while also acquiring new skills and competencies when introducing new practices, techniques, and technology (Edmondson, 2002). In addition to striving to achieve a level of shared cognition and mindfulness that facilitates smooth teamwork and short-term performance during action phases, these teams also have to be able to modify their existing mental models and shared frameworks of understanding to allow for the introduction of new routines and approaches. Gersick and Hackman's work (1990) on the liabilities of habitual routines makes clear the inherent tensions that these sorts of teams face between striving for conditions that facilitate coordinated performance versus those that contribute to long-term viability through change and adaptability.

In these settings team leaders must strive to both contribute to norms that support short-term performance and encourage a continual reexamination of group activities and patterns of behavior (e.g., working to help establish metanorms) (Hackman, 1990) that are necessary for long-term learning and adaptability. Recent work by Edmondson (2002) has begun to identify leader behaviors that are important for team learning and the creation of a climate of psychological safety. These include framing situations as opportunities for learning, encouraging team members to speak up, and communicating an openness to different views. What remains to be seen is whether leadership that promotes team adaptability and long-term functioning through the implementation of new practices and development of competencies can be done without sacrificing conditions that facilitate short-term coordination and performance.

A Delicate Balancing Act: Contingency Issues with Respect to Team Safety and Health

Throughout this chapter, we have attempted to note some of the issues underlying the inherent tensions that exist when considering group and normative influences on safety and health in teams. There may not be a single appropriate level of balance between short- versus long-term demands and between intrinsic versus extrinsic demands; a healthy balance may actually depend on factors such as the type of task involved, the stage of the team's development, the level of interdependence of team members, or the overall type of team. For example, a number of different types of teams exist: advice and involvement teams, production and service teams, project and development teams, and action and negotiation teams (Sundstrom et al., 1990). Each of these different team types may have different optimal levels of balance associated with them; for example, advice and involvement teams may require less focus on intrinsic, long-term aspects of team effectiveness or health and instead need to maximize extrinsic, short-term aspects of team effectiveness. The overarching point is that researchers need to consider different contingency factors when they study the balancing competing demands of team safety and health. In other words, instead of the notion that groups must strike a single, perfect balance between short- and long-term demands and intrinsic and extrinsic demands, we suggest that groups can strike multiple optimum balance points in order to be healthy and effective. The location of these balance points on Hoffman and Tetrick's Figure 1.1 for any given team depends on factors such as the type of team, task, and the nature of the interdependencies among work-group members.

Conclusion

In this chapter we both summarized the current state of research on group and normative influences on safety and health in team settings as well as suggested necessary extensions for future study. Arguing that safety and health in team settings can be considered from three primary pressure points (proactive orientations, reactive and adaptive responses, and team and individual learning), we suggested that a dynamic perspective focusing on team processes and different forms of emergent states will help us better understand the factors and conditions that promote safety and health. Finally, we attempted to outline how task and team design and team leadership serve as promising avenues for future research on safety and health in team settings. Work groups and teams are clearly a central fixture in the modern workplace. Understanding how group and normative influences affect safety and health outcomes from a broad-based perspective will add an important perspective to safety and health research and will contribute significantly to improving safety and health outcomes in the workplace. We hope that some of the ideas in this chapter will stimulate future research toward this end.

References

Ancona, D. G. (1990). Outward bound: Strategies for team survival in an organization. *Academy of Management Journal, 33,* 334–365.

Becker, T. E. (1992). Foci and bases of commitment: Are they distinctions worth making? *Academy of Management Journal, 35,* 232–244.

Bowers, C. A., Braun, C. C., & Morgan, B. B. (1997). Team workload: Its meaning and measurement. In M. T. Brannick, E. Salas, & C. Prince (Eds.), *Team performance assessment and measurement* (pp. 85–108). Hillsdale, NJ: Erlbaum.

Burke, M. J., Sarpy, S. A., Tesluk, P. E., & Smith-Crowe, K. (2002). General safety performance: The hazardous waste worker domain. *Personnel Psychology, 55,* 429–457.

Campion, M. A., Medsker, G. J., & Higgs, A. C. (1993). Relations between work group characteristics and effectiveness: Implications for designing effective work groups. *Personnel Psychology, 46,* 823–850.

Cannon, M. D., & Edmondson, A. C. (2001). Confronting failure: Antecedents and consequences of shared beliefs about failure in organizational work groups. *Journal of Organizational Behavior, 22,* 161–177.

Cannon-Bowers, J. A., Salas, S., & Converse, S. A. (1993). Shared mental models in expert team decision making. In M. T. Brannick, E. Salas, & C. Prince (Eds.), *Current issues in individual and group decision making* (pp. 221–246). Hillsdale, NJ: Erlbaum.

Cascio, W. F. (2000). Managing the virtual workplace. *Academy of Management Executive, 14*(3), 81–90.

Cohen, H. H., & Jensen, R. C. (1984). Measuring effectiveness of an industrial lift truck safety program. *Journal of Safety Research, 15,* 125–135.

Cohen, S. G., & Bailey, D. E. (1997). What makes teams work? Group effectiveness research from the shop floor to the executive suite. *Journal of Management, 23,* 239–290.

Cohen, S. G., & Ledford, G. E. (1994). The effectiveness of self managing teams: A quasi-experiment. *Human Relations, 47,* 13–43.

DeJoy, D. M. (1996). Theoretical models of health behavior and workplace self-protective behavior. *Journal of Safety Research, 27,* 61–72.

Denison, D. R., Hart, S. L., & Khan, J. A. (1996). From chimneys to crossfunctional teams: Developing and validating a diagnostic model. *Academy of Management Journal, 39,* 1005–1023.

Edmondson, A. C. (1996). Learning from mistakes is easier said than done: Group and organizational influences on the detection and correction of human error. *Journal of Applied Behavioral Science, 32,* 5–28.

Edmondson, A. C. (1999). Psychological safety and learning behavior in work teams. *Administrative Science Quarterly, 44,* 350–383.

Edmondson, A. C. (2002). *Leading for learning: How team leaders promote speaking up and learning in interdisciplinary action teams.* Manuscript submitted for publication.

Emory, F. E., & Trist, E. L. (1960). Socio-technical systems. In C. W. Churchman & M. Verhulst (Eds.), *Management sciences, models, and techniques* (Vol. 2, pp. 103–137). New York: Pergamon Press.

Feldman, D. C. (1984). The development and enforcement of group norms. *Academy of Management Review, 9,* 47–53.

Festinger, L. (1950). Informal social communication. *Psychological Review, 57,* 271–282.

Foushee, H. C., Lauber, J. K., Baetge, M. M., & Acomb, D. B. (1986). *Crew factors in flight operations III: The operational significance of exposure to short-haul air transport operations* (Technical Memorandum No. 88342). Moffett Field, CA: NASA–Ames Research Center.

Gaddy, C. D., & Wachtel, J. A. (1992). Team skills in nuclear power plant operations. In R. W. Swezey & E. Salas (Eds.), *Teams: Their training and performance* (pp. 379–396). Norwood, NJ: Ablex.

Gersick, C.J.G. (1988). Time and transition in work teams: Toward a new model of group development. *Academy of Management Journal, 31,* 9–41.

Gersick, C.J.G., & Hackman, J. R. (1990). Habitual routines in task-performing groups. *Organizational Behavior and Human Decision Processes, 47,* 65–97.

Gibson, C. B. (1999). Do they do what they believe they can? Group efficacy and group effectiveness across task and cultures. *Academy of Management Journal, 42,* 138–152.

Gist, M. E., Locke, E. A., & Taylor, M. S. (1987). Organizational behavior: Group structure, processes, and effectiveness. *Journal of Management, 13,* 237–257.

Gladstein, D. (1984). Groups in context: A model of task group effectiveness. *Administrative Science Quarterly, 29,* 499–517.

Goodman, P. S. (1979). *Assessing organizational change: The Rushton quality of work experiment.* New York: Wiley-Interscience.

Goodman, P. S., Devedas, R., & Hughson, T. L. (1988). Groups and productivity: Analyzing the effectiveness of self-managing teams. In J. P. Campbell, R. J. Campbell, & Associates (Eds.), *Productivity in organizations* (pp. 295–327). San Francisco: Jossey-Bass.

Goodman, P. S., & Leyden, D. P. (1991). Familiarity and group productivity. *Journal of Applied Psychology, 76,* 578–586.

Goodman, P. S., & Shah, S. (1991). Familiarity and work group outsomes. In S. Worchel & W. Wood (Eds.), *Group process and productivity* (pp. 276–298). Newbury Park, CA: Sage.

Guzzo, R. A., & Dickson, M. W. (1996). Teams in organizations: Recent research on performance and effectiveness. *Annual Review of Psychology, 47,* 307–338.

Guzzo, R. A., & Shea, G. P. (1992). Group performance and intergroup relations in organizations. In M. Dunnett & L. Hough (Eds.), *Handbook for industrial and organizational psychology* (2nd ed., Vol. 3, pp. 269–313). Palo Alto, CA: Consulting Psychologists Press.

Guzzo, R. A., Yost, P. R., Campbell, R. J., & Shea, G. P. (1993). Potency in groups: Articulating a construct. *British Journal of Social Psychology, 32,* 87–106.

Hackman, J. R. (1987). The design of effective work teams. In J. W. Lorsch (Ed.), *Handbook of organizational behavior.* Englewood Cliffs, NJ: Prentice Hall.

Hackman, J. R. (1990). Groups that work (and those that don't). San Francisco: Jossey-Bass.

Hackman, J. R. (1992). Group influences on individuals in organizations. In M. Dunnett & L. Hough (Eds.), *Handbook for industrial and organizational psychology* (2nd ed., Vol. 3, pp. 199–267). Palo Alto, CA: Consulting Psychologists Press.

Hackman, J. R. (1993). Teams, leaders, and organizations: New directions for crew-oriented flight training. In E. L. Wiener, B. G. Kanki, & R. L. Hendricks (Eds.), *Cockpit resource management.* Orlando, FL: Academic Press.

Hackman, J. R. (1998). Why teams don't work. In R. S. Tindall, J. Edwards, & E. J. Posavac (Eds.), *Theory and research on small groups* (pp. 277–301). New York: Plenum Press.

Hackman, J. R., & Morris, C. G. (1975). Group tasks, group interaction processes, and group performance effectiveness: A review and proposed integration. In L. Berkowitz (Ed.), *Advances in experimental social psychology* (Vol. 8, pp. 45–99). Orlando, FL: Academic Press.

Hackman, J. R., & Walton, R. E. (1986). Leading groups in organizations. In P. S. Goodman (Ed.), *Designing effective work groups.* San Francisco: Jossey-Bass.

Helmreich, R. L., & Foushee, H. C. (1993). Why crew resource management? Empirical and theoretical bases of human factors training in aviation. In E. L. Weiner & B. G. Klanki (Eds.), *Cockpit resource management* (pp. 3–45). Orlando, FL: Academic Press.

Hofmann, D. A., Jacobs, R., & Landy, F. L. (1995). High reliability process industries: Individual, micro, and macro organizational influences on safety. *Journal of Safety Research, 26,* 131–149.

Hofmann, D. A., & Morgeson, F. P. (1999). Safety-related behavior as a social exchange: The role of perceived organizational support and leader-member exchange. *Journal of Applied Psychology, 84,* 286–296.

Hofmann, D. A., Morgeson, F. P., & Gerras, S. J. (in press). Climate as a moderator of the relationship between LMX and content specific citizenship behavior: Safety climate as an exemplar. *Journal of Applied Psychology.*

Hofmann, D. A., & Stetzer, A. (1996). A cross-level investigation of the factors influencing unsafe behaviors and accidents. *Personnel Psychology, 49,* 307–339.

Hofmann, D. A., & Stetzer, A. (1998). The role of safety climate and communication in accident interpretation: Implications for learning from negative events. *Academy of Management Journal, 41,* 644–657.

Hurst, N. W., Bellamy, L. J., Geyer, T. A., & Astley, J. A. (1991). A classification scheme for pipework failures to include human and sociotechnical errors and their contributions to pipework failure frequencies. *Journal of Hazardous Materials, 26,* 159–186.

Hyatt, D. E., & Ruddy, T. M. (1997). An examination of the relationship between work group characteristics and performance: Once more into the breech. *Personnel Psychology, 50,* 553–585.

Ilgen, D. R. (1999). Teams embedded in organizations: Some implications. *American Psychologist, 54,* 129–142.

Ilgen, D. R., & Pulakos, E. D. (1999). Introduction: Employee performance in today's organizations. In D. R. Ilgen & E. D. Pulakos (Eds.), *The changing nature of performance: Implications for staffing, motivation, and development* (pp. 1–21). San Francisco: Jossey-Bass.

Jehn, K. A. (1995). A multimethod examination of the benefits and detriments of intragroup conflict. *Administrative Science Quarterly, 40,* 256–282.

Jex, S. M., & Bliese, P. D. (1999). Efficacy beliefs as a moderator of the impact of work-related stressors: A multi-level study. *Journal of Applied Psychology, 84,* 349–361.

Kirkman, B. L., & Rosen, B. (1999). Beyond self-management: Antecedents and consequences of team empowerment. *Academy of Management Journal, 42,* 58–74.

Kirkman, B. L., Rosen, B., Gibson, C. B., Tesluk, P., & McPherson, S. (in press). Five challenges to virtual team success: Lessons from Sabre, Inc. *Academy of Management Executive.*

Kletz, T. A. (1985). *An engineer's view of human error.* Warwickshire: Institution of Chemical Engineers.

Klimoski, R., & Mohammed, S. (1994). Team mental model: Construct or metaphor? *Journal of Management, 20,* 403–437.

Komaki, J., Barwick, K. D., & Scott, L. R. (1978). A behavioral approach to occupational safety: Pinpointing and reinforcing safe performance in a food processing plant. *Journal of Applied Psychology, 63,* 434–445.

Komaki, J., Heinzmann, A. T., & Lawson, L. (1980). Effect of training and feedback: Component analysis of a behavioral safety program. *Journal of Applied Psychology, 65,* 261–270.

Kozlowski, S.W.J., & Doherty, M. L. (1989). Integration of climate and leadership: Examination of a neglected issue. *Journal of Applied Psychology, 74,* 546–553.

Kozlowski, S.W.J., Gully, S. M., Brown, K. G., Salas, E., Smith, E. M., & Nason, E. (2001). Effects of training goals and goal orientation traits on multi-dimensional training outcomes and performance adaptability. *Organizational Behavior and Human Decision Processes, 85,* 1–31.

Kozlowski, S.W.J., Gully, S. M., Nason, E., & Smith, E. M. (1999). Developing adapting teams: A theory of compilation and performance across levels and time. In D. R. Ilgen & E. D. Pulakos (Eds.), *The changing nature of performance: Implications for staffing, motivation, and development* (pp. 240–292). San Francisco: Jossey-Bass.

Lawler, E. E., III, & Finegold, D. (2000). Individualizing the organization: Past, present, and future. *Organizational Dynamics, 29*(1), 1–15.

Lawler, E. E., III, Mohrman, S. A., & Ledford, G. E., Jr. (1995). *Employee involvement and Total Quality Management: Practices and results in Fortune 1000 companies.* San Francisco: Jossey-Bass.

Lawler, E. J. (1992). Affective attachments to nested groups: A choice process theory. *American Sociological review, 57,* 327–339.

Liang, D. W., Moreland, R., & Argote, L. (1995). Group vs. individual training and group performance: The mediating role of transactive memory. *Personality and Social Psychology Bulletin, 21,* 384–393.

Lindsley, D. H., Brass, D. J., & Thomas, J. B. (1995). Efficacy-performance spirals: A multilevel perspective. *Academy of Management Review, 20,* 645–678.

Marks, M. A., Mathieu, J. E., & Zaccaro, S. J. (2001). A temporally based framework and taxonomy of team processes. *Academy of Management Review, 26,* 356–376.

Marks, M. A., Zaccaro, S. J., & Mathieu, J. E. (2000). Performance implications of leader briefings and team-interactions training for team adaptation to novel environments. *Journal of Applied Psychology, 85,* 971–986.

Martocchio, J. J. (1994). The effects of absence culture on individual absence. *Human Relations, 47,* 243–262.

Mathieu, J. E. (1991). A cross-level nonrecursive model of the antecedents of organizational commitment and job satisfaction. *Journal of Applied Psychology, 76,* 607–618.

Mathieu, J. E., Heffner, T. S., Goodwin, G. F., Salas, E., & Cannon-Bowers, J. A. (2000). The influence of shared mental models on team processes and performance. *Journal of Applied Psychology, 85,* 272–283.

Mathieu, J. E., & Kohler, S. S. (1990). A cross-level examination of group absence influences on individual absence. *Journal of Applied Psychology, 75,* 217–220.

McAfee, R. B., & Winn, A. R. (1989). The use of incentives/feedback to enhance workplace safety: A critique of the literature. *Journal of Safety Research, 20,* 7–19.

McGrath, J. E. (1991). Time, interaction, and performance (TIP): A theory of groups: *Small Group Research, 22,* 147–174.

McIntyre, R. M., & Salas, E. (1995). Measuring and managing for team performance: Lessons from complex organizations. In R. A. Guzzo, E. Salas, & Associates (Eds.), *Team effectiveness and decision making* (pp. 9–45). San Francisco: Jossey-Bass.

Mohrman, S. A., Cohen, S. G., & Mohrman, A. M. (1995). *Designing team-based organizations: New forms for knowledge work.* San Francisco: Jossey-Bass.

Moreland, R., Argote, L., & Krishnan, R. (1996). Socially shared cognition at work: Transactive memory and group performance. In J. L. Nye & A. M. Brower (Eds.), *What's social about social cognition? Research on socially shared cognition in small groups* (pp. 57–84). Thousand Oaks, CA: Sage.

Moreland, R. L., & Levine, J. M. (2001). Socialization in organizations and work groups. In M. E. Turner (Ed.), *Groups at work: Theory and research* (pp. 69–112). Hillsdale, NJ: Erlbaum.

National Transportation Safety Board (NTSB). (1994). A review of flight-crew-involved, major accidents of U.S. air carriers, 1978 through 1990. Washington, D.C.:National Transportation Safety Board.

Nelson, D. L., Quick, J. C., & Eakin, M. E. (1988). A longitudinal study of newcomer role adjustment in organizations. *Work and Stress, 2,* 239–253.

Nicholson, N., & Johns, G. (1985). The absence culture and the psychological contract: Who's in control of absence? *Academy of Management Review, 10,* 397–407.

Nonaka, I., & Takeuchi, H. (1995). *The knowledge creating company.* New York: Oxford University Press.

Parker, S. K., Wall, T. D., & Cordery, J. L. (in press). Future of work design research and practice: Towards an elaborated model of work design. *Journal of Occupational and Organizational Psychology.*

Pearson, C.A.L. (1992). Autonomous workgroups: An evaluation at an industrial site. *Human Relations, 45,* 905–936.

Reason, J. (1990). *Human Error.* Cambridge: Cambridge University Press.

Reber, R. A., & Wallin, J. A. (1984). The effects of training, goal setting, and knowledge of results on safe behavior: A component analysis. *Academy of Management Journal, 27,* 544–560.

Reber, R. A., & Wallin, J. A. (1994). Utilizing performance management to improve offshore oilfield diving safety. *International Journal of Organizational Analysis, 2,* 88–98.

Reber, R. A., Wallin, J. A., & Chhokar, J. S. (1990). Improving safety performance with goal setting and feedback. *Human Performance, 3,* 51–61.

Riordan, C. M., & Griffith, R. W. (1995). The opportunity for friendship in the workplace: An unexplored construct. *Journal of Business and Psychology, 10,* 141–154.

Robbins, H., & Finley, M. (2000). *The new why teams don't work: What goes wrong and how to make it right.* San Francisco: Berrett-Koehler.

Roethlisberger, F. J., & Dickson, W. J. (1939). *Management and the worker.* Cambridge, MA: Harvard University Press.

Simard, M., & Marchand, A. (1995). A multilevel analysis of organizational factors related to the taking of safety initiatives by work groups. *Safety Science, 21,* 113–129.

Simard, M., & Marchand, A. (1997). Workgroups' propensity to comply with safety rules: The influence of micro-macro organizational factors. *Ergonomics, 40,* 172–188.

Sonnentag, S. (1996). Work group factors and individual well-being. In M. A. West (Ed.), *Handbook of work group psychology* (pp. 345–367). New York: Wiley.

Sonnentag, S., Brodbeck, F. C., Heinbokel, T., & Slolte, W. (1994). Stressor-burnout relationship in software development teams. *Journal of Occupational and Organizational Psychology, 67,* 327–341.

Sulzer-Azaroff, B., Harris, T. C., & McCann, K. B. (1994). Beyond training: Organizational performance management techniques. *Occupational Medicine: State of the Art Reviews, 9,* 321–339.

Sundstrom, E., DeMeuse, K. P., & Futrell, D. (1990). Work teams: Applications and effectiveness. *American Psychologist, 45,* 120–133.

Tesluk, P. E., & Mathieu, J. E. (1999). Overcoming road blocks to effectiveness: Incorporating management of performance barriers into models of work group effectiveness. *Journal of Applied Psychology, 84,* 200–217.

Tesluk, P., Mathieu, J., Zaccaro, S. J., & Marks, M. (1997). Task and aggregation issues in the analysis and assessment of team performance. In M. Brannick, E. Salas, & C. Prince (Eds.), *Assessment and measurement of team performance: Theory, research, and applications* (pp. 197–224). Greenwich, CT: JAI Press.

Thompson, L. (2000). *Making the team: A guide for managers.* Englewood Cliffs, NJ: Prentice Hall.

Trist, E. L., & Bamforth, K. W. (1951). Some social and psychological consequences of the Longwall method of coal-getting. *Human Relations, 4,* 3–38.

Trist, E. L., Higgins, G. W., Murray, H., & Pollack, A. B. (1963). *Organizational choice.* New York: Tavistock.

Turner, N., & Parker, S. K. (in press). Seeking safety with others: The effects of team working on safety. In J. Barling & M. R. Frone (Eds.), *The psychology of workplace safety.* Washington, DC: American Psychological Association.

Wegner, D. M. (1986). Transactive memory: A contemporary analysis of the group mind. In B. Mullen & G. R. Goethals (Eds.), *Theories of group behavior* (pp. 185–208). New York: Springer-Verlag.

Wright, C. (1986). Routine deaths: Fatal accidents in the oil industry. *Sociological Review, 4,* 265–289.

Zaccaro, S. J., & Dobbins, G. H. (1989). Contrasting group and organizational commitment: Evidence for differences among multilevel attachments. *Journal of Organizational Behavior, 10,* 267–273.

Zaccaro, S. J., & Lowe, C. A. (1988). Cohesiveness and performance on an additive task: Evidence for multidimensionality. *Journal of Social Psychology, 128,* 547–558.

Zaccaro, S. J., & Marks, M. A. (1999). The role of leaders in high-performance teams. In E. Sundstrom & Associates (Eds.), *Supporting work team effectiveness* (pp. 95–125). San Francisco: Jossey-Bass.

Zaccaro, S. J., & Riley, A. W. (1987). Stress, coping, and organizational effectiveness. In A. E. Riley & S. J. Zaccaro (Eds.), *Occupational stress and organizational effectiveness* (pp. 1–28). New York: Praeger.

Zohar, D. (2000). A group-level model of safety climate: Testing the effect of group climate on microaccidents in manufacturing jobs. *Journal of Applied Psychology, 85,* 587–596.

Zohar, D. (2002). Modifying supervisory practices to improve sub-unit safety: A leadership-based intervention model. *Journal of Applied Psychology, 87,* 156–163.

CHAPTER 6

Antisocial Work Behavior and Individual and Organizational Health

Michelle K. Duffy
Anne M. O'Leary-Kelly
Daniel C. Ganster

With good reason, Americans have become concerned with the violence that pervades our society. In recent years spheres of life that traditionally have been perceived as safe havens—our schools, churches, medical clinics, and workplaces—have been the setting for high profile incidents of violence. In regard to the workplace, many Americans fear that coworkers will "go postal," that disgruntled subordinates will resolve conflicts with weapons, or that they face great jeopardy when their jobs involve contact with the public. Certainly, violent behavior at work is not a new phenomenon, but the amount of fear and concern that the general public feels about this issue is perhaps higher than ever before.

The enhanced concern about workplace violence is reflected in an increase in the number of publications addressing this issue in the last decade (e.g., National Institute for Occupational Safety and Health [NIOSH], 1996, 1997). A recent article reported the results of a search of the PsycLIT database under key words related to workplace violence (O'Leary-Kelly, Duffy, & Griffin, 2000). This search identified 80 articles prior to 1990 and over 250 articles subsequent to 1990. Similar results were found for a search of the ABI database: 150 articles prior to 1990 and over 500 after 1990. A similarly dramatic increase can be found in the number of books written on the topic of workplace violence. A total of 43 books were published on the topic in the years 1994 to 1998, with a peak in 1996 and 1997. Because these databases include both academic and practitioner-oriented periodicals, the results of these searches suggest a broad interest in this topic among scholars and practitioners.

Although acts of workplace violence most often captured by the media involve the armed and disgruntled former employee or customer returning for an act of revenge, research indicates the majority of workplace violence incidents involve nonlethal forms of assault (Bulatao & VandenBos, 1996; National Crime Victimization Survey, 1998). Unfortunately, although these nonfatal forms of workplace violence occur more frequently, research has largely ignored them (Neuman & Baron, 1998). Perhaps for this reason, the definition of workplace violence delineated by Workplace Violence Research Institute includes both physical and verbal assaults, threats, coercion, intimidation, and all forms of harassment. Consistent with this approach, in the following review, we consider a broad range of antisocial behaviors.

The purposes of this chapter are threefold. First, we review the existing empirical literature on the relationship between forms of antisocial behavior in the workplace and assessments of employees' health and well-being. Our review of this literature is not designed to be exhaustive but rather descriptive of findings of extant research. It is designed to provide a foundation for the extension of antisocial work behavior research to broader definitions of organizational health. Second, we integrate the existing research base into the definitions of individual and organizational health reviewed by Hofmann and Tetrick (Chapter One of this volume). The integration involves a consideration of not only the consequences of this behavior but an assessment of the causes as well. Third, we outline an agenda for future research in this area based on the existing literature and the integration of organizational health definitions.

Antisocial Work Behavior and Well-Being: A Brief Review

The potential aftereffects of antisocial work behaviors are significant and broad. The negative impact might range from the individual level to the organizational level, where forms of antisocial behavior may substantially alter the bottom line (Bassman, 1992). Interestingly, nearly all of the empirical research that examines the consequences of antisocial work behavior focuses on individual-level outcomes. Although some estimates of the impact on organizational-level outcomes, bottom-line assessments like costs or profitability, are available, the base for such estimates has a shakier foundation than that found at the individual level. The following review is divided into two main sections—effects on individuals and effects on organizations—to reflect the research found at these levels. Moreover, as we noted earlier, the many types of antisocial work behavior range from the dramatically violent to the subtle and passive. The potential consequences to employees and organizations will clearly vary as a function of the type of exhibited behavior.

Antisocial Work Behavior and Individual Outcomes

The damaging consequences of the more severe forms of workplace violence have been well documented (Mack, Shannon, Quick, & Quick, 1998). Witnesses and survivors of workplace violence often report feelings of guilt, depression, vulnerability, and suicidal ideation, as well as alcohol and drug use (e.g., Alcorn, 1994). Depending on the severity of the violent incidents, effects can range from mild anxiety to full-blown posttraumatic stress disorder (Braverman, 1992). But such incidents are relatively rare compared to the prevalence of much nonphysical forms of antisocial behavior. These behaviors generally do not invoke feelings of direct physical threat (although that sometimes does occur), but rather they more often threaten outcomes such as one's position and security within the organization, one's self-esteem, and one's ability to maintain positive social relationships at work. Moreover, these milder forms of antisocial behavior tend to have a pernicious effect through chronic exposures over an extended time.

The most common theoretical lens for viewing such behaviors is the transactional, or cognitive appraisal, theory of stress (e.g., Lazarus & Folkman, 1984). This model has two types of appraisal. Primary appraisal occurs when the individual gives meaning to events and makes an evaluation of their potential to threaten his or her well-being. That is, it signals the individual that danger lurks and that vigilance must be increased. A primary appraisal of threat or harm triggers a secondary appraisal in which individuals take stock of the resources that they can use to cope with the appraised threat. This is a very general process model of stress, and it does not steer us to specific types of threats that might be generated by antisocial behaviors. But this perspective does alert us to the need to consider how targets (or bystanders) interpret such behaviors. Whereas most people perceive overt violent acts as threatening, we expect much greater diversity in the meanings assigned to milder forms of antisocial behavior. Consequently, we would expect to observe more variance in the outcomes that the targets of such behaviors experience. Compounding this variance in primary cognitive appraisals is further variance contributed by differing secondary appraisals. Some of this variance in primary appraisals might be linked to individual vulnerability factors, and specific coping resources such as social support might explain variance in secondary appraisals.

It is difficult, therefore, to identify or predict the effects of antisocial work behavior on employee well-being unless a specific type of antisocial behavior is designated. As this review is not intended to be exhaustive, we discuss in the remainder of this section the effects of a few specific forms of antisocial work behavior (sexual harassment and abusive-undermining-bullying behavior) on employee well-being.

Sexual Harassment

Researchers commonly view different forms of sexual harassment behavior as a job stressor (Fitzgerald, Hulin, & Drasgow, 1995), and as such they would be expected to be associated with measures of employee health and well-being. The evidence for this general association is relatively consistent. The work of Fitzgerald and Drasgow and their colleagues (e.g., Fitzgerald, Drasgow, Hulin, Gelfand, & Magley, 1997; Wasti, Bergman, Glomb, & Drasgow, 2000) present a direct effects model linking sexual harassment behaviors to both psychological and physical well-being outcomes as well as to job attitudes and employee withdrawal. Although empirical evidence for the direct effect on physical health is equivocal, sexual harassment correlates consistently with mental health and job satisfaction, with correlations in the magnitude of .20 to .30. Although the direct effects are specified, this model does not specify mediating pathways between sexual harassment behaviors and its mental and physical health consequences; hence, exploration of cognitive appraisal mediators is not part of the program. It is assumed that respondents appraise these various sex-related behaviors as posing some degree of threat, but this link is never explicitly modeled. We would expect, then, that individual differences exist in terms of how people appraise and react to such behaviors.

Cognitive mediating processes have received attention in other sexual harassment studies, however, most notably those concerned with the effects of labeling. For example, although approximately 50% of women in most samples report experiencing unwanted and offensive sex-related behaviors at work or school, fewer than 20% of these women label themselves as having experienced sexual harassment (Magley, Hulin, Fitzgerald, & DeNardo, 1999). From a cognitive appraisal perspective, one could hypothesize that labeling an incident as sexual harassment constitutes a primary appraisal that the event is threatening. If the event is not labeled as harassment, there may be no perceived threat and thus no sequelae of stressful responses. Alternatively, there are many reasons why someone might experience such behaviors as threatening and disturbing and yet fail to label them as sexual harassment. For example, ample evidence shows that individuals use different definitions of sexual harassment, typically defining sexually coercive behaviors as such but less consistently so for unwanted sexual attention and gender harassment (e.g., Frazier, Cochran, & Olson, 1995; Terpstra & Baker, 1987).

This latter explanation seems more likely, as support for the labeling hypothesis has been poor; that is, individuals report negative psychological and attitudinal outcomes from experiencing harassing behaviors whether or not they label them as such (e.g., Jacques, Sivasubramaniam, & Murry, 2000; Magley et al., 1999). But even if labeling does not significantly mediate the effects of harassing behaviors, other forms of stressful cognitive appraisals might. Jacques et al. (2000) found that employee outcomes could be explained by sexual harassment type, frequency, and duration of exposures. Individuals' cognitive appraisals of the behaviors (i.e., the extent to which they found them to be annoying, offensive, disturbing, or threatening) explained significant incremental variance over the direct exposure model. Labeling the exposures as sexual harassment explained variance in the cognitive appraisals but did not contribute any unique variance to the attitudinal outcomes.

In summary, ample evidence shows that experiencing sexually harassing behaviors at work is associated with poor job attitudes as well as mental and physical health outcomes. These associations hold whether or not individuals label their exposures as sexual harassment, but they are stronger for individuals who appraise the exposures as threatening. Unfortunately, we have little empirical evidence regarding the cognitive appraisal process in this domain, either in terms of which factors might explain why some individuals develop more negative cognitive appraisals or how differences in coping resources (e.g., social support) might moderate responses to harassment behaviors.

Abuse, Bullying, and Undermining

A variety of construct labels describe other nonphysical forms of antisocial behavior. Definitions of specific constructs vary, but most of these forms of harassing or socially abusive behavior in the workplace are seen as persisting over some period of time, most commonly six months to a year, and consisting of relatively mild acts (as opposed to more overt violent or aggressive acts). Labels for such behaviors vary across countries, with terms such as *workplace harassment* (Brodsky, 1976), *employee abuse* (Bassman, 1992), *emotional abuse* (Keashly, 1998), *abusive supervision* (Tepper, 2000), *passive and indirect workplace aggression* (Neuman & Baron, 1997, 1998), *workplace incivility* (Pearson, Andersson, & Porath, 2000; Pearson, 1998; Pearson, Andersson, & Wegner, 2001), *petty tyranny* (Ashforth, 1994), and *social undermining* (Duffy, Ganster, & Pagon, 2002) commonly used in North America; they are often referred to as *mobbing* in Northern Europe and *bullying* in the United Kingdom (Hoel, Rayner, & Cooper, 1999). Research in this area is characterized by survey studies that ask respondents to describe the frequency or duration of their exposures to various antisocial behaviors, sometimes using a label such as *bullying* and sometimes not. Such exposure measures are then linked to various self-reported outcomes that include mental and physical health symptoms, job attitudes, and behaviors.

European researchers, especially those in Scandinavia, have paid the most attention to the mental and physical health outcomes of abusive behavior. They have demonstrated consistent associations between bullying and a host of psychosomatic symptoms, musculoskeletal symptoms, anxiety, depression, and cognitive impairments such as concentration and attention problems (see reviews by Cortina, Magley, Williams, & Langhout, 2001; and Hoel, Rayner, & Cooper, 1999). In North America measures of abusive supervision (Tepper, 2000), incivility (Cortina et al., 2001), and workplace aggression (Neuman & Baron, 1997) have been linked to psychological distress, poor job attitudes, and low continuance commitment. For example, in one of the few studies with a prospective design, Tepper (2000) demonstrated an association between abusive supervision and subsequent turnover. In a follow-up study, longitudinal data collected from a sample of supervised employees supported a model in which abusive supervision led to less subordinate job satisfaction and affective commitment, lower performance, and greater depression and anxiety (Hoobler, Tepper, & Zellars, 2001). Furthermore, these authors also found that higher subordinate job satisfaction, affective commitment, and performance actually led to lower perceptions of abusive supervision. These findings are consistent with the characterization of abusive supervision as being both cause and consequence of important subordinate outcomes.

Duffy and her colleagues (2002) investigated another form of antisocial behavior, social undermining, which they defined as behavior intended to hinder over time the ability to establish and maintain interpersonal relationships, work-related success, and favorable reputation. Like workplace incivility and abusive supervision, undermining refers to low intensity behaviors that take their toll over some period of time. In a sample of national police officers in Slovenia, Duffy and her colleagues observed significant relationships between undermining from either supervisors or coworkers and reduced organizational commitment, lowered self-efficacy, and higher levels of somatic health complaints. This research also indicated that the effects of social undermining were uniformly stronger than the effects of social support, highlighting the salience of negative social behaviors on employee outcomes. Duffy et al. also examined the interaction between social support and undermining and found that receiving higher levels of social support from one's supervisor actually exacerbated the effects of undermining from the supervisor. Interestingly, social support is traditionally viewed as a buffer of work stressors (Ganster, Fusilier, & Mayes, 1986), but in this case it magnified the detrimental effects of undermining. Duffy et al. (2002) hypothesized this effect based on the presumed stress that would be caused by receiving inconsistent social behaviors from an important member of one's social environment.

Beyond physical and mental health, investigators have recently focused their attention on the effect of abusive and undermining work behaviors on employee behaviors. For example, Tepper, Duffy, and Shaw (2001) explored the relationships between abusive supervision and subordinates' resistance tactics. Results of a longitudinal study suggested that subordinates of abusive supervisors reported that they used both dysfunctional resistance tactics and constructive resistance tactics more frequently than did their nonabused counterparts. As the authors note, one way that subordinates can retaliate for the resentment and outrage that abusive supervision evokes is by resisting downward influence attempts dysfunctionally. Such dysfunctional resistance has clear implications for task and relational concerns that may ultimately interfere with organizational productivity. Likewise, Duffy et al. (2002), also found a link between perceptions of undermining and counterproductive behavior. Employees who reported higher levels of supervisor and coworker undermining were more likely to engage in active as well as passive counterproductive work behavior.

Tepper and his colleagues also examined the impact of abusive behavior on organizational citizenship behavior (OCB). In a longitudinal study examining the moderating role of abusive supervision, Tepper, Duffy, and Hoobler (2000) found that the relationship between coworkers' OCBs and job attitudes (job satisfaction) was negative when abusive supervision was high. As the authors note, these findings may indicate that "subordinates of abusive supervisors are more likely (than their non-abused counterparts) to perceive coworkers' OCBs as self-serving, which in turn produces low job satisfaction and attenuates what would otherwise be a positive relationship between coworkers' OCB and subordinates' commitment" (p. 16). The relationship between abusive and undermining supervision and subordinates' OCB was also explored among a sample of Air National Guard members and their military supervisors (Zellars, Tepper, & Duffy, in press). The results of this research suggest that subordinates of abusive supervisors perform fewer OCBs than do their nonabused counterparts.

Antisocial Work Behaviors and Organizational Outcomes

The costs to organizations of antisocial behavior are difficult to estimate. This difficulty may be reflected in the comparatively little research devoted to this issue at the organizational level. Moreover, the costs to organizations of antisocial work behavior tend to focus on the incidence of direct or severe forms rather than the costs of indirect, passive, or minor forms (e.g., bullying, verbal abuse, and undermining) of antisocial behavior. With these limitations in mind, it is still useful to survey the findings concerning the impact of antisocial behavior on organizational outcomes.

The direct costs to U.S. business of all consequences of workplace aggression are estimated to be more than $4 billion annually (Wilson, 1991). Moreover, the National Safe Workplace Institute estimated that a single act of workplace violence averages about $250,000 in lost work time and legal expenses (Blythe & Gardner, 1999). But these estimates must be considered in light of a study by the Department of Justice that reveals that fewer than half of all nonfatal workplace incidents were reported to the police (Wilson, 1991). Thus, estimates of costs must be adjusted for severe (and difficult to estimate) underreporting bias.

Attempts have been made to estimate the cumulative financial losses that organizations suffer in the form of diminished health and well-being, productivity losses, absenteeism, and turnover as a result of antisocial behavior. For example, according to the National Crime Victimization Survey (1994), half a million U.S. workers lost 1.75 million days of work (an average of 3.5 days) as well as

$55 million in wages (not including days covered by sick leave) due to various forms of antisocial behavior. In a recent survey, managers indicated that incidents of workplace violence resulted in decreased morale and productivity and increased security costs, litigation, and workers' compensation claims (American Management Association, 1994). Other estimates suggest that dealing with a single incident of workplace violence could cost as much as $250,000, given the high costs of alleviating the posttraumatic stress of victims and witnesses. Although few direct organizational-level tests of these relationships appear in the literature, anecdotal evidence suggests that antisocial behavior costs organizations in terms of what their employees do not do as a result of perceptions of workplace aggression and abuse (Bassman, 1992). Research and anecdotal evidence suggest that witnessing or being the target of antisocial behavior at work weakens relationships with customers, lessens product quality, redirects efforts away from goal commitment, and reduces positive discretionary behaviors among the workforce (Bassman).

A final cost to employers of antisocial behavior at work involves exposure to legal liability. As Paetzold (1998) notes, U.S. employers are generally liable for harms and claims of harms that occur at their worksites. The total legal costs of antisocial behaviors have yet to be isolated, but the direct costs of liability, in addition to the indirect legal liability costs of additional applicant screening, training of employees, and creation of crisis management teams, are no doubt considerable.

Toward Theoretical and Empirical Progress

In Chapter One of this volume, Hofmann and Tetrick reviewed and integrated several definitions and conceptualizations of organizational health (e.g., medical, wellness, and environmental models). As they pointed out, these perspectives on organizational health range from the narrow to the very broad. When applied to organizational health issues, the medical model is the most narrow and least value-laden; that is, organizational health is implied simply by the absence of illness. The wellness model is broader; it expands the issue of organizational health to include not only the absence of illness but also a consideration of the well-being of individuals in the organization, their view in terms of optimism, and also the positive functioning of the organization. The environmental model is the broadest. It encompasses the other issues but involves the additional facet of effective individual performance of tasks and roles demanded by the work environment.

We would argue that the existing empirical literature on the consequences of antisocial work behavior implicitly assumes a medical perspective on organizational health. The empirical literature we reviewed in the preceding section focuses on the incidences of antisocial behavior in the workplace and the negative consequences for individuals and organizations as a result of these behaviors. That is, the absence of antisocial behaviors in a given workplace implies a state of health. This approach to studying antisocial behavior at work is quite similar to what O'Leary-Kelly, Griffin, and Kilbourne (1999, p. 2) termed the "crime fighting" metaphor for the study of antisocial work behaviors. These authors argued that when fighting crime, police officers essentially scan the landscape, looking for criminal behaviors and arresting the offenders. In this approach law enforcement makes little effort to assess the process through which individuals progress on the road to committing crimes; the effort instead is on responding efficiently to crimes after they have been committed. If no crimes are committed, the landscape is deemed to be crime-free. According to the medical model approach, if no antisocial behaviors are committed in a given workplace, the workplace is deemed healthy. We are not trying to denigrate the empirical research to date in this area by using the crime-fighting metaphor. Indeed, given the perceived negative consequences of antisocial behavior, it was necessary to establish the links to the organizational outcomes we reviewed earlier.

Our point is that the existing approaches fail to capture the richer theoretical and empirical possibilities when a broader conceptualization of organizational health is considered. As noted, the crime-fighting approach has little to say about the causes of crimes or the problems lurking beneath the societal surface and even less to say about solving these fundamental problems. In a similar way, the medical model has little to say about the causes of antisocial behavior or the state of organizational health in the absence of antisocial behavior. To take steps in this direction, researchers must of necessity simultaneously consider the causes and consequences of antisocial work behavior and expand the scope of outcomes to include those subsumed under broader conceptualizations of organizational health, that is, the wellness and environmental models. In the following sections, we examine the possibilities associated with applying the study of antisocial work behaviors to the wellness and environmental models of health.

Antisocial Work Behavior and the Wellness Model of Organizational Health

The adoption of a wellness definition of organizational health has several exciting and important implications for organizational researchers. On a general level, the adoption of wellness as a framework would involve not only a consideration of the number of incidents of antisocial behavior exhibited by employees but also the state of employees in terms of their frustration, stress, optimism, and so on. So far, we have focused on the consequences of antisocial behavior in terms of individual and organizational health (the main focus of this chapter), but to extend the research base toward broader definitions of health, it is important to consider the antecedents or causes of such behavior as well. Researchers have focused on three general perspectives for predicting antisocial behavior in the workplace: frustration, social learning, and justice.

The earliest approach to studying antisocial work behavior suggested that these actions were the result of individuals' frustration with organizational practices (Spector, 1978). The organizational frustration model was based on the frustration-aggression hypothesis (Dollard, Doob, Miller, Mowrer, & Sears, 1939), which greatly influenced the study of general aggression in social psychology. This model suggested that certain environmental conditions trigger individual frustration that may be expressed through antisocial behavior. A more recent conceptualization of this model (Spector, 1997) emphasizes the importance of the individual's cognitive processing of environmental stimuli, an addition that addresses one of the major criticisms of the frustration-aggression hypothesis (Berkowitz, 1993; Tedeschi & Felson, 1994). Specifically, this most recent version of the frustration model suggests that environmental frustrators prompt a cognitive appraisal by the focal individual (Spector, 1997). Frustrators are events or situations that prevent or intrude on an individual's ability to achieve or maintain a personal goal at work. These situations may be significant (e.g., becoming company president) or comparatively inconsequential (e.g., leaving for lunch by noon). Previous research suggests that frustrators may include job-related information, tools and equipment, materials and supplies, budgetary support, required services and help from others, task preparation, time availability, and the work environment (Peters & O'Connor, 1980). In addition, several studies suggest that job stressors such as role ambiguity, role conflict, workload, and interpersonal conflict act as frustrators (Chen & Spector, 1992; Spector, Dwyer, & Jex, 1988).

Although the empirical evidence regarding frustration and antisocial behavior is probably best described as mixed, some is quite supportive. Neuman and Baron (1998) note that organizational frustration has been linked to interpersonal aggression, stealing, hostility, and sabotage in circumstances where the goal interference was perceived to be intentional and unjust (e.g., Martinko & Zellars, 1998; Spector, 1997). For example, Baron and Neuman (1996) found that frustrating changes in the workplace (e.g., downsizing, layoffs, and pay freezes) were positively related to aggression. Chen and Spector (1992) reported an association between frustration and aggression and hostility among a sample of 400 white-collar employees. Current research, then, indicates that frustration may be linked to antisocial behavior at work when employees perceive the cause of the frustration to be unjust, unfair, or unwarranted (e.g., Folger & Cropanzano, 1998).

A second approach to the study of antisocial behavior antecedents is social learning theory (Bandura, 1977), which emphasizes the role of several environmental factors—modeling, aversive treatment, incentive inducements, and conditions in the physical environment—in the promotion of antisocial behavior at work. Research generally confirms the importance of each of these dimensions as triggers for antisocial behavior. For example, in a study of work groups from a variety of industries, Robinson and O'Leary-Kelly (1998) found a positive relationship between the level of antisocial behavior exhibited by an individual group member and that exhibited by members of the work group. This positive relationship was enhanced as the individual's time in the group increased and as task interdependence increased. Although these findings do not directly test the modeling propositions from social learning theory, they are supportive.

Relatedly, the notion that incentive inducements in social learning theory, that is, aspects of organizational environments may actually support and encourage antisocial behavior is popular in the academic and practitioner literature alike (e.g., Bassman, 1992). To the extent that lack of punishment for antisocial behavior can be viewed as an inducement (e.g., see Trevino, 1992), studies examining employee theft offer empirical support for the role of organizational inducements in antisocial behavior. A review of the employee theft literature reveals that because employees are rarely prosecuted, most employees feel little guilt for stealing and thus may even feel rewarded for doing so (e.g., Carter, 1987). Greenberg and Scott (1996) report that fewer than 15% of employees caught stealing are prosecuted, as employers would rather fire the employees and have them repay the money.

A third important perspective on the antecedents of antisocial behavior is that offered by justice theory. Indeed, among the more consistent findings in the literature is that unfair or unjust treatment is often a catalyst of antisocial behavior at work. The range of behaviors associated with perceptions of unfair treatment is far-reaching. For example, perceptions of unfair organizational decisions or managerial actions elicit retribution among employees (e.g., Sheppard, Lewicki, & Minton, 1992). Moreover, unfair treatment also is associated with higher levels of interpersonal conflict in organizations (e.g., Mark & Folger, 1984), counterproductive work behaviors (Giacalone & Greenberg, 1997; Jermier, 1988), negative reactions to layoffs, and the spreading of untruthful rumors about the organization (e.g., DePaulo & DePaulo, 1989; Grover, 1997). Moreover, reactions to injustice in organizations often include severe forms of violence such as homicide. Perpetrators of workplace homicides often cite unjust treatment as the primary motive for their acts (e.g., Kinney, 1995; Mantell, 1994). In fact, the desire to even the score is cited as a justification for as many as 80% of workplace homicides (Weide & Abbott, 1994).

Finally, among the more consistent justice-based findings is that perceptions of injustice correlate with employee theft and retaliation in organizations. In an influential series of studies, Greenberg

(1990, 1993) demonstrated that theft behavior can be viewed as a reaction to perceived injustice. In both a field study (a plant in which workers were underpaid) and in a laboratory study (undergraduates being underpaid or equitably paid for a clerical task), he found that employees were more likely to steal when they were underpaid, suggesting that perceived distributive injustice motivated the theft. In addition, his findings suggested that when subjects perceive little procedural justice (i.e., payment is determined by the use of invalid information) or interactional justice (i.e., subjects are led to believe that the experimenter is unconcerned with their negative pay outcomes), then stealing behavior is greater compared to cases of high procedural and interactional justice. This research suggests that theft is an effort by an individual who believes that he or she has been unfairly treated to restore a state of equity (Greenberg & Scott, 1996).

What are the implications of these streams of research for future investigations of the relationship between antisocial behavior and organizational health? First, if researchers adopted a wellness-based approach to the study of antisocial work behavior, it would necessitate not only an assessment of how often these behaviors were occurring but would require organizational decision makers to assess the state of the organization in terms of employee frustration, social learning (e.g., modeling, aversive treatment, incentive inducements, and physical environment conditions), and justice. Such an approach would require that the organizations identify a priori whether conditions in the working environment warrant attention as potential triggers for antisocial behavior. That is, are there any perceived injustices, incentive inducements for antisocial behavior, or conditions in the physical environment that may serve as triggers for future antisocial behaviors? Or what is the state of perceived injustice in the organization, and could these perceptions possibly lead to employees engaging in antisocial behavior?

Let's return for a moment to the policing metaphor. As we discussed earlier, an organization that evaluates organizational health in terms of the absence of antisocial behavior is evaluating the issue with a crime-fighting approach. But if the underlying causes and the incidence of antisocial behavior are considered as only a single aspect of organizational health, the crime-fighting metaphor cannot be adequately applied. A new metaphor is required, one that O'Leary-Kelly et al. (1999) built from the community-policing model of law enforcement. The community-policing model is a participative technique between the community and law enforcement officials. The idea is for all interested groups to work together to examine problems and issues relevant to crime and to develop various innovative approaches to solve the problems (e.g., Carter, 1995; Crothers & Vinzant, 1996). Under the community-policing approach, the mandate of law enforcement officers is broader and farther reaching; that is, crime control is a single facet of a mandate that includes social and physical order, problem correction, and fear reduction (e.g., Moore & Trojanowicz, 1988).

Under the crime-fighting (medical model) approach, the absence of antisocial behavior would imply health at the individual and organizational levels, regardless of whether or not the underlying conditions in the organization (justice, aversive treatment, incentive inducements, physical conditions, etc.) were ripe for such behavior. When the issue is viewed through the lens of the broader wellness model, the absence of actual antisocial behavior is only a single indicator of organizational health. Under this view an organization could be considered unhealthy if working conditions or aspects of the organizational culture are perhaps incubating antisocial acts. This expansion would require the organization to be very proactive in preventing antisocial behavior, by seeking out and continually reevaluating the information on a broad range of issues that could possibly prompt various types of antisocial behavior. This approach also allows the organization to identify a priori whether any policies and practices in the organization—human resource management practices, work design issues, working conditions, and so on—are unjust or aversive or sending the wrong signals to employees. The organization could deal with these factors, as indications of unhealthy conditions, in such a way as to avert the incidents of antisocial work behaviors and restore a measure of organizational health. Adoption of the wellness view of organizational health has challenges; primarily, it requires organizational decision makers and employees to be very proactive in preventing antisocial behavior. It also has several advantages, not only in terms of the possible reductions of antisocial work behaviors but also in terms of positive feelings, optimism, and organizational attachment—all factors that are subsumed under the wellness model of organizational health.

Antisocial Work Behavior and the Environmental Model of Organizational Health

The environmental model of organizational health is another step up the ladder in terms of scope, complexity, and values associated with the health definitions. It subsumes many of the aspects of the medical and wellness models, including the absence of illness (in our case, antisocial behavior), well-being, higher-level functioning, and optimism, and other aspects including how well individuals are performing their tasks and roles in the organization. Beyond this, the environmental approach takes a view that is even more systemic than the wellness approach. Rather than considering individual and organizational health in terms of their separate antecedents, it sees them as part of a mutually reinforcing, reciprocal system. If researchers adopt an environmental definition of organizational health, a much broader and richer range of issues enters the realm of possibility in terms of antisocial work behavior research. For example, this definition could help researchers assess the impact of a given antisocial work behavior not just on the targeted employee or group of employees but on other employees (e.g., witnesses or other affected individuals) as well. The social learning approach to the study of antisocial behavior may hold the most promise in this regard. Robinson and O'Leary-Kelly's study (1998) of crossover effects of antisocial behavior from groups to individuals is a good example in this vein. More recently, Duffy, Ganster, & Pagon (2000) examined how the work-group context affected the impact of undermining. They found that the negative impact of social undermining on employee well-being was stronger when other members of the focal person's work group reported low levels of undermining from the supervisor, hence leaving the individual to feel singled out for the negative intention.

Beyond individual and cross-level group studies, the adoption of an environmental view of organizational health would open many doors of opportunity for organizational researchers. If they added the effective performance of tasks and roles to the health equation, researchers could consider not only how individuals are affected by antisocial behaviors directed at others but also how individuals and organizations cope with violent or deviant incidents both in the short and long term. As Hofmann and Tetrick (Chapter One) highlight, the definitions of organizational health differ in their time orientation. It is reasonable to assume that the environmental model incorporates the lengthiest conceptualization of the time continuum; that is, it views health not only from a short-term illness and wellness perspective but also in terms of adaptability, survivability, and other long-term consequences. Researchers could also study the creation of organizational violence or antisocial behavior cultures, as sociologists and others study the creation of violence cultures in society at large. Finally, the study of antisocial work behavior from an environmental view could also assess how incidents of antisocial behavior and programs and practices designed to strengthen the health of the organization by creating an environment that thwarts antisocial acts affects organizational-level outcomes like financial performance and public image or societal reputation.

In summary, we agree with Hofmann and Tetrick's assessment (Chapter One) of the literature's definitions of organizational health. The wellness and environmental models are more value-laden than the medical model in that they provide a conceptualization of health beyond the absence of illness. They imply that organizations must determine what they consider to be the most important aspects of individual and organizational health and well-being. A consideration of these values as they relate to the study of antisocial behavior in the workplace, we believe, will lead to a much stronger research base, stronger antisocial behavior prevention programs, and ultimately to more positive assessments of organizational health.

Challenges and Opportunities for Research on Antisocial Behavior and Organizational Health

In this section we outline several challenges for researchers in light of our discussion, as well as offer several suggestions for future antisocial work behavior research.

Challenges for Researchers

Although we devoted most of our review to the consequences of antisocial work behavior, more research is still needed on its antecedents. In particular, we challenge researchers to move beyond the current focus on antisocial work behavior as a reaction. Currently, influential theoretical models (e.g., the organizational frustration model and the justice model) depict antisocial actions as a response to some work situation that the individual perceives as frustrating, uncomfortable, or unjust. However, antisocial work behavior may at times likely be a proactive effort to achieve some valued goal rather than a reaction to some disliked circumstance. For example, individuals may use antisocial actions to gain something that they value (e.g., social undermining of a colleague in order to enhance one's chances for promotion) or to create a desired social identity (e.g., sexual harassment to establish one's place in a valued work group) (O'Leary-Kelly et al., 2000). Existing theoretical frameworks provide comparatively little insight into these proactive antisocial actions, suggesting that additional theoretical perspectives (e.g., self-presentation theory) may be usefully applied.

We also challenge researchers to recognize the differences in severity in various types of antisocial work behavior. Antisocial actions might include very severe forms of conduct (e.g., homicide, physical assault), as well as those that have less extreme consequences (e.g., social undermining, sabotage of work). Clearly, these actions differ, and therefore their theoretical explanations also are likely to differ. Although we advocate an interactionist approach in which both situational and dispositional antecedents are considered, we do suspect that these antecedents may be differentially important, depending on the severity of the conduct. Specifically, we expect that the most fundamental antecedents of less severe forms of conduct (e.g., social undermining of a coworker) may be found in the social environment (i.e., work-group norms, modeling), whereas personality characteristics may be more critical to the prediction of severe forms of conduct (e.g., homicide or physical assault). Therefore, more careful matching of theoretical perspectives to the types of conduct that are being explained may be useful.

Relatedly, we encourage researchers to adopt a more complex approach to the study of perpetrators and victims. Specifically, we expect that the traits that characterize perpetrators will vary by the type of conduct under consideration. Simply put, the personality trait that encourages physical assault by one individual is likely to be very different from the trait that encourages another individual to spread gossip. Similarly, the characteristics that typify victims are likely to depend on the type of conduct considered. To be most predictive, then, research on perpetrators and victims must consider these individuals' traits within the context of a specific form of antisocial action.

Current research also is skewed toward understanding the conditions that prompt antisocial actions versus the responses of individuals or the organization to these actions. Although some research identifies the costs of antisocial actions for targets and the organization, surprisingly little is known about how individuals react when confronted with such behavior. Antisocial actions often occur within a larger social dynamic that involves ongoing interactions between the parties involved. Research that examines the target's reaction to negative actions and its effect on the perpetrator's subsequent behavior would provide useful insights into the destructive dynamics that contribute to antisocial actions and their prevention. Similarly, research that examines organizational actions following antisocial actions and their effects on subsequent perpetrator behavior would be useful.

Research on the Relationship Between Antisocial Work Behavior and Health

Beyond the general ideas we offered under the conceptualizations of organizational health, we would also offer several specific ideas for future research on the relationship between antisocial work behaviors and aspects of individual and organizational health. First, despite the increased research attention to antisocial work behavior in recent years, little effort has gone into delineating the mediating processes that determine their impact on employee outcomes (for exceptions see Jacques et al., 2000; Tepper, 2000). Thus, researchers still have abundant opportunity to explore different intervening pathways between exposure to antisocial behaviors and negative outcomes. In fact, multiple pathways could exist that represent different causal effects of the antisocial behaviors. For example, some antisocial behaviors might operate mostly through their damaging effects on the target's self-esteem, whereas others might be harmful because they encourage feelings of social isolation. Still others might directly threaten the individual's job security.

In addition to examining the impact of antisocial behaviors on the targets themselves, researchers should search for wider effects of such behaviors on the work group and the organization. As a point of departure, Andersson and Pearson (1999), for example, assert that such behaviors (incivility, specifically) can produce a spiral of negative events that can ultimately produce more extreme forms of antisocial behavior, including physical violence. A pattern of abusive behavior or undermining from a supervisor can in time elicit similar negative behaviors from the targets of the abuse or even from observers if they interpret such behavior as normative in the organization. Duffy, Tepper, and O'Leary-Kelly (2002), observed such a pattern in a longitudinal investigation of the role of moral disengagement at work. Individuals who witnessed undermining supervision were more likely to morally disengage and commit deviant acts at work. This relationship was more pronounced when there were vicarious reinforcements for engaging in antisocial behavior. This finding conforms to Tepper's observation (2000) of the effect of abusive supervision on organizational justice perceptions. Likewise, a relationship between abuse and dysfunctional resistance and OCBs has also been noted (e.g., Tepper et al., 2001). Although this research is encouraging, the impact of such behaviors on the subsequent antisocial behaviors of the targets as well as on the climate of the work group itself remains mostly unexplored, and it thus represents an area worthy of investigation.

Moderators of the effects of antisocial behaviors also have received scant attention. One class of moderators common in the occupational stress arena concerns individual vulnerability factors. Such factors are typically characteristics of the person or the person's situation that make him or her more likely to suffer negative consequences from exposure to the antisocial behaviors. One candidate is the gender of the target. From sexual harassment research and the broader bullying research, we note that females are more frequently targets of antisocial behavior, and this constitutes a form of vulnerability. But no evidence proves that females are more adversely affected by antisocial behaviors when they encounter them. Other vulnerability factors have been suggested, such as being in a minority group, being young, or being unmarried (Cortina et al., 2001), but little empirical evidence supports these hypotheses. Another factor that would make one more vulnerable to antisocial behaviors is an inability to escape from the situation because of low job mobility. Indeed, Tepper (2000) found significant interactions between low job mobility and abusive supervision that indicated that those with low perceived mobility suffered more negative consequences than did those with high mobility. This type of vulnerability factor is relevant to the secondary appraisal process in that it represents a kind of coping resource (i.e., an escape route) available to the target of antisocial behaviors. A number of other such resources merit attention, such as formal organizational mechanisms that give individuals a means of challenging or at least escaping the antisocial behaviors. Investigating such moderating factors can also lead us to the discovery of potential remedies for counteracting the effects of antisocial behaviors.

Finally, we must note that evidence pertaining to antisocial work behavior consists of the self-reports of targets concerning both their exposure to the behaviors and their effects. Generally, these associations are observed in cross-sectional designs (although see Hoobler et al., 2001; Tepper, 2000; and Tepper et al., 2000 for exceptions). We will repeat the shibboleth about the need for longitudinal designs but are reluctant to overemphasize the spurious effects of self-report measurement strategies. Longitudinal designs would be particularly productive when one could begin observation at an early point in a dyadic relationship or in the formative stages of a work group. In these situations one has the possibility of tracing the development of work-group norms or climate and assessing the etiologic role of antisocial behaviors. One might observe, for example, that groups characterized by high levels of antisocial behaviors from key members (or leaders) show evidence of the snowballing effect that Andersson and Pearson (1999) described.

We are less sanguine about the potential for avoiding self-report exposure measures. Antisocial behaviors in general, and especially certain forms such as sexual coercion, are frequently private events that are not amenable to third-party observation. One strategy is to rely on the convergence of self-reports from several potential targets, which is particularly relevant to supervisory antisocial behavior. If there is high convergence among observers (i.e., high interrater reliability), then we might be more confident in the validity of the self-reports. But even this approach is ambiguous because low convergence of reports within a work group about a particular supervisor might just be reflective of the dyadic nature of the phenomenon rather than an indication that self-reports are not true. The experience of antisocial behaviors is largely a subjective event, but we should strive to measure the behaviors themselves as objectively as possible. This can be accomplished to some degree by relying on descriptive items rather than evaluative ones. For example, the abusive supervision item "[My supervisor] tells me I'm incompetent" (Tepper, 2000, p. 90) is an objective descriptor of behavior. In contrast, an item asking respondents whether they have been sexually harassed involves a considerable amount of cognitive interpretation (labeling) and is an ambiguous description of a particular behavior. Given the ethical constraints on conducting experimental studies in this area, investigators will undoubtedly continue to rely on respondent reports of antisocial behaviors. But we still have much to learn by the careful and thoughtful use of this research strategy.

Conclusion

In this chapter we reviewed the empirical literature concerning the relationship between antisocial behavior at work and its consequences for individual and organizational health and well-being. Empirical literature generally supports the prediction that antisocial behavior at work is associated with a broad array of potentially negative consequences to both individuals and organizations. Drawing on the existing literature base, we incorporated the findings into broader definitions of organizational health, including those that focus on the presence of health rather than simply the presence of negative outcomes. We believe that research efforts overall have created a significant body of knowledge in a relatively short period. Valuable frameworks and findings now may assist practicing managers or guide researchers in their efforts to further understand antisocial work actions. Although we find much here to be enthusiastic about, much remains to be done and much yet to be learned. We hope the agenda and the challenges associated with antisocial work behavior will encourage future researchers to tackle these issues in their research.

References

Alcorn, S. (1994). *Anger in the workplace*. Westport, CT: Quorum/Greenwood.

American Management Association (1994). *Workplace violence: Policies, procedures, and incidents*. New York: Publisher.

Andersson, L. M., & Pearson, C. M. (1999). Tit for tat? The spiraling effect of incivility in the workplace. *Academy of Management Review, 24,* 452–471.

Ashforth, B. (1994). Petty tyranny in the workplace. *Human Relations, 47,* 755–778.

Bandura, A. (1977). *Social learning theory.* Englewood Cliffs, NJ: Prentice Hall.

Baron, R. A., & Neuman, J. H. (1996). Workplace violence and workplace aggression: Evidence on their relative frequency and potential causes. *Aggressive Behavior, 22,* 161–173.

Bassman, E. S. (1992). *Abuse in the workplace: Management remedies and bottom line impact.* Westport, CT: Quorum/Greenwood.

Berkowitz, L. (1993). *Aggression: Its causes, consequences, and control.* New York: McGraw-Hill.

Blythe, B., & Gardner, R. (1999). Eye on workplace violence. *Risk Management, 46,* 46–49.

Braverman, M. (1992). Post trauma crisis intervention in the workplace. In J. Quick, L. Murphy, & J. Hurrell (Eds.), *Stress and well-being at work.* Washington, DC: American Psychological Association.

Brodsky, C. M. (1976). *The harassed worker.* Lexington, MA: Heath.

Bulatao, E., & VandenBos, G. (1996). Workplace violence: Its scope and the issues. In G. R. VandenBos & E. Bulatao (Eds.), *Violence on the job: Identifying risks and developing solutions* (pp. 1–23). Washington, DC: American Psychological Association.

Carter, D. L. (1995). Politics and community policing: Variables of changes in the political environment. *Public Administration Quarterly, 19*(1), 6–25.

Carter, R. (1987). Employee theft often appears legitimate. *Accountancy, 100,* 75–77.

Chen, P. Y., & Spector, P. E. (1992). Relationships of work stressors with aggression, withdrawal, theft, and substance abuse: An exploratory study. *Journal of Occupational and Organizational Psychology, 65,* 177–184.

Cortina, L. M., Magley, V. J., Williams, J. H., & Langhout, R. D. (2001). Incivility in the workplace. *Journal of Occupational Health Psychology, 6,* 64–80.

Crothers, L., & Vinzant, J. (1996). Cops and community: Street-level leadership in community-based policing. *International Journal of Public Administration, 19*(7), 1167–1191.

DePaulo, P. J., & DePaulo, B. M. (1989). Can deception by salespersons and customers be detected through nonverbal behavioral cues? *Journal of Applied Social Psychology, 19,* 1552–1577.

Dollard, J., Doob, L. W., Miller, N. E., Mowrer, O. H., & Sears, R. R. (1939). *Frustration and aggression.* New Haven, CT: Yale University Press.

Duffy, M. K., Ganster, D. C., & Pagon, M. (2000, August). The influence of the social context on individual reactions to perceived supervisory undermining. Paper presented at the annual meeting of the Academy of Management, Toronto.

Duffy, M. K., Ganster, D. C., & Pagon, M. (2002). Social undermining in the workplace. *Academy of Management Journal, 45,* 331–351.

Duffy, M., Tepper, B., & O'Leary-Kelly, A. (2002, November). Moral disengagement and antisocial behavior at work. Paper presented at the annual Southern Management Association meeting, Atlanta.

Fitzgerald, L. F., Drasgow, F., Hulin, C. L., Gelfand, M. J., & Magley, V. J. (1997). The antecedents and consequences of sexual harassment in organizations: A test of an integrated model. *Journal of Applied Psychology, 82,* 578–589.

Fitzgerald, L. F., Hulin, C. L., & Drasgow, F. (1995). The antecedents and consequences of sexual harassment in organizations. In G. Keita & J. J. Hurrell Jr. (Eds.), *Job stress in a changing workforce: Investigating gender, diversity, and family issues* (pp. 55–73). Washington, DC: American Psychological Association.

Folger, R., & Cropanzano, R. (1998). *Organizational justice and human resource management.* Thousand Oaks, CA: Sage.

Frazier, P. A., Cochran, C. C., & Olson, A. M. (1995). Social science research on lay definitions of sexual harassment. *Journal of Social Issues, 51,* 21–37.

Ganster, D. C., Fusilier, M. R., & Mayes, B. T. (1986). Role of social support in the experience of stress at work. *Journal of Applied Psychology, 71,* 102–110.

Giacalone, R. A., & Greenberg, J. (1997). *Antisocial behavior in organizations.* Thousand Oaks, CA: Sage.

Greenberg, J. (1990). Employee theft as a reaction to underpayment inequity: The hidden cost of pay cuts. *Journal of Applied Psychology, 75,* 561–568.

Greenberg, J. (1993). Stealing in the name of justice: Informational and interpersonal moderators of theft reactions to underpayment inequity. *Organizational Behavior and Human Decision Processes, 54,* 81–103.

Greenberg, J., & Scott, K. S. (1996). Why do workers bite the hands that feed them? Employee theft as social exchange process. In B. M. Staw & L. L. Cummings (Eds.), *Research in organizational behavior* (Vol. 18, pp. 111–156). Greenwich, CT: JAI Press.

Grover, S. L. (1997). Lying in organizations: Theory, research, and future directions. In R. A. Giacalone & J. Greenberg (Eds.), *Antisocial behavior in organizations* (pp. 68–84). Thousand Oaks, CA: Sage.

Hoel, H., Rayner, C., & Cooper, C. L. (1999). Workplace bullying. In C. L. Cooper & I. T. Robertson (Eds.), *International review of industrial and organizational psychology* (Vol. 14, pp. 195–230). New York: Wiley.

Hoobler, J. M., Tepper, B. J., & Zellars, K. (2001, August). *Causal connections among abusive supervision and subordinates' attitudes, psychological distress, and performance.* Paper presented at the Academy of Management meetings, Washington, DC.

Jacques, P. H., Sivasubramaniam, N., & Murry, W. D. (2000, August). *Sexual harassment labeling v. cognitive appraisal: Employing a stress-strain model to resolve the theoretical tension.* Paper presented at the annual meeting of the Academy of Management, Toronto.

Jermier, J. M. (1988). Sabotage at work: The rational view. In S. B. Bachrach (Ed.), *Research in the sociology of organizations* (Vol. 6, pp. 101–134). Greenwich, CT: JAI Press.

Keashly, L. (1998). Emotional abuse in the workplace: Conceptual and empirical issues. *Journal of Emotional Abuse, 1,* 85–117.

Kinney, J. A. (1995). *Violence at work: How to make your company safer for employees and customers.* Englewood Cliffs, NJ: Prentice Hall.

Lazarus, R. S., & Folkman, S. (1984). *Stress, appraisal, and coping.* New York: Springer.

Mack, D. A., Shannon, C., Quick, J. D., & Quick, J. C. (1998). Stress and the preventative management of workplace violence. In R. W. Griffin, A. M. O'Leary-Kelly, & J. M. Collins (Eds.), *Dysfunctional behavior in organizations: Violent and deviant behavior* (pp. 119–142). Greenwich, CT: JAI Press.

Magley, V., Hulin, C., Fitzgerald, L., & DeNardo, M. (1999). Outcomes of self-labeling sexual harassment. *Journal of Applied Psychology, 84,* 390–402.

Mantell, M. R. (1994). *Ticking bombs: Defusing violence in the workplace.* Burr Ridge, IL: Irwin.

Mark, M. M., & Folger, R. (1984). Responses to relative deprivation: A conceptual framework. In P. Shaver (Ed.), *Review of personality and social psychology* (Vol. 5, pp. 192–218). Thousand Oaks, CA: Sage.

Martinko, M. J., & Zellars, K. L. (1998). Toward a theory of workplace violence and aggression: A cognitive appraisal perspective. In R. W. Griffin, A. M. O'Leary-Kelly, & J. M. Collins (Eds.), *Dysfunctional behavior in organizations: Violent and deviant behavior* (pp. 1–42). Greenwich, CT: JAI Press.

Moore, M. H., & Trojanowicz, R. C. (1988). Corporate strategies for policing. *Perspectives on policing* (No. 6, pp. 1–15). Washington, DC: National Institute of Justice, U.S. Department of Justice.

National Crime Victimization Survey. (1994). Bureau of Justice Statistics, United States Department of Justice. (www.ojp.usdoj.gov)

National Crime Victimization Survey. (1998). Bureau of Justice Statistics, United States Department of Justice. (www.ojp.usdoj.gov)

National Institute for Occupational Safety and Health (NIOSH). (1996). *Violence in the workplace: Risk factors and prevention strategies* [Current Intelligence Bulletin No. 96–100). Cincinnati, OH: Author.

National Institute for Occupational Safety and Health (NIOSH). (1997). *Violence in the workplace* [NIOSH Fact Sheet]. Cincinnati, OH: Author.

Neuman, J. H., & Baron, R. A. (1997). Aggression in the workplace. In R. A. Giacalone & J. Greenberg (Eds.), *Antisocial behavior in organizations* (pp. 37–67). Thousand Oaks, CA: Sage.

Neuman, J. H., & Baron, R. A. (1998). Workplace violence and workplace aggression: Evidence concerning specific forms of, potential causes, and preferred targets. *Journal of Management, 24,* 391–412.

O'Leary-Kelly, A. M., Duffy, M. K., & Griffin, R. W. (2000). Construct confusion in the study of antisocial work behavior. In G. R. Ferris (Ed.), *Research in personnel and human resource management* (Vol. 18, pp. 275–303). Greenwich, CT: JAI Press.

O'Leary-Kelly, A. M., Griffin, R. W., & Kilbourne, L. M. (1999). Community policing as an aggression-prevention philosophy in business organizations. Unpublished manuscript.

Paetzold, R. (1998). Workplace violence and employer liability: Implications for organizations. In R. W. Griffin, A. M. O'Leary-Kelly, & J. M. Collins (Eds.), *Dysfunctional behavior in organizations: Violent and deviant behavior* (pp. 143–165). Greenwich, CT: JAI Press.

Pearson, C. (1998). Organizations as targets and triggers of aggression and violence: Framing rational explanations for dramatic organizational deviance. *Research in the sociology of organizations: Vol. 15. Deviance in and of organizations* (pp. 197–223). Greenwich, CT: JAI Press.

Pearson, C., Andersson, L., & Porath, C. (2000). Assessing and attacking workplace incivility. *Organizational Dynamics,* Fall, pp. 123–137.

Pearson, C., Andersson, L., & Wegner, J. (2001). When workers flout convention: A study of workplace incivility. *Human Relations, 54,* 1387–1419.

Peters, L. H., & O'Connor, E. J. (1980). Situational constraints and work outcomes: The influences of a frequently overlooked construct. *Academy of Management Review, 5,* 391–397.

Robinson, S. L., & O'Leary-Kelly, A. M. (1998). Monkey see, monkey do: The influence of work groups on the antisocial behavior of employees. *Academy of Management Journal, 41,* 658–672.

Sheppard, B. H., Lewicki, R. J., & Minton, J. W. (1992). *Organizational justice: The search for fairness in the workplace.* New York: Lexington Books/Macmillan.

Spector, P. E. (1978). Organizational frustration: A model and review of the literature. *Personnel Psychology, 31,* 815–829.

Spector, P. E. (1997). The role of frustration in antisocial behavior at work. In R. A. Giacalone & J. Greenberg (Eds.), *Antisocial behavior in organizations* (pp. 1–17). Thousand Oaks, CA: Sage.

Spector, P. E., Dwyer, D. J., & Jex, S. M. (1988). The relationship of job stressors to affective, health, and performance outcomes: A comparison of multiple data sources. *Journal of Applied Psychology, 73,* 11–19.

Tedeschi, J. T., & Felson, R. B. (1994). *Violence, aggression, and coercive actions.* Washington, DC: American Psychological Association.

Tepper, B. J. (2000). Consequences of abusive supervision. *Academy of Management Journal, 43,* 178–190.

Tepper, B., Duffy, M. K., & Hoobler, J. (2000, August). Moderating effects of abusive supervision on relationships between coworkers' organizational citizenship behavior and fellow employees' attitudes. Paper presented at the Academy of Management meetings, San Diego, CA.

Tepper, B., Duffy, M. K., & Shaw, J. D. (2001). Personality moderators of the relationship between abusive supervision and subordinates' resistance. *Journal of Applied Psychology, 86,* 974–983.

Terpstra, D. E., & Baker, D. D. (1987). A hierarchy of sexual harassment. *Journal of Psychology, 121,* 599–605.

Trevino, L. (1992). The social effects of punishment in organizations: A justice perspective. *Academy of Management Journal, 17,* 646–674.

Wasti, S. A., Bergman, M. E., Glomb, T. M., & Drasgow, F. (2000). Test of the cross-cultural generalizability of a model of sexual harassment. *Journal of Applied Psychology, 85,* 766–778.

Weide, S., & Abbott, G. (1994). Management on the hot seat. *Employment Relations, 21,* 23–34.

Wilson, C. B. (1991). U.S. businesses suffer from workplace trauma. *Personnel Journal, 70,* 47–50.

Workplace Violence Research Institute. (www.workviolence.com)

Zellars, K., Tepper, B., & Duffy, M. K. (in press). Abusive supervision and subordinates' organizational citizenship behavior. *Journal of Applied Psychology.*

CHAPTER 7

The Influence of Leadership and Climate on Occupational Health and Safety

Dov Zohar

Occupational health and safety pose major managerial challenges in terms of societal and financial costs. Work accidents alone cost the U.S. economy about $108.4 billion a year (National Safety Council, 1999) in addition to human suffering and loss of life. Occupational diseases are estimated as more than doubling those costs (Leigh, Markowitz, Fahs, Shin, & Landrigan, 1997). Whereas occupational health issues have received much attention in the medical and social-psychological literature (the latter focusing mainly on occupational stress and burnout), safety issues have received only cursory attention by management scholars (Fahlbruch & Wilpert, 1999; Marcus, Nichols, & McAvoy, 1992; Shannon, Mayr, & Haines, 1997). Yet a fatal work injury occurs every two hours, and a disabling injury every eight seconds (National Safety Council, '999). Two exceptions are the growing body of research on safety imate/culture and the more recent research on leadership in-f. ences on safety outcomes. Both issues are receiving increasing at ntion since the inquiry into the Chernobyl disaster identified inadequate safety climate/culture stemming from faulty managerial practices as the major underlying factor (International Atomic Energy Agency [IAEA], 1986, 1991).

Although scholars have long recognized that "leaders create climates" through their customary modes of action (Lewin, Lippitt, & White, 1939, p. 272), few have explored this link. The purpose of the present chapter is to identify variables underlying this relationship and to offer an integrated model of leadership and climate influences on safety and health. A primary implication, discussed in the concluding section, concerns the distinction between absence of illness and presence of health, highlighted in Chapter One of this volume and in the organizational health approach (Rochlin, 1999). It draws on a related distinction between (1) mostly latent organizational pathogens such as deficient procedures and faulty equipment and (2) organizational salutogens (health-enhancing factors) associated with health-promoting practices and development-oriented management (Reason, 1997). Jointly, these distinctions imply that, in the context of occupational safety, absence of accidents and control of currently known safety and health hazards are insufficient indicators of operational safety unless accompanied by health-promoting resources such as positive organizational climate and high quality leadership. Given the prevailing ambiguity of safety climate research and the indiscriminate use of *safety climate* and *safety culture* as interchangeable terms (Cox & Flin, 1998), I consider several relevant issues concerning organizational climate in general and safety climate in particular before attempting integration with leadership. In this chapter I refer specifically to organizational climate, reserving discussion of differences between climate and culture for another occasion.

Motivational Welfare as a Referent of Organizational Climate

Organizational climate has been associated with employees' welfare since its inception in Lewin's pioneering work (Lewin, Lippitt, & White, 1939). Lewin and his colleagues coined the term *climate* to describe (leader-induced) social contexts that vary along the autocratic-democratic–laissez-faire continuum. Their description of group members' behavior, attitudes, and emotions leaves little doubt that climate influences psychological welfare. The subsequent formulation of global organizational climate as a measure of employees' perceptions of their work environment has turned psychological welfare into a primary albeit tacit referent of climate perceptions (for a brief outline of climate research, see Schneider, Bowen, Ehrhart, & Holcombe, 2000). Available taxonomies of global climate dimensions (Ostroff, 1993; Schmidt & DeShon, 2001), intended to clarify the core meaning of this construct, identified three primary attributes by which the work environment is assessed, namely, social attributes (participation, cooperation, social rewards), personal growth (autonomy, skill improvement, challenge), and instrumental attributes (formalization, hierarchy, extrinsic rewards). These attributes can be mapped onto available need taxonomies (e.g., Alderfer, 1972), indicating that they have been selected because of their relevance to employees' motivational welfare. Apparently for this reason, five out of the six main outcome categories in global climate research include direct or indirect indicators of motivational welfare: job satisfaction, organizational commitment, absenteeism and turnover, psychological well-being, and job performance (Ostroff, 1993). A meta-analysis of climate-outcome relationships indicated significant correlations between all three climate dimensions and job satisfaction and commitment, with the latter predicting psychological welfare, absenteeism, and job performance criteria (Schmidt & DeShon, 2001).

Current approaches in climate research are associated with facet-specific rather than global climates (i.e., "climate for something" such as service quality or employee safety; see Schneider et al., 2000). Specific climates provide summary measures of employees' perceptions of focal facets of the work environment, primarily the relevant (formal and informal) policies, procedures, and practices (Reichers & Schneider, 1990). Thus, the policies and procedures pertaining to service or safety as executed by managers across the organizational hierarchy provide the referent object of climate perceptions for respective climates. However, employees are assumed to assess policies and procedures primarily in psychologically meaningful terms, that is, as indicators of the kinds of role behavior likely to be rewarded and encouraged rather than in terms of some objective, motive-irrelevant criterion. Motivational welfare thus constitutes, as before, a cognitive referent or assessment criterion in terms of which work environments are being assessed.

These ideas suggest a hierarchical appraisal model in which climate perceptions are associated with external and internal cognitive referents. A body of facet-specific policies and procedures must first be assessed in terms of emphasis on particular strategic goals (e.g., How important is service quality or operational safety in this company?). Secondly, the expected consequences of acting in accordance with these goals must also be assessed, in order to determine their motivational relevance for employees (e.g., Is service- or safety-oriented role behavior likely to be recognized and rewarded?). The climate for service or safety is the integrated outcome of both appraisals; that is, it refers to shared assessments of service or safety policies and procedures where assessments convey the importance of these facets for the organization at one level of analysis and the importance of congruent role behavior at the job-level of analysis. This process derives from a collective effort of employees' making sense of the organization, resulting in a consensual, socially validated assessment (Ashforth, 1985; Schneider & Reichers, 1983). It is due to this hierarchical appraisal that climate influences role definitions (Hofmann, Morgeson, & Gerras, in press) and role behavior (Neal, Griffin, & Hart, 2000), thus providing the necessary condition for the empirically supported climate-outcomes relationship. Operationally, climate perceptions inform organization- or group-level behavior-outcome expectancies (Zohar, 1980, 2000) and provide the requisite expectancy-based motivation for congruent role behavior (Bandura, 1986; Lawler, 1971; Vroom, 1964). This leads in turn to organizational outcomes such as increased customer satisfaction (Schneider, White, & Paul, 1998) or decreased injury rates (Hofmann & Morgeson, 1999; Zohar, 2000).

Climate thus represents employees' socially shared interpretations of the organizational environment and informs the kinds of role behavior likely to be recognized and rewarded. In other words, facet-specific climates reveal which role facets will best improve motivational welfare by means of recognition and other (formal and informal) rewards. By default, if the leadership of the organization or of its subunits is deficient in terms of recognition or contingent rewarding, amounting to corrective management by exception (Bass, 1990), poor climate will result, regardless of formal policies. This climate infers poor motivational welfare akin to the autocratic climate described 70 years ago by Lewin and colleagues (1939). Thus, the theme common to earlier and more recent conceptualizations of climate, as Schneider and colleagues (1998) noted, concerns its core meaning as a shared perception of focal aspects of the work environment, assessed in terms of its implications for motivational welfare.

Safety climate is unique in that its external referents include policies and procedures concerning physical well-being. In a positive climate, expected consequences include higher frequency of safety behavior leading to improved physical well-being, which is accompanied by improved motivational well-being associated with supervisory recognition and reward. Thus, in organizations where safety is a relevant aspect of work, employees focus on safety policies and procedures so that if there is sufficient consensus concerning its importance relative to other aspects of work, then the aggregate score represents the level of safety climate (i.e., lack of consensus implies the absence of climate; see James, 1982). For the remainder of this chapter, I refer to safety climate as an example of facet-specific climates.

Cognitive Challenges in Climate Formation

Assessment of organizational and motivational meanings of policies, procedures, and practices can be quite complex, requiring, among other things, establishing the differences between formally declared policies and procedures and their enforced or enacted counterparts (i.e., managerial practices). Formal policy is explicit, relating to overt statements and official procedures, whereas enacted policy, which relates to managerial practice, is tacit, derived from observing senior, middle, and lower management patterns of action concerning key policy issues. (This distinction is akin to that of Argyris and Schön, 1996, between formally espoused theories of action or policies and theories or policies in use.) From a functional perspective, climate perceptions refer only to policies in use or enacted policies, rather than to their formal counterparts, because they inform employees of the importance and probable consequences of safety behavior. Thus, a consensus should occur when management displays an internally consistent pattern of action concerning safety, even if this differs from formally declared policy. For example, despite official claims to the contrary, site managers might expect workers to bend company safety rules, except in life-threatening situations, whenever production falls behind schedule. If this is done consistently, workers will conclude that formal declarations have little substance in terms of motivational outcomes. This promotes a low safety climate, as described by Pate-Cornell (1990) and Wright (1986) with regard to managerial practices on offshore oil platforms, turning into a contributing factor to the high incidence of risk taking in this industry (Flin, Mearns, Gordon, & Fleming, 1996), identified as a major factor underlying the Alpha Piper disaster, in which 167 workers lost their lives (Cullen, 1990; Pate-Cornell, 1993).

Given that safety issues are inherent to any manufacturing process, competing with other primary issues such as speed and profitability, it follows that (enforced) safety policies and procedures can be construed in terms of relative priorities of safety versus production goals. Because relative priorities are a parsimonious way of interpreting motivational implications of enforced policies for company employees, safety climate perceptions are likely to refer to those policy attributes that indicate the priority of safety. The focal issue for safety climate perceptions is, therefore, the (true) priority of safety, so that climate level reflects this consensual priority rather than numerous procedures considered individually. In other words, climate perceptions involve pattern-recognition processes relating to "procedures-as-pattern" rather than to individual procedures. In this respect safety climate is a social cognitive construct (Rochlin, 1999), part of an active process of organizational sense-making (Drazin, Glynn, & Kazanjian, 1999; Weick, 1995) as opposed to passive observation of isolated safety procedures. Hence, when safety issues are ignored or become contingent on production pressures, workers will infer low safety priority. All that is required for such a policy to become a source of (low) climate perceptions is that it remains unequivocal and stable.

Level-Specific Attributes of Safety Climate

Organizational climate can be investigated at different hierarchical levels, most notably the organizational and the subunit or group level (Glick, 1988; Patterson, Payne, & West, 1996; Rousseau, 1988; Zohar, 2000). This is because policies and procedures must be implemented by upper management as well as by unit managers lower in the organizational hierarchy. Top managers are concerned with policy making and the establishment of company-wide procedures to facilitate policy implementation, whereas unit managers at lower hierarchical levels execute these policies and associated procedures in the subunits for which they are responsible. Given its parsimony, individual employees intuitively use relative priority as their metric to compare and decide whether unit managers assign low or high priority to safety issues relative to that of top management. Employees can discriminate between procedures instituted by top management and those executed by unit managers by using two main sources of information. One is the degree of difference between subunits, detected through social comparisons among employees in different subunits (Schneider & Reichers, 1983). For example, members of one subunit may conclude that their immediate superior is more lenient regarding use of protective gear than are other superiors, assigning it lower priority. Thus, whereas typical or modal supervisory pressures and expectations identify company-level policy regarding use of protective gear, the discrepancy between what most supervisors do and what a specific supervisor does identifies subunit practices. Another source of information concerns the degree to which company management backs supervisory behavior during interaction with group members. For example, if a supervisor initiates disciplinary action in response to a safety violation, this indicates supervisory emphasis on safety, that is, high priority. The degree to which higher-level managers are willing to back this action is indicative of organizationwide priorities. Similarly, the extent to which safety considerations influence promotion prospects in a company is indicative of company-level priorities. Together, this information helps employees to discriminate between company-level and group-level emphasis on safety.

Note that level of analysis defines both the unit of aggregation and the referent object of climate perceptions (Chan, 1998). At the organization level, climate perceptions are aggregated across the company, with company-level emphasis on safety as the referent. At the group level, perceptions are aggregated within subunits, with supervisory emphasis (and attendant motivational consequences) as the referent. By adjusting the referent of perceptions and assuming that individuals do discriminate between procedural and supervisory emphasis on safety, it becomes possible to measure safety climate at different organizational levels (Zohar, 2000).

Antecedents of Safety Climate

The remainder of this chapter presupposes that appreciation of the antecedents of climate in organizational subunits requires a multilevel perspective. As I noted above, a multilevel interpretation suggests that organization-level procedures differ from group-level practices in that unit managers must turn company-level procedures into situation- and task-specific action directives appropriate to their immediate environment. This process requires discretionary decision making because procedures cannot cover all possible tasks, work environments, and situational contingencies in every subunit (except in highly standardized technologies and processes, such as classic assembly-line manufacturing). Hence, unit managers must continually decide how to implement procedures, while taking into account situation-specific factors. This cross-level process provides the "space" where climate antecedents largely reside.

Supervisory roles entail considerable discretion because the implementation of procedures depends inextricably on technological and human considerations; that is, supervisors manage production through other people rather than dealing directly with production technology. For example, a supervisor must decide whether to put more or less emphasis on certain safety procedures in a situation where increased workload may fatigue or irritate some of the workers (i.e., increased supervisory pressure could induce even greater fatigue or irritation). The discretionary power inherent in supervisory roles creates a potential for group-level variation, which should result in corresponding climate variation based on the proposition that leaders create climates. A recent study of group-level safety climate supported this, using 53 work groups in a single manufacturing company (Zohar, 2000). In this study safety climate perceptions revealed sufficiently high within-group homogeneity and between-group variance to warrant group-level aggregation. Climate levels were also shown to predict minor injury records in work groups for a five-month period following climate measurement, after controlling for departmental risk levels. Similar results were obtained in military field units, using 61 platoons in several brigades (Zohar & Luria, 2001).

Given discretionary power, supervisory execution of safety procedures is likely to be influenced by three factors acting as antecedents of (group-level) climate: production technology, personal beliefs, and leadership quality. Production technology often differs across subunits, creating different risk levels. Most manufacturing companies have some subunits where risk is substantially greater than in others (e.g., operating handheld power tools is more dangerous than operating stationary machinery; vaporous raw materials pose greater health hazards than do nonvaporous or solid materials). Supervisors in high risk units are likely to implement safety procedures more rigorously and attach greater motivational consequences to safety behavior than are supervisors in low risk units, resulting in higher safety climates. By default, if the safety procedures are not executed more rigorously in riskier departments, they will have lower group climate scores than the safer departments.

Safety beliefs and attributions of subunit supervisors have been shown to exhibit significant individual variation (Cox & Cox, 1991; Williamson, Feyer, Cairns, & Biancotti, 1997) regardless of risk levels, thus affecting the execution of company-level procedures. Individuals attribute accidents either to external or internal factors, adopt fatalistic or optimistic beliefs, and differ as to whether employees' safety is considered the responsibility of management or of the individual worker. Supervisors are likely, therefore, to execute instituted procedures according to their own bias. For example, some supervisors expect subordinates to assume greater responsibility for their own safety, which should result in greater emphasis on safety behavior and consequently higher climate. Individual differences also characterize managers in higher hierarchical positions, thus influencing larger units (e.g., whole departments or divisions). However, safety attributions are also influenced by climate and communication factors, which limit individual-level variations (Hofmann & Stetzer, 1998). This suggests a cross-level relationship whereby safety attributions of key individuals influence climate in respective subunits, influencing in turn the attributions of members in those subunits. As concerns the third factor, leadership quality, here too it is presumed that between-group variation stems from idiosyncratic modes of policy execution in subunits. This factor, the antecedent of main interest for this chapter, is the subject of the next section.

Leadership Quality as Antecedent of Safety Climate

Where job performance has direct safety implications, the quality of leader-member interaction is likely to influence the leader's concern for members' physical welfare. Higher quality interactions will result in greater emphasis on safety, which will in turn influence safety-climate perceptions in the group. This premise is based on evidence that closer and higher quality relationships increase leaders' concern for members' emotional welfare (Bass, 1990; Fairhurst, 1993, 2000). High quality relationships are characterized by a balanced social exchange (Blau, 1964) in which each party facilitates achievement of the other party's personal goals. Because goal achievement is instrumental in motivational welfare (Austin & Vancouver, 1996), superiors and subordinates alike will feel responsible for each other's welfare. In situations involving heightened risk of injury, this also applies to physical well-being (Hofmann & Morgeson, 1999). Reciprocity is based on such values as trust, openness, and loyalty that result in the value-based interactions characteristic of high LMX and transformational leadership (Yukl, 1998). These values encourage leaders to resist short-term production pressures, which workers often meet at the cost of safety (Pate-Cornell, 1990), and encourage open safety communication (Hofmann & Morgeson, 1999), including discussion of nonroutine problems conducive to accidents (Fairhurst, 1993). Practices generated by closer leader-member interaction should therefore promote shared perceptions among the group concerning safety priority, that is, higher safety climate (Zohar, 2002a).

The full-range leadership model (Bass, 1990; Bass & Avolio, 1997) offers a comprehensive framework for discussing the effects of leadership quality. It covers several dimensions of leaders' behavior that can be arguably arranged along a continuum of quality of leader-member interaction and, consequently, concern for (emotional and physical) welfare (see Figure 7.1). As noted, transformational leadership is characterized by value-based and individualized interaction, resulting in better exchange quality and

Figure 7.1. A Conceptual Description of Relationships Between Leadership Dimensions and Safety Climate.

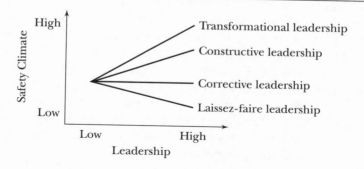

greater concern for welfare. Transactional leadership, the other global dimension, can be subdivided into constructive, corrective, and laissez-faire dimensions (Bass & Avolio, 1997). All of these are based to some extent on nonindividualized exchanges in hierarchical rather than egalitarian relationships. Reciprocity in this case does not relate to goal achievement of either party but to position-based expectations, in which organizational positions override those of the individuals occupying them (Katz & Kahn, 1978).

Constructive leadership (i.e., the contingent-reward dimension in the full-range model) implies an intermediate level of concern for members' welfare because, although it is based on hierarchical relationships (i.e., reward-for-effort exchanges), leaders have to identify needs, desires, and individual capabilities in order to offer motivationally relevant rewards (Bass, 1990; Yukl, 1998). This decreases the power distance between hierarchical positions, resulting in closer relationships and more individualized consideration. It is probably due to such individualized consideration that the transformational and constructive dimensions often merge into a single factor (Avolio, Bass, & Jung, 1999; Carless, 1998).

Corrective leadership (i.e., management by exception) mainly includes error detection and correction based on active or passive monitoring of subordinates' performance in relation to required standards. This results in poorer nonindividualized interaction, because superiors see subordinates as interchangeable, with little concern for their welfare beyond that required "by the book." Concern for physical well-being thus depends on the extent to which the required standards include safety as a criterion. Whenever production speed and costs are the primary criteria, as is often the case, corrective leaders maintain a threshold level of concern for members' welfare, that is, the minimal emphasis required for preventing major injury. This is represented in Figure 7.1 as a zero-slope line. Given the distinction between active and passive corrective leadership (Bass, 1990), concern for members' welfare is expected to decrease accordingly; that is, passive supervisors exhibit less concern.

Finally, laissez-faire leadership offers the lowest level of concern for welfare: nonleadership or disregard for supervisory responsibility despite rank in the hierarchy. Consequently, it should be negatively correlated with supervisory concerns (see Figure 7.1). Given that leadership dimensions assume such an (ordinal) order as regards concern for members' welfare, they should result in similarly ordered safety climate levels. This rationale was supported in several recent studies, using different measures of leader-member relationships, i.e., LMX scores (Hofmann & Morgeson, 1999), transformational scores (Barling, Loughlin, & Kelloway, 2002), and scores across the entire leadership range (Zohar, 2002a). In these studies higher LMX and transformational scores were associated with higher safety climate scores, and transactional and laissez-faire leadership scores were negatively related with climate.

Moderating Effects of Assigned Safety Priority

Although supervisory roles entail considerable personal discretion, the expectations communicated by an immediate superior will influence supervisory practice, resulting in an equilibrium reflecting the settled preferences of both parties (Bacharach, Bamberger, & Sonnenstuhl, 1996). This suggests that supervisory emphasis on safety issues (and consequently safety-climate perceptions) is an interactive function of concern for members' welfare derived from the quality of interaction with group members and externally assigned safety priorities. Altogether, this should result in a moderated relationship between leadership quality and safety climate. However, in the full-range leadership model, the moderation effect of assigned safety priority should differ for each leadership dimension due to the qualitative shift from reciprocal, value-based interaction to hierarchical, utility-based interaction between leader and group members. A recent study by Zohar (2002a) tested these differences and documented the different interaction shapes between leadership dimensions and assigned safety priority. Figure 7.2 displays the shape of these interactions.

As Figure 7.2a indicates, transformational leaders consider safety as a moral obligation, especially under adverse conditions indicated by low assigned priority. Safety climate level is thus positively related to transformational level even under low priority conditions. Because assigned priority makes a difference for low transformational leaders, resulting in lower climate under low assigned priority, the result

Figure 7.2. Leadership Dimensions and Safety Priority as Predictors of Safety Climate.

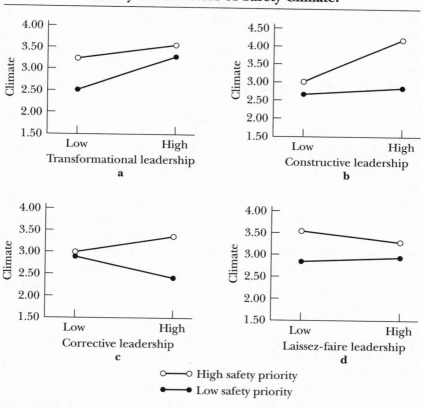

Note: Adapted from Zohar (2002a). © John Wiley & Sons Limited. Reproduced with permission.

is a stronger positive relationship (a steeper slope), indicative of the compensatory effect of increasingly transformational leadership under adverse conditions. Figure 7.2b indicates that constructive leaders monitor and reward aspects of subordinates' performance that are assigned higher priority by immediate superiors (e.g., speed, safety). Under high priority conditions, safety is a relevant performance goal, resulting in higher safety climate for increasingly constructive leadership. However, because of their commitment to safety as a consequence of closer interaction with subordinates, the superiors do not disregard safety completely, even when it has low priority. This results in a near-zero increase in safety climate level as a function of leadership score, that is, no relationship.

Conversely, because corrective leadership derives from nonindividualized relationships, it implies less concern for members' welfare. If the assigned safety priority is low, corrective leaders may disregard safety deviations (apart from imminent danger) in favor of higher priorities. This results in a negative leadership-climate relationship. Some studies suggest that under such conditions corrective supervisors will even disregard imminent danger in favor of singled-out production targets (Pate-Cornell, 1990; Wright, 1986). However, under high priority conditions, safety behavior is a referent standard for monitoring subordinates' performance, resulting in a positive relationship between leadership and safety levels (Figure 7.2c). Laissez-faire leadership involves little individualized consideration and only limited concern over any aspect of performance. Consequently, the effect of assigned priorities diminishes as laissez-faire scores increase. High assigned safety priority results in a negative relationship between this leadership dimension and safety climate. Low assigned priority results in no relationship (i.e., near-zero slope) because disregard for safety (due to low priority) remains unchanged as passivity increases (Figure 7.2d).

A Leadership—Climate Injury Mediation Model

Given that leadership quality influences safety climate in work groups and that safety climate influences safety behavior of group members, safety climate thus becomes a mediator. Furthermore, because human behavior plays a dominant role in occupational accidents (National Safety Council, 1999), there should be a mediated relationship between leadership, climate, and behavior or injury. The influence of climate on safety behavior and injury rate is supported by substantial evidence (e.g., Hofmann & Stetzer, 1996; Neal, Griffin, & Hart, 2000; O'Dea & Flin, 2000; Zohar, 2000). Such a relationship is inherent in the definition of climate presented earlier; that is, climate is the collective assessment of supervisory practices that indicate the true priorities in key task facets, thus informing behavior-outcome expectancies.

Zohar (2002a) tested the mediation model, using within-group split-sample analysis of 42 work groups in a manufacturing plant. This methodology was designed to eliminate single-source bias by randomly splitting each team in half, then obtaining climate variables from one half and leadership variables from the other. Results indicated that transformational and constructive leadership predicted medically recorded injury rates, with the effect being fully mediated by commensurate climate variables. Corrective leadership provided indirect, conditional prediction. Similar results were reported in a study of restaurant employees, using a modified scale of transformational and constructive leadership referring specifically to safety leadership (Barling et al., 2002). Using structural equations modeling, Barling et al. found that leadership scores predicted climate levels directly and indirectly, due to their effect on group members' level of safety consciousness. Climate level in turn predicted self-reported injury data. Yet another study tested safety climate as a cross-level moderator of the relationship between LMX and safety citizenship behavior (Hofmann et al., in press). In other words, instead of studying leadership and climate at the same level of analysis, these scholars considered climate as a higher order context variable. Thus, LMX and safety behavior were measured at the individual level, whereas climate was measured at the group level of analysis, that is, as a social-environmental variable. Results indicated that under high safety climate the quality of dyadic leader-member exchanges is positively related to safety citizenship behavior of respective group members, whereas this relationship disappears under low climate. Overall, therefore, these studies indicate that group climate influences the leadership-safety relationship and is thus a proximal social-cognitive antecedent of facet-specific role behavior.

Direct Effects of Leadership on Safety

In addition to their mediated effects, leadership variables are also likely to have direct (i.e., unmediated) effects on safety records, merely as a result of better management. As before, these effects can be best appreciated from the perspective of the full-range leadership model and the attendant distinction between leadership and supervision (House, 1996; Yukl, 1998). Transactional supervision has to do with organizing tasks and getting people to perform them more reliably and efficiently. Transformational leadership has to do with development and with encouraging people to commit themselves to more challenging goals. As noted by several authors (Bass & Avolio, 1997; Waldman, Bass, & Yammarino, 1990), the relationship between the two roles is that of augmentation in that transactional supervision provides reliability and predictability (expected performance), whereas transformational leadership provides heightened motivation and development orientation (performance beyond expectations). Effective managers employ elements of both strategies.

Transactional supervision can influence safety directly because effective monitoring and contingent rewarding practices will maintain reliable performance based on Skinnerian (i.e., operant) principles regardless of climate perceptions (Komaki, 1998). Supervisory reliability in and of itself plays an important role in occupational safety (Reason, 1990). This is supported by the fact that failure to use protective gear provided at the workplace (indicative of poor supervision) accounts for about 40% of work accidents (National Safety Council, 1999). Disorganization often results in unauthorized or uncoordinated acts leading to accidents (Lowe, Kroeck, & Sivasubramaniam, 1996; Feyer, Williamson, & Cairns, 1997). Removing safety violators from risky jobs or inducing them to leave the group (Schneider, 1987) will also have unmediated effects on safety records. However, effectiveness of supervision-based safety will depend on priorities assigned by upper management, as demonstrated in Figure 7.2.

Transformational leadership also influences safety behavior directly due to several of its inherent characteristics. Transformational leaders encourage open, egalitarian communication within and across group boundaries (Yukl, 1998). Openness of communication, especially concerning nonroutine problems, is relevant to safety because such problems are conducive to accidents (Rasmussen, 1990; Wright, 1986). A study conducted by Parker, Axtell, and Turner (in press) at two manufacturing sites of a large company indicated positive relationships between supportive leadership (a proxy for transformational leadership) and safety communication, with a lagged effect on safety-compliance behavior 18 months later. A second transformational attribute, development orientation, is also relevant because emphasizing continuous development of work-group members' knowledge, skills, and abilities will improve transfer of learning from formal training or compensate for inadequate training. Training reduces accident rates, especially when coupled with congruent contingencies (Krispin & Hantula, 1996). Furthermore, in little-routinized work, situations often arise in which standard procedures are not applicable, requiring employees to adjust their actions and solve problems based on a broad base of technical and decision-making skills (Bacharach & Conley, 1989). Investment in members' development would be especially effective in such situations.

Inspirational motivation and intellectual stimulation, two other attributes of transformational leaders, may also exert influence when leaders challenge group members to go beyond individual needs for the collective good (Barling et al., 2002). The challenge should encourage a shift from a passive behavioral mode identified as safety compliance (i.e., mere adherence to relevant procedures) to a proactive mode reminiscent of organizational citizenship behavior, which includes helping coworkers, demonstrating initiative, and making efforts to improve safety in the workplace, identified as safety participation (Neal et al., 2000). A study involving military transportation teams provided empirical support for this contention by showing that high LMX scores were associated with safety participation, identified in that study as safety citizenship behavior (Hofmann et al., in press). Williams, Turner, and Parker (2000) reported similar results, based on surveys completed by team members in a chemical processing plant. Transformational leadership was positively associated with both safety compliance and participation, and it exerted a stronger effect on individuals who had been less committed to safety. Finally, the idealized influence attribute of transformational leadership is likely to induce inclusion of safety as a core value in situations involving heightened risk of injury (Barling et al., 2002). Leaders providing idealized influence convey safety as social responsibility and safety behavior as the moral or right thing to do. Thus, safety considerations receive higher priority than do speed or production, serving as a counterbalance to short-term production pressures. The extent of these theoretical links and the dearth of empirical studies (whose results are nevertheless encouraging) offer an extensive agenda for research.

A Contingency Model for Leadership and Safety

A key proposition in leadership research is that leader effectiveness depends on a fit between leadership style and situational characteristics (Fiedler & Chemers, 1982; House, 1996). Occupational safety as the relevant criterion for effectiveness should prove no exception. If transformational leadership and transactional supervision exert complementary influences on safety behavior, one of them may be more influential than the other under some conditions (Komaki, 1998). Rule formalization and activity routinization as defined in the routinization-formalization model (Hage & Aiken, 1969; Perrow, 1967, 1979) are especially relevant contingency factors in the context of occupational safety. Routinization is determined by the degree of variation in problem types encountered during work and the level of difficulty in problem solving. As Perrow (1967) noted, both parameters can be collapsed into a continuum of routinization. For example, production workers engaged in batch processing encounter more exceptions and a wider variety of problems than do assembly-line workers. Similarly, maintenance workers or those involved in mobile work (i.e., away from stationary workstations) encounter more exceptions than do production workers in stationary work. Formalization is defined and measured in terms of job codification, relating to the degree to which job descriptions are specified (Hage & Aiken, 1967). High formalization implies an increase in the number and specificity of rules and procedures. Not only are there more procedures to cover possible contingencies, but the procedures themselves are more specific and rigid. This, of course, reduces decision latitude or autonomy at the lower levels of the organizational hierarchy, because the range of variation that is tolerated within the rules is reduced (Hage, 1974). According to the routinization-formalization model, the two elements are closely linked, in that organizations with routine work have greater formalization of organizational roles (Hage & Aiken, 1969).

It should be noted that some jobs may be more codified than others, so that formalization in some subunits may be higher than in others within the same organization, depending on characteristics of the main activity that each performs. Bacharach and Aiken (1976) provided empirical evidence for this in their study of local government organizations. In an industrial context, it is easy to demonstrate a wide range of subunit routinization levels. For example, activity performed with chemically unstable raw materials is less routinized than that performed with solid materials, and older machinery in some subunits often requires significantly more problem solving, resulting in less routinization than in units with newer, automated machinery.

The classification of transactional versus transformational leadership can now be considered from the perspective of routinization. As Komaki (1998) noted, although transformational leaders may excel at innovating and inducing extraordinary effort, effective transactional leaders are needed to maintain regular and reliable activity. House, Spangler, and Woycke (1991, p. 391) suggested accordingly that in situations "requiring routine but reliable performance in the pursuit of pragmatic goals, charismatic leadership . . . may even be dysfunctional." This implies that in subunits where activity is more routinized, so that work procedures (including safety procedures) are more formalized, "routine but reliable" performance is required to improve safety outcomes. Under such conditions, therefore, transactional supervision will prove superior to transformational leadership (see Table 7.1). In that case the best supervisory practices include performance-based monitoring and timely communication of contingent consequences (Komaki, 1998). Furthermore, safety compliance should be considered more important than safety participation because the available procedures guide job performance under all or nearly all circumstances.

Table 7.1. A Contingency Approach for Safety Leadership.

	Routinization Level	
	High	Low
Transactional leadership	Compliance fit	Knowledge-based misfit
Transformational leadership	Rule-based misfit	Participation fit

The situation is quite different in subunits where routinization is low. In this case workers encounter problems for which no procedures are available, requiring problem solving and decision making of varying complexity. Under such conditions transformational leadership that induces heightened safety participation (in addition to compliance with relevant rules) should provide better safety outcomes. In other words, because procedures do not cover all situations and contingencies, behavior must be based on broad technical knowledge and problem-solving skills rather than on adherence to specific rules. Such situations call for the development of "line professionals" (Bacharach & Conley, 1989, p. 312) who learn to solve problems through constructive dialogue, open communication, and self-regulation backed by continuous learning (Weick & McDaniel, 1989). Here, transformational leadership can provide its greatest benefit based on attributes such as intellectual stimulation, idealized influence, and development orientation, as described earlier.

As noted in Table 7.1, this classification also implies two types of misfit. Transformational leaders who manage workers performing highly routinized work are likely to have lower safety results than could have been expected based on their concern for well-being and open communication. This stems from undue emphasis on the challenge of performance beyond expectations through intellectual stimulation (e.g., suggesting new ways of doing things, encouraging nontraditional thinking to deal with traditional problems), whereas routine but reliable performance is all that is required. Based on the categorization of human performance as skill-based, rule-based, and knowledge-based (Rasmussen, 1982, 1990; Reason, 1990, 1997), this misfit is categorized as a rule-based misfit. In this case the likelihood of errors of commission, where workers take unnecessary action, is increased because group members are encouraged to improve performance beyond expectations (e.g., improve productivity, reduce costs) by initiating new ways of doing things instead of emphasizing predictable and reliable performance (i.e., safety compliance behavior). In effect, a situation where transformational leaders supervise highly routinized work creates a conflict between their tendency to excel by challenging subordinates to consider new ways of doing things and that of caring for their well-being through routine but reliable performance. Assuming that such leaders strike a balance between the two, this results in lower-than-expected safety records.

The other misfit, where a transactional leader is in charge of a work group performing little-routinized work is identified as a knowledge-based misfit (see Table 7.1). In this case supervisory practices are likely to focus on safety compliance rather than on participation, despite the fact that existing rules and procedures do not match the complexities involved. In this case workers are not given sufficient decision latitude or allowed to solve problems on their own. Rather, they are expected to refer all decisions to their supervisor even when problems need to be solved on the spot (e.g., stopping a leak of vaporous raw material). They are expected to act according to standard procedures that may not be relevant or appropriate, resulting in "incorrect compliance" (Reason, Parker, & Lawton, 1998, p. 296). This increases the risk of omission errors because appropriate action is not taken due to overly restrictive practices. Overall, the contingency model suggests that, contrary to available studies in which transformational leadership proved superior to transactional supervision, apparently due to the fact that workers in these studies performed little-routinized work (e.g., Barling et al., 2002; Zohar, 2000), the reverse might apply when routinization is high. More research is needed here, including searching for other contingency factors.

Improving Organizational Health

One of the most important implications of the integrated leadership-climate model presented here relates to the proactive organizational health perspective highlighted in Chapter One (see also Rochlin,

1999). This perspective is based on the distinction between absence of illness (e.g., lack of accidents, control of known risks) and presence of health (e.g., safety citizenship behavior, well-being as a core value). Health, distinguished from lack of illness, is associated with development and maintenance of (physical, mental, and social) salutogenic resources. The present chapter deals with two primary resources, high quality leadership and positive organizational climate, with safety climate as the exemplar. Hofmann and Tetrick's wellness model provides two goal parameters with which these resources can be characterized: temporal orientation (short-term versus long-term goals) and spatial orientation (extrinsic versus intrinsic goals). The reactive, pathogen-centered perspective that currently dominates the occupational safety and health literature focuses on short-term external goals, epitomized by the idea that "Hazard identification, risk assessment, and risk control are at the heart of a successful occupational safety and health (OS&H) management system and should be reflected in the organization's OS&H policy" (British Standards Institute [BSI], 2000, p. 7). My contention is that effective, contingency-adaptive leadership that nourishes the development of positive safety climates in organizational subunits is an important resource for organizational health, augmenting the imperative efforts toward accident reduction by means of hazard identification, risk assessment, and control. This complementary approach shifts attention to a proactive, salutogen-centered perspective that adopts long-term, intrinsic goals. Thus, in addition to focusing on technical procedures associated with risk assessment and control, this perspective draws attention to health implications of the development-orientation and moral-obligation attributes of high quality leadership. It also draws attention to the health implications of a positive safety climate, especially in departments and companies characterized by little routinization, where internal behavioral standards rather than externally imposed rules provide answers for maintaining error- and accident-free performance.

This suggests a dual agenda for future research. First, mediated relationships involving leadership and safety climate, as well as unmediated effects on safety, should be explored. The relatively small body of research on this subject suggests that relevant relationships require explication. For example, Zohar's study (2002a) demonstrated strong moderating effects of assigned safety priority, to the point where leadership-climate relationships assume opposite signs under low assigned priority depending on leadership dimension (see Figure 7.2). Other moderators, such as the level of supervisory discretion or supervisory beliefs and attributions, may have equally strong effects. Thus, leadership by supervisors given greater discretion due to seniority or physical separation is likely to have a stronger effect on climate, and this may also be true of supervisors holding safety-congruent beliefs (e.g., "safety is part of my job as supervisor"). The same study suggested that the mediation effect of safety climate is only provided by climate variables commensurate with the relevant leadership dimension. For example, the effect of transformational leadership was mediated by the only climate variable concerned with dynamic supervisory action involving open communication. Thus, different leadership dimensions seem to be differentially related to congruent climate dimensions. This issue requires further exploration. Other issues include the proposition that transformational leadership influences safety outcomes mainly through the safety-participation behavior of group members, whereas transactional leadership exerts influence mainly through safety-compliance behavior. Finally, the contingency approach discussed earlier should first be tested with routinization as the relevant situational factor and subsequently expanded to include other factors. This should provide a rich and diverse research agenda of theoretical and practical significance.

A second line of research concerns development and evaluation of intervention methods for developing and sustaining health resources such as leadership and climate. Improvement of both leadership and climate involves modification of supervisory practices. However, the former relates to a more general change, relating to characteristic patterns of interaction with subordinates regardless of the issue at hand, whereas the latter involves a more specific change, relating to the priority of safety implicated during such interaction. Zohar recently reported one attempt in this direction (2002b), aimed at modifying supervisory monitoring and rewarding of subordinates' safety behavior. Line supervisors in a metal processing plant received weekly feedback based on repeated episodic interviews with subordinates concerning the cumulative frequency of their safety-oriented interactions. This information identified the true priority of safety over competing goals such as speed or schedules. Section managers received the same information and used it to communicate high safety priority. Safety-oriented interaction increased significantly in the experimental groups but remained unchanged in the control groups. This change was accompanied by a significant increase in safety climate in the experimental groups; safety climate remained unchanged in the control groups. Furthermore, minor-injury rate and earplug use exhibited significant improvement in the experimental groups, which remained stable during the five-month postintervention period. Because much literature concerning leadership development exists (see review in Gist & McDonald-Mann, 2000), it would be interesting to examine the effect of other available leadership development programs on safety behavior, including transformational leadership improvement as leverage for improving safety behavior in units performing little-routinized work.

Given the sustained emphasis on administrative, engineering, and process-control solutions for occupational safety and health (BSI, 2000), characterized by short-term, external goals associated with the control of organizational pathogens, this chapter suggests that this perspective should be augmented by longer-term, internal goals associated with development and maintenance of health resources. The main lesson here is that if leadership style is inappropriate, resulting in poor leadership, and if safety climate is also poor, resulting in situations where workers commonly perceive speed and schedules to override safety considerations, then engineering solutions will be of limited (although vital) utility. As noted by Shannon et al. (1997), this is apparently why, despite the last two decades' sustained efforts regarding occupational safety and health, safety has remained a major problem, costing the U.S. economy in excess of $100 billion annually, in addition to immeasurable human suffering. The fact that a fatal work injury occurs in the United States every two hours and a disabling injury every eight second seconds indicates that occupational safety and health are still major managerial challenges. I hope this chapter has also illuminated the scientific challenge.

References

Alderfer, C. P. (1972). *Existence, relatedness, and growth: Human needs in organizational settings.* New York: Free Press.

Argyris, C., & Schön, D. A. (1996). *Organizational learning II: Theory, method, and practice*. Reading, MA: Addison-Wesley.

Ashforth, B. E. (1985). Climate formation: Issues and extensions. *Academy of Management Review, 10,* 837–847.

Austin, J. T., & Vancouver, J. B. (1996). Goal constructs in psychology: Structure, process, and content. *Psychological Bulletin, 120,* 338–375.

Avolio, B. J., Bass, B. M., & Jung, D. I. (1999). Re-examining the components of transformational and transactional leadership using the MLQ. *Journal of Occupational and Organizational Psychology, 72,* 441–462.

Bacharach, S. B., & Aiken, M. (1976). Structural and process constraints on influence in organizations: A level-specific analysis. *Administrative Science Quarterly, 21,* 623–642.

Bacharach, S. B., Bamberger, P., & Sonnenstuhl, W. J. (1996). The organizational transformation process: The micropolitics of dissonance reduction and the alignment of logics of action. *Administrative Science Quarterly, 41,* 477–506.

Bacharach, S. B., & Conley, S. C. (1989). Uncertainty and decision making in teaching: Implications for managing line professionals. In T. J. Sergiovanni & J. H. Moore (Eds.), *Schooling for tomorrow* (pp. 311–329). Needham Heights, MA: Allyn & Bacon.

Bandura, A. (1986). *Social foundations of thought and action*. Englewood Cliffs, NJ: Prentice Hall.

Barling, J., Loughlin, C., & Kelloway, K. (2002). Development and test of a model linking transformational leadership and occupational safety. *Journal of Applied Psychology, 87,* 488–496.

Bass, B. M. (1990). *Bass and Stogdill's handbook of leadership*. New York: Free Press.

Bass, B. M., & Avolio, B. J. (1997). *Full range leadership development: Manual for the MLQ*. Palo Alto, CA: Mind Garden.

Blau, P. (1964). *Exchange and power in social life*. New York: Wiley.

British Standards Institute (BSI). (2000). *Occupational health and safety management systems: Guidelines for the implementation of OHSAS 18001* (BSI-02-2000). London: Author.

Carless, S. A. (1998). Assessing the discriminant validity of transformational leadership behavior as measured by the MLQ. *Journal of Occupational and Organizational Psychology, 71,* 353–358.

Chan, D. (1998). Functional relations among constructs in the same content domain at different levels of analysis: A typology of composition models. *Journal of Applied Psychology, 83,* 234–246.

Cox, S. J., & Cox, T. (1991). The structure of employee attitudes to safety: A European example. *Work and Stress, 5,* 93–106.

Cox, S. J., & Flin, R. (1998). Safety culture: Philosopher's stone or man of straw? *Work and Stress, 12,* 189–201.

Cullen, D. (1990). *The public inquiry into the Piper Alpha disaster* (Vols. 1–2, Cm 1310). London: Her Majesty's Stationery Office.

Drazin, R., Glynn, M. A., & Kazanjian, R. K. (1999). Multilevel theorizing about creativity in organizations: A sensemaking perspective. *Academy of Management Review, 24,* 286–307.

Fahlbruch, B., & Wilpert, B. (1999). System safety: An emerging field for I/O psychology. In C. L. Cooper & I. T. Robertson (Eds.), *International review of industrial and organizational psychology* (Vol. 14, pp. 55–93). New York: Wiley.

Fairhurst, G. T. (1993). The leader-member exchange patterns of women leaders in industry: A discourse analysis. *Communication Monographs, 60,* 321–351.

Fairhurst, G. T. (2000). The leader-follower communication. In F. Jablin & L. Putnam (Eds.), *Handbook of organizational communication* (2nd ed.). Thousand Oaks, CA: Sage.

Feyer, A., Williamson, A. M., & Cairns, D. R. (1997). The involvement of human behavior in occupational accidents: Errors in context. *Safety Science, 25,* 55–65.

Fiedler, F. E., & Chemers, M. M. (1982). *Improving leadership effectiveness: The leader match concept* (2nd ed.). New York: Wiley.

Flin, R., Mearns, K., Gordon, R., & Fleming, M. (1996). Risk perception by offshore workers on U.K. oil and gas platforms. *Safety Science, 22,* 131–145.

Gist, M. E., & McDonald-Mann, D. (2000). Advances in leadership training and development. In C. L. Cooper & E. A. Locke (Eds.), *Industrial and organizational psychology: Linking theory with practice* (pp. 52–71). Oxford: Blackwell.

Glick, W. H. (1988). Organizations are not central tendencies: Shadowboxing in the dark, round 2. *Academy of Management Review, 13,* 133–137.

Hage, J. (1974). *Communications and organizational control.* New York: Wiley.

Hage, J., & Aiken, M. (1967). Relationship of centralization to other structural properties. *Administrative Science Quarterly, 12,* 72–91.

Hage, J., & Aiken, M. (1969). Routine technology, social structure, and organizational goals. *Administrative Science Quarterly, 14,* 366–378.

Hofmann, D. A., & Morgeson, F. P. (1999). Safety-related behavior as a social exchange: The role of perceived organizational support and leader-member exchange. *Journal of Applied Psychology, 84,* 286–296.

Hofmann, D. A., Morgeson, F. P., & Gerras, S. J. (in press). Climate as a moderator of the relationship between LMX and content-specific citizenship behavior: Safety climate as an exemplar. *Journal of Applied Psychology.*

Hofmann, D. A., & Stetzer, A. (1996). A cross-level investigation of factors influencing unsafe behaviors and accidents. *Personnel Psychology, 49,* 307–339.

Hofmann, D. A., & Stetzer, A. (1998). The influence of safety climate and defensive communication on attributions about accidents. *Academy of Management Journal, 78,* 644–657.

House, R. J. (1996). Path-goal theory of leadership: Lessons, legacy, and a reformulated theory. *Leadership Quarterly, 7,* 323–352.

House, R. J., Spangler, W. D., & Woycke, J. (1991). Personality and charisma in the U.S. presidency: A psychological theory of leadership effectiveness. *Administrative Science Quarterly, 36,* 364–396.

International Atomic Energy Agency (IAEA). (1986). *Summary report on the post-accident review meeting on the Chernobyl accident* (IAEA Safety Series 75-INSAG-1). Vienna: Author.

International Atomic Energy Agency (IAEA). (1991). *Safety culture* (IAEA Safety Series 75-INSAG-4). Vienna: Author.

James, L. R. (1982). Aggregation bias in estimates of perceptual agreement. *Journal of Applied Psychology, 67,* 219–229.

Katz, D., & Kahn, R. L. (1978). *The social psychology of organizations* (2nd ed.). New York: Wiley.

Komaki, J. L. (1998). *Leadership from an operant perspective.* New York: Routledge.

Krispin, J., & Hantula, D. (1996). A meta-analysis of behavioral safety interventions in organizations. *Proceedings of the 1996 Annual Meeting of the Eastern Academy of Management, 15,* 435–455.

Lawler, E. E. (1971). *Pay and organizational effectiveness: A psychological view.* New York: McGraw-Hill.

Leigh, J. P., Markowitz, S., Fahs, M., Shin, C., & Landrigan, P. (1997). Occupational injury and illness in the U.S. *Archives of Internal Medicine, 157,* 1557–1568.

Lewin, K., Lippitt, R., & White, R. K. (1939). Patterns of aggressive behavior in experimentally created social climates. *Journal of Social Psychology, 10,* 271–299.

Lowe, B. K., Kroeck, K. G., & Sivasubramaniam, N. (1996). Effectiveness correlates of transformational and transactional leadership: A meta-analytic review of the MLQ literature. *Leadership Quarterly, 7,* 385–425.

Marcus, A., Nichols, M. L., & McAvoy, G. E. (1992). Economic and behavioral perspectives on safety. *Research in Organizational Behavior, 15,* 323–355.

National Safety Council. (1999). *Injury facts.* Itasca, IL: Author.

Neal, A., Griffin, M. A., & Hart, P. M. (2000). The impact of organizational climate on safety climate and individual behavior. *Safety Science, 34,* 99–110.

O'Dea, A., & Flin, R. H. (2000). *Safety leadership in the offshore oil and gas industry* (Society of Petroleum Engineers Technical Report SPE-61158). Richardson, TX: SPE-International.

Ostroff, C. (1993). The effects of climate and personal influences on individual behavior and attitudes in organizations. *Organizational Behavior and Human Decision Processes, 56,* 56–90.

Parker, S. K., Axtell, C. M., & Turner, N. (in press). Designing a safer workplace: Importance of job autonomy, communication quality, and supportive supervisors. *Journal of Occupational Health Psychology.*

Pate-Cornell, M. E. (1990). Organizational aspects of engineering system safety: The case of offshore platforms. *Science, 250,* 1210–1217.

Pate-Cornell, M. E. (1993). Learning from the Alpha Piper accident: A post-mortem analysis of technical and organizational factors. *Risk Analysis, 13,* 215–232.

Patterson, M., Payne, R., & West, M. (1996). Collective climates: A test of their socio-psychological significance. *Academy of Management Journal, 39,* 1675–1691.

Perrow, C. (1967). A framework for the comparative analysis of organizations. *American Sociological Review, 32,* 194–208.

Perrow, C. (1979). *Complex organizations: A critical essay* (2nd ed.). Glenview, IL: Scott, Foresman.

Rasmussen, J. (1982). Human errors: A taxonomy for describing human malfunction in industrial installations. *Journal of Occupational Accidents, 4,* 311–333.

Rasmussen, J. (1990). The role of error in organizing behavior. *Ergonomics, 33,* 1185–1199.

Reason, J. (1990). *Human error.* New York: Cambridge University Press.

Reason, J. (1997). Managing the risks of organizational accidents. Aldershot, England: Ashgate.

Reason, J., Parker, D., & Lawton, R. (1998). Organizational controls and safety: The varieties of rule-related behavior. *Journal of Occupational and Organizational Psychology, 71,* 289–304.

Reichers, A. E., & Schneider, B. (1990). Climate and culture: An evolution of constructs. In B. Schneider (Ed.), *Organizational climate and culture* (pp. 5–39). San Francisco: Jossey-Bass.

Rochlin, G. I. (1999). Safe operations as a social construct. *Ergonomics, 42,* 1549–1560.

Rousseau, D. M. (1988). The construction of climate in organizational research. In C. L. Cooper & I. T. Robertson (Eds.), *International review of industrial and organizational psychology* (Vol. 3, pp. 139–158). New York: Wiley.

Schmidt, A. M., & DeShon, R. P. (2001, April). *The impact of workplace climates on work outcomes: A meta-analytic perspective.* Paper presented at the annual conference of the Society of Industrial and Organizational Psychology, San Diego, CA.

Schneider, B. (1987). The people make the place. *Personnel Psychology, 40,* 437–453.

Schneider, B., Bowen, D. E., Ehrhart, M. G., & Holcombe, K. M. (2000). The climate for service: Evolution of a construct. In N. M. Ashkanasy, C. P. Wilderom, M. F. Peterson (Eds.), *Handbook of organizational culture and climate* (pp. 21–36). Thousand Oaks, CA: Sage.

Schneider, B., & Reichers, A. E. (1983). On the etiology of climates. *Personnel Psychology, 36,* 19–39.

Schneider, B., White, S., & Paul, M. C. (1998). Linking service climate and customer perceptions of service quality: Test of a causal model. *Journal of Applied Psychology, 83,* 150–163.

Shannon, H. S., Mayr, J., & Haines, T. (1997). Overview of the relationship between organizational and workplace factors and injury rates. *Safety Science, 26,* 201–217.

Vroom, V. H. (1964). *Work and motivation.* New York: Wiley.

Waldman, D. A., Bass, B. M., & Yammarino, F. J. (1990). Adding to contingent-reward behavior: The augmenting effect of charismatic leadership. *Group and Organizational Studies, 15,* 381–394.

Weick, K. E. (1995). *Sensemaking in organizations.* Thousand Oaks, CA: Sage.

Weick, K. E., & McDaniel, R. R. (1989). How professional organizations work: Implications for school organization and management. In T. J. Sergiovanni & J. H. Moore (Eds.), *Schooling for tomorrow* (pp. 330–355). Needham Heights, MA: Allyn & Bacon.

Williams, H., Turner, N., & Parker, S. (2000, August). *The compensatory role of transformational leadership in promoting safety behaviors.* Paper presented at the annual meeting of the Academy of Management, Toronto.

Williamson, A. M., Feyer, A. M., Cairns, D., & Biancotti, D. (1997). The development of safety climate: The role of safety perceptions and attitudes. *Safety Science, 25,* 15–27.

Wright, C. (1986). Routine deaths: Fatal accidents in the oil industry. *Sociological Review, 4,* 265–289.

Yukl, G. (1998). *Leadership in organizations* (4th ed.). Englewood Cliffs, NJ: Prentice Hall.

Zohar, D. (1980). Safety climate in industrial organizations: Theoretical and applied implications. *Journal of Applied Psychology, 65,* 96–102.

Zohar, D. (2000). A group-level model of safety climate: Testing the effect of group climate on microaccidents in manufacturing jobs. *Journal of Applied Psychology, 85,* 587–596.

Zohar, D. (2002a). The effects of leadership dimensions, safety climate, and assigned priorities on minor injuries in work groups. *Journal of Organizational Behavior, 23,* 75–92.

Zohar, D. (2002b). Modifying supervisory practices to improve sub-unit safety: A leadership-based intervention model. *Journal of Applied Psychology, 87,* 156–163.

Zohar, D., & Luria, G. (2001, April). *Climate strength: Identifying boundary conditions for organizational climate.* Paper presented at the 16th annual conference of the Society for Industrial and Organizational Psychology, San Diego, CA.

PART 2

Strategy and Policy

CHAPTER 8

Strategic HRM and Organizational Health

Jason D. Shaw
John E. Delery

The study of the links among business strategy, human resource management (HRM) systems, and organizational performance is increasingly popular in the organizational sciences (e.g., Arthur, 1992, 1994; Delery & Doty, 1996; Huselid, 1995; Huselid, Jackson, & Schuler, 1997; Lee & Miller, 1999; Tsui, Pearce, Porter, & Tripoli, 1997). This line of research is commonly referred to as strategic HRM (SHRM) or the strategic management of people in work organizations (Delery & Shaw, 2001). The area of SHRM is relatively new; its roots can be traced to a few conceptual papers from the 1980s that attempted to directly integrate the role of HRM and business strategy (e.g., Dyer, 1984; Schuler & Jackson, 1987).

In this chapter we attempt to broaden the landscape of SHRM research by viewing facets of organizational health as important consequences of an organization's HRM strategy or array of HRM practices. The vast majority of SHRM research focuses solely on the financial performance (e.g., profitability) of organizations as the dependent variable of interest. This singular focus is logical: the proposition that HRM has a strategic role in the organization required that the scholar establish theoretical and empirical links to performance. Interestingly, given the broad and encompassing definitions of organizational health reviewed by Hofmann and Tetrick

(Chapter One, this volume), one could argue that some of the links between HRM strategy and organizational health have been well explored. But organizational health includes many more facets than distal measures of bottom-line performance, including many intermediate performance dimensions such as safety, productivity, efficiency, and assessment of employee health and well-being. Indeed, in several calls for additional SHRM research (e.g., Delery & Shaw, 2001), the issue of intermediate links between HRM systems and financial performance is central. For example, after reviewing the SHRM literature, Becker and Gerhart (1996, p. 793) argued the following:

> Future work on the strategic perspective must elaborate on the black box between a firm's HR system and the firm's bottom line. Unless and until researchers are able to elaborate and test more complete structural models—for example, models including key intervening variables—it will be difficult to rule out alternative causal models that explain observed associations between HR systems and firm performance. Without intervening variables, one is hard pressed both to explain *how* HR influences firm performance and to rule out an alternative explanation.

Some progress has been made in this regard (e.g., see Guthrie, 2001; Shaw, Delery, Jenkins, & Gupta, 1998), but much remains to be accomplished. Among the neglected intermediate outcomes in the relationship between HRM strategy and financial performance are assessments of facets germane to the issue of organizational health. Our aim is to establish a point of departure for researchers to begin exploring this vast unexplained "black box" between HRM strategy, the use of various HRM systems, and the financial performance of the organization. Specifically, this chapter is divided into six main sections. First, we define the landscape of SHRM, distinguishing it from the technical or functional perspective on HRM. Second, we examine the concept of HRM strategy, including definitional issues and a survey of how HRM strategy has been operationalized in the literature. Third, we briefly review the existing empirical literature on the relationship between strategic HRM and the most distal assessment of organizational health: financial performance. Fourth, we examine the concept of systems or bundles of HRM practices and their expected relationships with some of the more intermediate facets of organizational health proposed by Hofmann and Tetrick (Chapter One). In this section we draw on previously identified systems or clusters of HRM practices and discuss how these systems may relate to facets of organizational health in the short and long term. Fifth, we integrate the HRM strategy literature with the idea of modes of employment or the expected relationship between employee and employers, and with the conceptualization of organizational health offered by Hofmann and Tetrick (Chapter One). We examine the issue of the internal consistency or fit of strategic HRM systems and conceptual and empirical links to aspects of organizational health. This section also includes arguments that link each employment mode–HRM strategy combination with aspects of individual and organizational health. Sixth, we provide a brief discussion of the research implications of adding health to the SHRM literature. This will entail additional theoretical development and more creative research designs. The chapter concludes with an assessment and summary of lessons learned.

SHRM: The Landscape and Definitional Issues

The Landscape of SHRM

In this section, we briefly review the history of SHRM and distinguish it from other areas of HRM research. Next, we describe the various perspectives on HRM strategy that organizational researchers have used in theoretical and empirical investigations of these issues.

Although SHRM has many facets in common with long-standing areas of research in selection (e.g., Taylor & Russell, 1939), job and task design (e.g., Turner & Lawrence, 1965; Hackman & Oldham, 1980), and participatory management (McGregor, 1960), there are three primary differences between research labeled SHRM and other research simply labeled HRM. First, almost by definition, SHRM research seeks to explicate the role that HRM can play in the functioning of the entire organization. As Wright, Dunford, and Snell (2001) point out, human resource professionals have historically struggled to justify the function's value in terms of how well the organization performs. Linking the HRM function and business strategy and investigating how combinations or systems of

HRM practices may affect performance paves the way for more specific estimates of the value of HRM. This integrative or systems view of HRM contrasts with a functionalist perspective on HRM that seeks to explain the antecedents or consequences of a single facet of HRM. A functionalist view of HRM research generally either investigates how an HRM practice (e.g., a selection technique, a method of pay, or a performance appraisal method) affects the attitudes or behavior of individuals or explores how well or poorly HRM practices work when implemented. The functionalist focus describes nearly all of the HRM research before the 1980s and a still sizeable portion of HRM research today. Second, following logically from the systems approach, the level of analysis of SHRM research tends to be the organizational level, whereas functional HRM research is generally conducted at the level of the individual. Third, the strategic view generally assumes a bundles or systems approach to the study of HRM practices (Delery & Shaw, 2001). This assumption can be traced in part to resource-based theory that posits that a firm's resources (including human resources) are the key to sustainable competitive advantage (Barney, 1991; Wright & McMahan, 1992). The resource-based perspective assumes that resources lead to sustainable advantage only when they are rare, valuable, and inimitable. As such, a single HRM practice in isolation (e.g., a particular compensation or selection practice) may be easily imitated by a rival organization, but the key to success of a system or bundle of interrelated practices may not be as transparent. Thus, although the operationalization of systems of HRM practices varies considerably across the empirical literature, the idea of combinations of practices is a distinguishing characteristic of SHRM research. Although these distinctions are rather simple, they constituted a major shift in the research landscape, one that ultimately led to the consideration of the role of HRM as a strategic business partner in the organization (Delery & Shaw, 2001).

Perspectives on HRM Strategy

The SHRM literature contains some debate concerning the definition of HRM strategy. Indeed, researchers have taken several paths to explain the SHRM and organizational performance linkages. One vein of this research might be labeled the direct integration approach. According to this conceptualization, SHRM is captured by the integration or fit between the array of HRM practices that the organization uses and the business strategy it pursues. One can trace the history of the direct integration approach from a call for these linkages by Walker (1978) to several conceptual works published in the 1980s (e.g., Devanna, Fombrun, & Tichy, 1984; Dyer, 1984; Schuler & Jackson, 1987) and finally to often-cited empirical tests in the 1990s (e.g., Delery & Doty, 1996; Wright, Smart, & McMahan, 1995; Youndt, Snell, Dean, & Lepak, 1996). The basic theoretical foundation explored in these studies is a contingency or configurational theory; that is, HRM practices must be consistent with the pursued business strategy in order for the organization to be maximally effective.

The more popular approach, in our opinion, pursues this basic theme but without a direct integration of strategy and HRM practices (e.g., Arthur, 1992, 1994; Guthrie, 2001; Huselid, 1995; MacDuffie, 1995). For example, Wright and McMahan (1992, p. 298) defined SHRM as the "pattern of planned human resource deployments and activities intended to enable an organization to achieve its goals." The idea of HRM strategy as a pattern of decisions that relates to HRM practices in terms of goal achievement has several advantages over the direct integration approach. First, it provides a simpler framework from which to view the strategic impact of HRM on organizational outcomes. Second, it allows researchers to assume that the practices in use by the organizations are essentially the operationalization of strategy. The conceptual idea is that the HRM systems in place at a given time signal an organization's strategic approach; that is, an investment in a certain array of HRM practices indicates that the organization is pursuing a certain type of strategy, whereas another type of investment signals a different strategy (Arthur, 1994; see also MacDuffie, 1995). Third, it focuses attention on the issue of the internal consistency or congruence among the HRM practices. That is, in addition to the overall level of investment in HRM systems, another critical factor in the level goal achievement is whether or not the practices are mutually reinforcing (e.g., Delery, Gupta, & Shaw, 1997).

We follow the latter view in this chapter, as it offers the greater utility in examining the relationship between SHRM and organizational health. In the following sections, we briefly review the empirical literature that attempts to link HRM systems to measures of organizational performance. We view a distal performance measure, such as financial performance, as an indication, albeit of only a single facet, of an organization's health. Next, we expand the current conceptual base by linking HRM systems to facets under broader definitions of organizational health, including consequences to internal functioning, social systems, safety, and individual strain. We also take steps in the direction of identifying short- and long-term consequences of different HRM systems. Following this, we address the issues of HRM system congruence and organizational health, drawing on some recent initial empirical examinations in this area.

SHRM and Organizational Health

Several authors have proposed conceptual definitions or frameworks of organizational health that can be directly applied to the SHRM research arena (e.g., Bennis, 1962; Jaffe, 1995; Miles, 1965). In particular, organizational-level conceptualizations of organizational health found in the organizational literature (e.g., Jaffe, 1995; Miles, 1965) are easily synthesized with the view of organizational performance and effectiveness found in the SHRM literature. Moreover, some aspects of micro- or individual-level conceptualizations of organizational health (e.g., role theory perspective from Katz & Kahn, 1978) seem to be amenable to a vertical synthesis to the organizational level. The common theme among the various definitions is the multifaceted nature of organizational health (Chapter One of this volume). Jaffe (1995), for example, argued that organizational effectiveness includes aspects such as profitability or financial performance (the outcomes of choice in the SHRM literature) but also the way the organization treats employees, presumably through the system of HRM and other work practices. Miles (1965) proffered 10 dimensions of health under the categories of goals (focus, communication adequacy, power equalization), internal system status (resource use, cohesiveness, morale), and growth and change (innovativeness, autonomy, adaptation, problem solving). Finally, although framed from the individual perspective, Katz and Kahn's role theory perspective (1978) highlights two criteria necessary for organizational health: (1) Does the performance of the individual enhance the effectiveness of the organization? (2) Does role enactment enhance individual well-being? A common thread among these conceptualizations is that organizational performance, typically defined as profitability or productivity, is but a single, albeit important, facet of the overall health of the organization.

After reviewing the existing empirical SHRM literature, we base our analysis on Hofmann and Tetrick's integration of the organizational health definitions (Chapter One of this volume), drawing heavily on the idea of competing demands. In the following section, we discuss the relationship between HRM systems on the one hand and measures of financial performance and productivity on the other. We then proceed to extend the literature and point future researchers in the direction of exploring a broader and richer conceptual domain of organizational health.

SHRM, Financial Performance, and Productivity

Much of the theorizing in SHRM strongly suggests that no one best way exists for organizations to manage their human resources. Rather than a single set of best practices, configurational or contingency approaches (specific sets of practices matched to particular situations) are what scholars argue will lead to superior organizational performance (e.g., Delery, 1998; Delery & Shaw, 2001; Schuler & Jackson, 1987). But the operationalizations of these sets often amount only to an assessment of the level of investment an organization makes in a specific set of practices deemed to be superior (e.g., training, performance appraisal, high pay and benefits, participation in decision making, etc.) to their polar opposites (e.g., low training, centralized decision making, low pay, etc.). That is, organizations that invest more heavily in their HRM systems are likely to perform more effectively than are organizations with lesser investments. As Becker and Huselid (1998, p. 55) point out, the appropriate level of analysis for SHRM research is a set or system of HRM practices because "they more accurately reflect the multiple paths through which HRM policies will influence successful strategy implementation." Indeed, this set of practices (although slightly different across various studies) is often labeled high performance (Becker & Huselid, 1996; Huselid, 1995), high involvement (Pil & MacDuffie, 1996), or high commitment (Arthur,

1992, 1994; Guthrie, 2001), implying that they are indeed superior to other sets of practices in terms of enhancing organizational performance.

Thus, SHRM does differ from the technical or functional literature in terms of number of practices considered, but the overall picture is generally the same. Some practices are simply more effective than others, and organizations that invest more heavily in them will be more effective. The assumed theoretical mechanism from this view is a fairly straightforward mediation model: employees of organizations that invest heavily (in terms of superior HRM practices) will outperform employees of less heavily invested organizations, and higher productivity and profitability will result. A corollary theoretical perspective, the resource-based view, suggests that competitive advantage results from the inimitable human capital found in the high investment organizations. As Delery and Shaw (2001) point out, existing empirical research generally confirms that human capital accumulations as a function of an organizational HRM system have a substantial impact on an organization's ability to sustain competitive advantage; it also consistently demonstrates that HRM practices, and combinations of these practices, are proximal precursors of these human capital accumulations.

We should mention several exemplars of this general statement. In a pioneering series of studies, Huselid and colleagues (e.g., Becker & Huselid, 1996; Becker, Huselid, Pickus, & Spratt, 1997; Huselid, 1995; Huselid et al., 1997) consistently demonstrate that a set of HRM practices designed to enhance employee skills (e.g., training, formal grievance and dispute resolution procedures, valid selection practices, participation programs) and motivation (e.g., performance appraisals, qualified applicants per position) relate to various measures of corporate financial performance (return on assets), market value, productivity, and turnover rates. Huselid's studies include data collections from national surveys of U.S. organizations with more than 100 employees, including biannual surveys from initial samples of more than 4,000 organizations. The results have set the standard for SHRM studies. For example, Huselid (1995) estimated that a one standard-deviation upward change in HRM practices resulted in implied economic effects of more than $27,000 in sales per employee (a standard measure of productivity) and more than $18,000 in the market value per employee among the studied organizations. Using improved and more extensive measures of HRM practices (high performance work practices, using the semantics of Huselid and colleagues) among a similar cross-industry sample in 1996, Huselid and Becker (1997) reported that a one standard-deviation positive change in HRM practices was associated with 24% higher market value of shareholder equity and 25% higher accounting or financial profits. Although the series of studies by Huselid and colleagues provides the most compelling evidence with regard to effect sizes, several others studies also demonstrate a link between HRM practice and financial performance. For example, Ichniowski, Shaw, and Prennushi (1997) found that increases in the use of a set of innovative HRM practices was associated with a 3.5% increase in production uptime, resulting in a minimum increase of $105,000 in operating profits per month. In addition, in a study of HRM systems in commercial banks, Delery and Doty (1996) reported that measures of HRM systems explained between 6% and 11% of variation in measures of financial performance (e.g., return on assets).

The results of SHRM studies also extend beyond general measures of financial performance or accounting profits. Indeed, the relationship between HRM systems and organizational outcomes is arguably as strong when measures of workforce productivity or performance are the outcomes of interest. In an ambitious study of automotive manufacturing facilities worldwide, MacDuffie (1995) found that plants using team-based and high commitment HRM practices consistently outperformed plants in terms of labor hours worked per vehicle (a standard measure of productivity) and product quality. The HRM indices in MacDuffie's equations explained from 3% to 7% of the variation in labor hours per vehicle and 9% to 37% of the variation in product quality across the manufacturing plants.

Several other studies that examine the link between HRM systems and productivity are also worthy of note. First, although the results were not as striking as the financial performance estimates, Huselid (1995) found significant direct relationships between measures of HRM systems and workforce productivity. Second, Arthur (1994) clustered a sample of steel minimill plants in terms of their HRM systems. He labeled organizations from the resulting two-cluster solution as either high commitment or control facilities. Organizations included in the high commitment cluster were characterized by higher levels of training, employee involvement and decision-making participation, social integration, and other practices associated with enriched and enlarged jobs. Organizations included in the control cluster used only low levels of training, participation, and so on; generally defined jobs narrowly; and had hierarchical and centralized decision-making structures. Arthur's results (1994) were generally consistent with those of MacDuffie (1995) and others. High commitment HRM system facilities required fewer labor hours to produce a given quantity of output (i.e., higher productivity) and experienced lower employee turnover. The opposite results were found among those facilities characterized by control (or low commitment systems). Third, testing the external validity of Arthur's narrow, single-industry study, Guthrie (2001) explored the external validity of Arthur's commitment and control categorization among a cross-industry sample of Australian organizations. Rather than clustering organizations by HRM practices, Guthrie employed a single continuum approach ranging from high (commitment-oriented HRM systems) to low (control oriented). Guthrie's results (2001) tended to parallel those of Arthur (1994): organizations employing commitment-oriented systems tended to outperform control-oriented organizations in terms of labor productivity.

Evaluation of the Empirical Research Base

This series of productivity studies, viewed in light of the results from those that examine profitability, paint a rather convincing picture of the relationship between HRM systems and important organizational outcomes. Although the specific operationalizations of HRM systems vary somewhat across empirical studies, the basic formulation has remained fairly simple and fairly constant; that is, those organizations that invest more heavily in their HRM systems are expected to perform better across the board. On the one hand, it is difficult to argue with some of the extant empirical research findings in this regard. Recent reviews (e.g., Becker & Huselid, 1998; Delery & Shaw, 2001) show clear and convincing evidence of the SHRM–organizational performance relationship. On the other hand, the current literature base has several limitations.

This stream of research clearly carries with it two primary assumptions flowing from the belief that people are the most important asset. These two assumptions are that (1) heavy use of an array of sophisticated HRM practices is better than the alternative low investment approach and (2) organizations should strive to increase performance, defined as profitability or productivity, via this array of practices. Although these assumptions have served the SHRM literature quite well, they carry with them advantages and disadvantages. Importantly, these assumptions have the advantage of relative simplicity, given that attempting to conceptualize and operationalize all possible combinations of practices that might yield functional and healthy organizations in different circumstances is a daunting task. Much of the conceptual work emphasizes combinations, synergies, fit, or congruence, but the strongest findings to date employ more simple assessments of HRM indices, factor scores, or organizations clustered by their HRM profile.

But the common empirical approaches also have several limitations or downsides that, in our opinion, limit scientific progress and our understanding of the consequences of SHRM systems. In particular, the omnibus view that a high HRM investment is superior fails to capture the richer and more complex notion of balanced or competing goals that can be addressed by viewing organizational effectiveness more broadly, that is, in terms of organizational health. Although heavy investment in HRM practices may generally be a positive step for organizations, it also likely comes with costs beyond that required by the design and implementation of these systems. Kaminski (2001) argues that although nearly all adopted HRM practices are intended to improve organizational functioning, managers often downplay the side effects of their implementation. As an example, any practice that increases the expected contributions of the workforce (e.g., developmental performance appraisal, incentive pay, participation in decision making) may improve organizational performance in terms of workforce productivity (and perhaps profitability) but also places greater burdens on the employees themselves. In the terminology of the work stress literature, high investment HRM systems increase job demands and may threaten other aspects of organizational health. Although job demands per se are not necessarily harmful, adding job demands may increase competition between demands (e.g., see Parker, Turner, & Griffin, Chapter Four of this volume). This notion of competing demands has not been adequately addressed in the SHRM literature.

A final related limitation of the existing empirical base is a failure to fully explore the notion of incongruence among the HRM practices that organizations use. The idea of a logically implemented system of mutually supportive practices is deeply ingrained in the conceptual literature, but only recently have researchers begun to explore the consequences of system incongruence (e.g., Kaminski, 2001; Shaw, Gupta, & Delery, 2001). We address the notion of competing demands and system incongruence in turn later in this chapter.

A Broader Conceptualization of SHRM and Organizational Health: HRM Strategies

As we noted, SHRM researchers generally categorize a cluster or bundle of HRM practices under a single moniker (e.g., commitment, skills and motivation, inducements, investments, or control). Among the most cited of the conceptualizations are Arthur's commitment and control HRM systems (1992, 1994). According to Arthur (1994, p. 672), the goal of "control human resource systems is to reduce direct labor costs, or improve efficiency, by enforcing employee compliance with specified rules and procedures and basing employee reward on some measurable output criteria" (see also Eisenhardt, 1985). The practices that characterize control HRM systems include centralized (or low employee participation in) decision making, little training, high span of control, low justice or due process mechanisms, low pay and benefits, and high use of individual incentives. By contrast, commitment systems are designed to shape behaviors through psychological mechanisms (e.g., trust, goal alignment, etc.). Although Arthur (1992, 1994) clustered organizations according to their HRM profiles, the commitment and control systems can be thought of as simply opposite ends of a single continuum (e.g., see Guthrie, 2001). Commitment-oriented HRM systems include high employee participation, training, due process, base pay and benefit, and low use of individual incentives.

Others, although differing in the specific practices included, offer similar conceptualizations. Huselid (1995) categorized practices as those reinforcing employee skills and organizational structures (e.g., information sharing, job analysis, employee participation programs, organizational incentive plans, and formal due process procedures) and those that relate to employee motivation (e.g., performance appraisals, merit pay, applicant quality). Shaw, Delery, Jenkins, and Gupta (1998) classified practices as those that serve as inducements and investments (e.g., pay, benefits, job stability, training, and justice) and those that related to expected or enhanced employee contributions (e.g., monitoring, high job demands, etc.).

Tsui et al. (1997) offered the most integrated conceptualization of HRM strategy by combining the approach offered by Arthur (1994) and others with the employee perspective of the HRM strategy. The result is essentially a description of the broad modes of employment found in organizations. These authors identified four distinct types of employee-employer relationships, characterized by the array of HRM practices used and the expectations of the individual employees within each. They argued for two balanced employer-employee relationships (quasi-spot and mutual investment) and two unbalanced approaches (underinvestment and overinvestment). The four approaches basically represent a two-by-two grid that crosses economic or social exchange by short- or open-ended time orientation. The quasi-spot relationship is considered balanced because the employer offers simple economic inducements for a specified well-defined set of activities from the employee; the expected duration is short. Mutual investment relationships in terms of HRM systems and employee expectations are similar to the commitment system. Under this type the employer provides high training, participation, organizational incentives (e.g., profit sharing), and so on and expects the employee to complete broad ranging, loosely defined tasks. The time orientation in the mutual investment approach is the long term.

The two other approaches cross the level of investment with the time frame and are considered unbalanced. In underinvestment situations, employees are expected to complete tasks and fulfill the long-term obligations as are those in mutual investment organizations, but the organization reciprocates with short-term, purely economic HRM practices (e.g., low training, participation, and high individual incentives). In the overinvestment approach, HRM practices are high commitment in style, but the expectations of employees in terms of their job duties concern only well-specified, narrow job tasks.

Tsui et al. (1997) thus offer an integration of HRM strategy with the expectations of employees under each of these strategies. We would argue that interesting implications for organizational health can be derived for each of the strategies in this framework. The integration of Tsui et al.'s modes of employment and HRM strategy highlights, in our opinion, offers one of the most obvious paths for future researchers to explore. We will discuss the health implications of this integration.

HRM Strategy, Modes of Employment, and Organizational Health

Table 8.1 contains a depiction of the integration of HRM strategy, modes of employment, and organizational health. Using the framework from Tsui et al. (1997), the table depicts the four HRM strategies along the two dimensions of employer-expected employee contributions and employer-offered inducements. The upper-left quadrant contains organizations using the mutual investment mode of employment and the commitment-oriented HRM strategy. The lower-right quadrant contains the quasi-spot employment mode and the control HRM strategy. These quadrants are distinguishable from the other quadrants in that they represent balanced employer-employee approaches; that is, the expected contributions of employees match the type of investments (e.g., HRM practices) that the organization uses. By contrast, the alternative quadrants contain unbalanced approaches to the employee-employer relationships. The lower-left quadrant includes organizations using the overinvestment employment mode and the commitment HRM strategy. The upper-right quadrant contains organizations in the underinvestment mode using a control HRM approach. As the table shows, some initial consequences in terms of macro-level and individual-level (employee) organizational health are outlined in each quadrant. Also, the notion of competing demands in terms

Table 8.1. Employment Mode, HRM Strategy, Employee Expectations, and Organizational Health Consequences.

	Employer-Offered Inducements	
	High	**Low**
Employer-Expected Employee Contributions — High	Employment mode: Mutual investment HRM strategy: Commitment Employee expectations: Broad job definitions, open-ended orientation, loosely-defined tasks Macro-organizational health consequences: • Low short-run productivity and profitability • High long-run productivity and profitability • High long-run adaptability • Low turnover and absenteeism • High citizenship behavior • Low safety records Individual health consequences: Strain outcomes associated with high decision-making, responsibility, role ambiguity and conflict; higher well-being as a result of job security, safety, and social integration.	Employment mode: Underinvestment HRM strategy: Control Employee expectations: Broad job definitions, open-ended orientation, loosely-defined tasks Macro-organizational health consequences: • Low short- and long-term productivity and profitability • Very high turnover and absenteeism • High accident rates • Low citizenship and prosocial behavior Individual health consequences: Strain outcomes associated with HRM strategy-employee expectation incongruence, role ambiguity and conflict; lower well-being as a result of lower safety.
Employer-Expected Employee Contributions — Low	Employment mode: Overinvestment HRM strategy: Commitment Employee expectations: Narrow job definitions, short-term employment orientation, clearly defined tasks Macro-organizational health consequences: • Questionable short-term and long-term productivity and profitability • Low turnover and absenteeism • High work force stagnation Individual health consequences: Strain outcomes associated with HRM strategy-employee expectation incongruence, work monotony, low task meaningfulness, low task significance, low social integration, and low job security	Employment mode: Quasi-spot HRM strategy: Control Employee expectations: Narrow job definitions, short-term employment orientation, clearly defined tasks Macro-organizational health consequences: • High short-term productivity and profitability • Low long-term productivity and profitability • High turnover and absenteeism • Low citizenship and prosocial behavior Individual health consequences: Strain outcomes associated with work monotony, low task meaningfulness, low task significance, low social integration, and low job security

of organizational health is incorporated into each quadrant; that is, each employment mode–HRM strategy combination is likely to yield some positive and some negative consequences in terms of individual and macro-level organizational health. We will detail each of the quadrants and some initial arguments about the consequences for organizational health.

Upper-Left Quadrant

The mutual investment approach balances HRM practices designed to enhance skills through training, selection, organizational incentive systems and other practices oriented toward the long term with jobs that are broad, loosely defined, and open-ended in time. As Tsui et al. (1997) point out, this approach is designed for high individual employee performance, high levels of morale, and a general focus on high workforce productivity. Indeed, this approach is what the SHRM literature implies to be superior across the board. Table 8.1 lists the likely positive consequences in terms of individual and organizational health for this relationship. Osterman (1987) refers to such high investment organizations as those that are seeking competitive advantage through high individual performance and ultimately high productivity. Heavy investment in the training and development as well as other practices supportive of broadly defined work responsibilities and high levels of participation should result in high levels of productivity and adaptability, low turnover, low accident rates, and other positive macro-outcomes. Indeed, much of the SHRM research is probably picking up these effects. Aside from productivity and other macro-level effects, the positive consequences should extend to individuals employed in mutual investment systems in terms of job security, opportunity to redress grievances, safety, and social integration. The emphasis on the enhancement of the workforce should also increase the frequency of citizenship and other forms of prosocial behaviors in the organization (Tsui et al., 1997).

In line with the competing-values notion of organizational health, this quadrant also lists several negative consequences of the mutual investment approach. First, as Table 8.1 shows, the organizational-level positive effects may not be immediately observable. The advantages of mutual investment employment arrangements and the commitment HRM strategy in general may take time to produce positive results. Moreover, as Arthur (1994) and Guthrie (2001) demonstrate, higher turnover rates among commitment-oriented organizations also have a more powerful negative effect on productivity than do the corresponding level of turnover among organizations pursuing a control HRM strategy. The other negative consequences are those that result from the higher job demands on several dimensions including higher decision-making authority, increased responsibility, and two common stressors resulting from loosely defined tasks: role ambiguity and conflict. The work stress research provides some convincing evidence in this regard. For example, Schaubroeck, Ganster, and Kemmerer (1994) found that highly complex jobs were associated with physical health problems including cardiovascular disease.

Lower-Right Quadrant

The outcomes in terms of organizational health in this quadrant are likely to be in evidence as a result of the combination of a quasi-spot employment mode and the control HRM strategy. Again, this approach is considered balanced because the HRM practices are consistent with the assumed expectations in the employer-employee relationship. As Osterman (1987) notes, organizations pursuing this type of low investment approach attempt to satisfy stakeholder demands (i.e., high profitability) through cost reduction rather than through high workforce productivity. Thus, the organizational health consequences listed in this quadrant reflect the possibility that these organizations may be highly profitable in the short run, although their long-term success and perhaps their survival is an open question. As Shaw, Gupta, and Delery (2002a) argue, organizations pursuing this type of employment mode and HRM strategy should also be able to tolerate high levels of employee turnover and remain profitable in the short run. These authors found the highest levels of profitability among organizations that either invested heavily in their HRM systems and experienced low turnover rates or among organizations that pursued a low-road HRM strategy and experienced high levels of turnover. Osterman (1987) referred to this effect as the HRM practice–turnover trade-off. Thus, we expect higher levels of turnover and absenteeism among organizations in this quadrant as well as lower social integration and citizenship, given that the organizations invest little in the employees and expect only sharply defined, narrow contributions from them in return. In terms of individual-level health outcomes, the expectations are generally negative. Strain outcomes should be those associated with the types of job demands resulting from this employment mode and HRM strategy, that is, specialized tasks, little decision-making authority, and so on. From a job design perspective, these tasks should have little meaning and significance; they should produce little integration in the social network; and the jobs should be fairly monotonous. Moreover, because the nature of the exchange relationship is purely economic, negative health consequences as a result of low job security would also be expected. Probst and Brubaker (2001), for example, found that employees who reported that their jobs were insecure reported lower motivation to comply with safety procedures and had higher levels of workplace injuries and accidents.

Upper-Right Quadrant
Tsui et al. (1997) labeled the mode of employment contained in this quadrant underinvestment. It combines an HRM strategy focused on the stakeholders and designed to lower internal costs with an open-ended mode of employment focused on the individual. Very few positive organizational-level health outcomes are expected to result from this unbalanced approach. Because employees will not be appropriately trained or oriented to open-ended jobs with high levels of authority and responsibility, productivity and ultimately profitability are likely to be low. Moreover, because employees are not adequately rewarded (in a broad sense) for their contributions, high levels of turnover, absenteeism, and accidents should be observed over time. The individual-level health outcomes should parallel the organizational-level outcomes, including risk of injuries and the strain outcomes associated with role ambiguity and conflict.

The negative health outcomes are also expected to be affected by the mixed signals that organizations using a control-oriented HRM strategy and an intrinsically oriented mode of employment send. In addition to inadequately preparing employees for the tasks at hand, the inconsistent messages may contribute to negative individual and organizational-health problems among employees through their effect on the safety climate in the organization (e.g., see Zohar, Chapter Seven of this volume). Interestingly, some recent empirical research involving organizational health outcomes (primarily safety) can be applied directly to this quadrant. Tsui et al. (1997, p. 1115) found that the underinvestment approach yielded the worst organizational outcomes because "evidently, employees managed by this type of employer-defined employee-organization relationship reacted by reducing their performance on core tasks, by refraining from engaging in citizenship behavior, and by being absent more often than others."

Shaw and his colleagues (2001) examined the consequences (including a measure of accident rates) of combinations of compensation systems and dimensions of integrated manufacturing among a sample of production facilities in the concrete pipe industry. They argued that the dimensions of integrated manufacturing (Total Quality Management [TQM] and advanced manufacturing technology) change the nature of the employment relationship in an open-ended, highly integrated, highly empowered manner. For example, integrated manufacturing increases employee responsibility levels, task interdependence, and the frequency of complex tasks while reducing supervision (Dean & Snell, 1991).

Shaw et al. (2001) argued that certain compensation systems (i.e., team- and skill-based pay) are logically congruent for use with integrated manufacturing and that their concurrent use should be associated with higher safety levels; other systems (i.e., individual incentives and seniority-based pay) are incongruent, and their use with integrated manufacturing should be associated with poorer safety records. Skill-based pay rewards encourages up-skilling and knowledge acquisition (e.g., Gupta & Shaw, 2001) that are needed in demanding integrated manufacturing environments and therefore lessen the likelihood that accident and safety problems will occur. By contrast, Shaw et al. (2001) argued further that integrated manufacturing techniques blur lines of distinction among individuals and encourage teamwork, whereas effective individual incentives hinge on the identification and measurement of individual performance and encourage intraorganizational competition. As such, the simultaneous use of integrated manufacturing (characteristics similar to open-ended, enriched employment relationships) and individual incentives (characteristic of control-oriented HRM systems) provides a mixed message to employees that may lead to poorer safety records. The findings of Shaw et al. supported these predictions; that is, five-year accident rates were higher among facilities using TQM systems and individual incentives and lower among facilities concurrently using TQM and skill-based pay. Supporting these empirical findings, Landsbergis, Cahill, and Schnall (1999) reviewed the literature on the health consequences of TQM, finding that the positive effects of empowerment were mitigated by higher job demands including more rapid work pace, ergonomic stressors, and other adverse health effects. Although they did not present empirical evidence that parallels that of Shaw et al. (2001), the implications of the review would be that these negative effects would be more apparent in organizations that pursue this employer-employee exchange relationship in the absence of a supporting HRM strategy.

In a second study, Shaw, Gupta, and Delery (2002b) examined how compensation structure (pay dispersion), task design (work interdependence), and the use of individual incentives interacted to predict workforce performance outcomes (including measures of workforce safety records). Drawing on justice, motivation, and institutional theories, Shaw and his colleagues argued that highly dispersed pay would negatively relate to accident rates only when a normatively accepted dispersion-creating practice like individual incentives was also in place. Of particular interest to the topic at hand, they also argued that highly dispersed pay (a characteristic of individual incentive systems in control-oriented HRM systems) would result in higher accident rates when it was coupled with high levels of work interdependence (a characteristic of an intrinsic or social exchange view of the employment relationship). The results were generally consistent with the congruence predictions in two large single-industry studies, a trucking industry study and the concrete pipe industry study (Shaw et al., 2001).

Finally, in a recent individual-level study, Probst (2002) argued that the negative effects of job insecurity (common in control-oriented HRM strategies) would be exacerbated by a production- or economic-oriented organizational climate. Supporting her contentions, she found that the relationship between job insecurity and an array of safety outcomes (safety knowledge, safety compliance, near miss accidents, and physical health problems) was stronger when the organizational climate emphasized production goals rather than safety.

Lower-Left Quadrant

This health quadrant contains the overinvestment mode of employment described by Tsui et al. (1997). This unbalanced approach combines a high investment HRM strategy with narrowly defined jobs, little decision-making responsibility, and pure economic orientation in terms of the employee-employer relationship. Among the four quadrants, research evidence on this type of HRM strategy–employment mode combination is the scarcest. Also scarce is individual-level research that can be used in a vertical synthesis to provide evidence about the outcomes of the incongruence. Indeed, Tsui et al. found that the hypothesized relationships between the overinvestment mode of employment and organizational outcomes (e.g., individual performance, job attitudes, etc.) failed to materialize as they expected. These authors reasoned that one explanation for the inconsistent pattern was that few organizations fit the requirements of this cell, making tests of the hypotheses and conclusions drawn from them difficult.

We can, however, make some general speculations about the likely health consequences of this unbalanced approach. First, the short- and long-term productivity of these organizations is questionable, given that they are applying HRM practices designed for broadly defined, highly empowered jobs to a narrowly defined, centralized situation. The employment mode is designed for an economic exchange, whereas the HRM strategy is designed for a social exchange. According to Osterman (1987), such organizations are unlikely to be successful financially, given the mismatch of investments and expectations. Turnover and absenteeism rates are expected to be low under such systems, because the high level of inducements and investments offered by the organization may bind employees to the organization (Becker, 1960), but these rates, combined with the unfulfilling nature of the tasks, may result in a stagnated workforce. To some degree the expected health consequences for the individual employee should match those of the other control HRM strategy quadrant, that is, strain outcomes as a result of low task meaningfulness, autonomy, and significance, as well as strain resulting from work monotony. More important, additional negative health consequences may result from the mixed signals that the organization gives by the incongruence between employment mode and HRM strategy (Shaw et al., 2001).

Research Implications

One of the most obvious conclusions from this review is that very little research addresses health issues at the organizational level. The body of work, however, is growing rapidly. In fact, a number of studies are now investigating HRM systems and one measure of health, financial performance. We strongly urge researchers in this area to widen the choice of dependent variables to include some of the many health and safety factors discussed in this chapter. This will take additional work in both theoretical development and research design. Such work, we expect, will not be easily accomplished.

As for theoretical development, the area of SHRM in general still needs significantly better theory. As we have mentioned, several general notions prevail in the literature; however, scholars have made relatively few attempts to clarify them and bring them into a more comprehensive theoretical framework (Delery & Shaw, 2001, Lepak & Snell, 1999). Arguably, this theoretical development is essential prior to the beginning of empirical work. The problems of ill-defined constructs and frameworks are prevalent in the SHRM literature. Although many scholars discuss HRM systems and various components of those systems but with such little specificity that empirical measures of what those scholars argue to be the same construct rarely are similar in empirical studies.

One of the first constructs in need of greater conceptual clarity is that of organizational health. Hofmann and Tetrick (Chapter One of this volume) do a nice job of pulling together notions of this construct, but we argue that it is still somewhat vaguely defined. Conceptual development is needed to clarify this construct and, just as important to empirical research, point to measurable indicators. In the SHRM literature to date, even when the constitutive definition of constructs may have been clear, the methods needed to measure the construct are not. Many researchers have relied on measurement techniques developed to measure individual-level constructs to measure organizational-level constructs, with poor results (Gerhart, Wright, McMahan, & Snell, 2000; Wright, Gardner, et al., 2001).

Beyond conceptual development, researchers will have to look to innovative methods to collect empirical data. No consensus exists as to the features of the ideal study, nor should one. Many different perspectives and approaches are needed. That said, we do have a few suggestions for readers to keep in mind. First, given the multilevel nature of the issues surrounding organizational health, we suggest collecting data at multiple levels. This would entail collecting data on organizational, unit, and individual variables. The methods used to collect these data, however, may be different across levels. Second, researchers should conduct studies across time, something rarely done in SHRM research to date. The basic ideas are causal in nature, and only longitudinal designs can really address them. Much more, the use of longitudinal designs may enable researchers to use fewer organizations in their samples, thus minimizing the number of responding companies one must get to participate. Third, we highly recommend studying changes in HRM systems over time. Ideally, the researcher could gain access to an organization prior to a major change in HRM strategy. Then the researcher could study this over a period of time. Qualitative in addition to quantitative methods could prove very useful in such a case. Through additional qualitative studies, we may be able to refine our theories.

In the end it will take many different research designs to gain the hoped-for insight into organizational health. We encourage researchers to use their creative capacity to design new methods for measuring organizational characteristics. Methods designed to measure individual-level constructs may be ill suited to measure the organizational-level constructs discussed in this chapter.

Conclusions

Despite the conceptual and empirical progress evident from two decades of research, work that attempts to address the unexplored black box between SHRM systems and the ultimate financial performance of organizations is still sorely needed. In particular, the impact of HRM strategy on organizational health and safety issues is among the issues that scholars barely touch. Thus, with the extant field of SHRM as a baseline, we began the process of addressing the implications of SHRM research to individual and organizational health. As our review and synthesis indicates, the speculations about the possible consequences outweigh solid theoretical foundations and the existing research base in many areas. We argue for SHRM research to extend the existing paradigm by considering the notion of competing demands. The limited empirical research base (e.g., Shaw et al., 2001, 2002b; Probst, 2002) provides encouraging, although preliminary, evidence demonstrating the potential deleterious nature of mismatches between SHRM systems and modes of employment on the safety and health of the workforce. In particular, these studies demonstrate that mismatches or a lack of internal congruence between organization systems can create situations that negatively affect organizational health through injustice, perceived illegitimacy of organizational systems, or the creation of situations in which employees receive mixed messages.

Investigations that leave previously held assumptions behind (e.g., commitment-oriented HRM strategies are superior in all cases, and financial performance is the most important outcome), will, in our opinion, lead to a more interesting, more fundamentally sound, and more useful scientific research base. We are excited about the possibilities for theoretical and empirical research on organizational health that the integration of HRM strategy and modes of employment offers, as well as about the research streams possible from considering both the short- and long-term aspects of individual and organizational health. We hope that other researchers share our enthusiasm and that reviews in the decades to come will include analyses of such studies.

References

Arthur, J. B. (1992). The link between business strategy and industrial relations systems in American steel minimills. *Industrial and Labor Relations Review, 45,* 488–506.

Arthur, J. B. (1994). Effects of human resource systems on manufacturing performance and turnover. *Academy of Management Journal, 37,* 670–687.

Barney, J. (1991). Firm resources and sustained competitive advantage. *Journal of Management, 17,* 99–120.

Becker, H. S. (1960). Notes on the concept of commitment. *American Journal of Sociology, 66,* 32–40.

Becker, B., & Gerhart, B. (1996). The impact of human resource management on organizational performance: Progress and prospects. *Academy of Management Journal, 39,* 779–801.

Becker, B., & Huselid, M. A. (1996). Methodological issues in cross-sectional and panel estimates of the HR-firm performance link. *Industrial Relations, 35,* 400–422.

Becker, B., & Huselid, M. A. (1998). High performance work systems and firm performance: A synthesis of research and managerial implications. *Research in Personnel and Human Resources Management, 16,* 53–101.

Becker, B. E., Huselid, M. A., Pickus, S. A., & Spratt, M. F. (1997). HR as a source of shareholder value: Research and recommendations. *Human Resource Management, 36,* 39–47.

Bennis, W. G. (1962). Towards a "truly" scientific management: The concept of organization health. *General Systems: Yearbook of the Society for General Systems Research, 7,* 269–282.

Dean, J. W., & Snell, S. A. (1991). Integrated manufacturing and job design: Moderating effects of organizational inertia. *Academy of Management Journal, 34,* 776–804.

Delery, J. E. (1998). Issues of fit in human resource management: Implications for research. *Human Resource Management Review, 8,* 289–309.

Delery, J. E., & Doty, D. H. (1996). Modes of theorizing in strategic human resource management: Test of universal, contingency, and configurational performance predictions. *Academy of Management Journal, 39,* 802–835.

Delery, J. E., Gupta, N., & Shaw, J. D. (1997, August). *Human resource management and firm performance: A systems perspective.* Paper presented at the Southern Management Association meeting, Atlanta, GA.

Delery, J. E., & Shaw, J. D. (2001). The strategic management of people in work organizations: Review, synthesis, and extension. *Research in Personnel and Human Resources Management, 20,* 165–197.

Devanna, M. A., Fombrun, C. J., & Tichy, N. M. (1984). *Strategic human resource management.* New York: Wiley.

Dyer, L. (1984). Studying human resource strategy: An approach and an agenda. *Industrial Relations, 23,* 156–169.

Eisenhardt, K. (1985). Control: Organizational and economic approaches. *Management Science, 31,* 134–149.

Gerhart, B., Wright, P. M., McMahan, G. C., & Snell, S. A. (2000). Measurement error in research on human resources and firm performance: How much error is there, and how does it influence effect size estimates? *Personnel Psychology. 53,* 803–834

Gupta, N., & Shaw, J. D. (2001). Successful skill-based pay plans. In C. H. Fay, D. Knight, & M. A. Thompson (Eds.), *The executive handbook on compensation: Linking strategic rewards to business performance* (pp. 513–526). New York: Free Press.

Guthrie, J. P. (2001). High involvement work practices, turnover, and productivity: Evidence from New Zealand. *Academy of Management Journal, 44,* 180–190.

Hackman, J. R., & Oldham, G. R. (1980). *Work redesign*. Reading, MA: Addison-Wesley.

Huselid, M. A. (1995). The impact of human resource management practices on turnover, productivity, and corporate financial performance. *Academy of Management Journal, 38,* 635–672.

Huselid, M. A., & Becker, B. E. (1997, August). The impact of high performance work systems, implementation effectiveness, and alignment with strategy on shareholder wealth. Paper presented at the annual meetings of the Academy of Management, Boston, MA.

Huselid, M. A., Jackson, S. E., & Schuler, R. S. (1997). Technical and strategic human resource management effectiveness as determinants of firm performance. *Academy of Management Journal, 40,* 171–188.

Ichniowski, C., Shaw, K., & Prennushi, G. (1997). The effects of human resource management systems on economic performance: An international comparison of U.S. and Japanese plants. *Management Science, 45,* 704–721.

Jaffe, D. T. (1995). The healthy company: Research paradigms for personal and organizational health. In S. L. Sauter, & L. R. Murphy (Eds.), *Organizational risk factors for job stress* (pp. 13–39). Washington, DC: American Psychological Association.

Kaminski, M. (2001). Unintended consequences: Organizational practices and their impact on workplace safety and productivity. *Journal of Occupational Health Psychology, 6,* 127–138.

Katz, D., & Kahn, R. L. (1978). *The social psychology of organizations*. New York: Wiley.

Landsbergis, P. A., Cahill, J., & Schnall, P. (1999). The impact of lean production and related new systems of work organization on worker health. *Journal of Occupational Health Psychology, 4,* 108–130.

Lee, J., & Miller, D. (1999). People matter: Commitment to employees, strategy, and performance in Korean firms. *Strategic Management Journal, 20,* 579–593.

Lepak, D. P., & Snell, S. A. (1999). The human resource architecture: Toward a theory of human capital allocation and development. *Academy of Management Review, 24,* 31–48.

MacDuffie, J. P. (1995). Human resource bundles and manufacturing performance: Organizational logic and flexible production systems in the world auto industry. *Industrial and Labor Relations Review, 48,* 197–221.

McGregor, D. (1960). *The human side of enterprise*. New York: McGraw-Hill.

Miles, M. B. (1965). Planned change and organizational health: Figure and ground. In F. D. Carver & T. J. Sergiovanni (Eds.), *Organizations and human behavior: Focus on schools*. New York: McGraw-Hill.

Osterman, P. (1987). Turnover, employment security, and the performance of the firm. In M. Kleiner (Ed.), *Human resources and the performance of the firm* (pp. 275–317), Madison, WI: Industrial Relations Research Association.

Pil, F. K., & MacDuffie, J. P. (1996). The adoption of high-involvement work practices. *Industrial Relations, 35,* 423–455.

Probst, T. M. (2002, August). Organizational climate: Moderating job insecurity's toll on employee safety. Paper presented at the annual meetings of the Academy of Management, Denver, CO.

Probst, T. M., & Brubaker, T. L. (2001). The effects of job insecurity on employee safety outcomes: Cross-sectional and longitudinal explorations. *Journal of Occupational Health Psychology, 6,* 139–159.

Schaubroeck, J., Ganster, D. C., & Kemmerer, B. E. (1994). Job complexity, "type A" behavior, and cardiovascular disorder: A prospective study. *Academy of Management Journal, 37,* 426–439.

Schuler, R. S., & Jackson, S. E. (1987). Linking competitive strategies with human resource management practices. *Academy of Management Executive, 1,* 207–219.

Shaw, J. D., Delery, J. E., Jenkins, G. D., Jr., & Gupta, N. (1998). An organizational-level analysis of voluntary and involuntary turnover. *Academy of Management Journal, 41,* 511–525.

Shaw, J. D., Gupta, N., & Delery, J. E. (2001). Congruence between technology and compensation systems: Implications for strategy implementation. *Strategic Management Journal, 22,* 379–386.

Shaw, J. D., Gupta, N., & Delery, J. E. (2002a, August). Voluntary turnover and work force performance. Paper presented at the annual meeting of the Academy of Management, Denver, CO.

Shaw, J. D., Gupta, N., & Delery, J. E. (2002b). Pay dispersion and work force performance: Moderating effects of incentives and interdependence. *Strategic Management Journal, 23,* 491–512.

Taylor, H. C., & Russell, J. T. (1939). The relationship of validity coefficients to the practical effectiveness of tests in selection. *Journal of Applied Psychology, 46,* 27–48.

Tsui, A. S., Pearce, J. L., Porter, L. W., & Tripoli, A. M. (1997). Alternative approaches to the employee-organization relationship: Does investment in employees pay off? *Academy of Management Journal, 40,* 1089–1121.

Turner, A. N., & Lawrence, P. R. (1965). *Industrial jobs and the worker.* Boston: Harvard Graduate School of Business Administration.

Walker, J. (1978). Linking human resource management and strategic planning. *Human Resource Planning, 1,* 1–18.

Wright, P. M., Dunford, B. B., & Snell, S. A. (2001). Human resources and the resource-based view of the firm. *Journal of Management, 27,* 701–722.

Wright, P. M., Gardner, T. M., Moynihan, L. M., Park, H. J., Gerhart, B., & Delery, J. E. (2001). Measurement error in research on human resources and firm performance: Additional data and suggestions for future research. *Personnel Psychology, 54,* 875–901.

Wright, P. M., & McMahan, G. C. (1992). Theoretical perspectives for strategic human resource management. *Journal of Management, 18,* 295–320.

Wright, P. M., Smart, D. L., & McMahan, G. C. (1995). Matches between human resources and strategy among NCAA basketball teams. *Academy of Management Journal, 38,* 1052–1074.

Youndt, M. A., Snell, S. A., Dean, J. W., Jr., & Lepak, D. P. (1996). Human resource management, manufacturing strategy, and firm performance. *Academy of Management Journal, 39,* 836–866.

CHAPTER 9

Work Arrangements
The Effects of Shiftwork, Telework, and Other Arrangements

Carlla S. Smith
Lorne M. Sulsky
Wayne E. Ormond

Shiftwork and telework are indisputable components of modern work organizations. Shiftwork changes the time and telework the place work is performed relative to standard organizational practices. For several reasons the use of atypical means of arranging work is increasing: the advent of modern industrial processes, the proliferation of information technology, the globalization of the economy, and the acceptance of alternative lifestyles and career paths. However, beyond the fact that both shiftwork and telework are atypical work arrangements, their goals, historical roots, and research literatures appear to differ greatly, a topic we explore later in this chapter.

Our purpose is not to argue for or against these alternative work arrangements. Rather, our goals are to define these constructs and selectively review their research literatures as they relate to health and safety. We then briefly discuss systematic attempts to improve their effectiveness or caveats about introducing them into work settings. Finally, we explore broader linkages between these arrangements and personal and organizational health. We conclude with an examination of future research needs.

Shiftwork

Of the work arrangements we discuss in this chapter, shiftwork has the longest formal history, arising with the advent of modern industrial technologies, which frequently require 24-hour maintenance. However, industrial processes have not been the only force driving atypical work arrangements. As the demand for 24-hour availability of goods and services rose over the past couple of decades due to globalization and new technologies, the prevalence of shiftwork has likewise increased. In the United States, approximately 20% of all nonagricultural workers experience some form of shiftwork, and 25% of these shiftworkers work at night (U.S. Congress, Office of Technology Assessment, 1991).

Shiftwork can be defined very generally as any arrangement of daily working hours that differs from the standard daylight hours (e.g., 8 A.M. to 5 P.M.). Organizations that use shiftwork typically extend the hours of work past eight hours by using successive teams of workers. Beyond this global definition, shiftwork varies widely and to some extent is unique to each organization's needs. Moreover, shifts may be fixed or rotating. Rotating shifts refer to alternating the time of the shift (e.g., day vs. night) for an individual worker. The number and length of shifts, the presence or absence of night work, the number and placement of rest days between shifts, and the direction and speed of shift rotation (or whether the shift rotates) are only a few of the factors that can describe any shift system. This diversity can be a potential confound in comparing shift effects across studies unless the shift systems are carefully described. As we discuss later in this chapter, researchers have discovered that various types of shift systems may exert differential effects on shiftworkers' health.

Brief Review of the Research: Health and Safety

The medical and scientific communities, in hundreds of research papers and in the popular media, have warned society of the dangers of shiftwork to personal health. Specifically, they maintain that persons who work atypical work hours, especially at night, are at increased risk for physical and psychological dysfunction or disease (Costa, Folkard, & Harrington, 2000; Smith, Folkard, & Fuller, in press). The risk is assumed to originate from the fact that shift-work, which requires sleeping at atypical times of the day or night, disrupts biological functions and sleep.

This biological disruption has very specific origins. Humans are day-oriented or diurnal animals whose biological processes are finely tuned to day activity and night rest. Many physiological processes (e.g., blood pressure, core body temperature), as well as some nonphysiological ones (e.g., mood), are directly influenced by this 24-hour cycle. These 24-hour body rhythms are called circadian rhythms. When the biological system is disrupted by changing sleep and wake periods, neither sleep nor biological processes function optimally. And chronic disruptions can lead to serious disease, the most common of which we discuss next.

Health

The disruption of sleep and the resulting fatigue are the most obvious and immediate health problems associated with shiftwork. Workers who must work and sleep at atypical times of the 24-hour day (e.g., night work and day sleep, rotating work and sleep schedules) frequently report that they have difficulty falling asleep and maintaining sleep for several hours. This disruption occurs because shiftworkers are attempting to function at odds with their circadian rhythms. Shiftworkers' circadian rhythms can adjust to new schedules (e.g., consider transcontinental or international travel); adaptation to an eight-hour shift change typically takes 10–14 days. However, given competing family and social demands, true biological adjustment rarely occurs (Smith et al., in press).

Disruption of the quantity and quality of sleep leads to waking fatigue, which in the longer term may result in the development of chronic fatigue, anxiety, nervousness, and depression, all of which may require medical intervention (Costa, 1996; Costa et al., 2000). It is therefore not surprising that researchers find that psychological and emotional distress often accompany shiftwork (Smith et al., in press). According to Costa (1996), this psychological distress may be the determining factor that persuades 20 to 50% of new shiftworkers to leave shiftwork.

Gastrointestinal disorders, in the form of heartburn, irregular bowel movements, constipation, gas, and appetite disturbances, are the most commonly cited health complaints of shiftworkers, affecting 20 to 75% of shift and night workers (Costa et al., 2000). These complaints may develop into chronic gastrointestinal disease, such as peptic ulcers and chronic gastritis (Costa, 1996). Well-designed longitudinal studies (e.g., Angersbach et al., 1980) and review articles (e.g., Costa, 1996) have indicated that the amount of night work, not just shiftwork, is the critical factor in the development of gastrointestinal disease.

Scientists speculated that gastrointestinal disorders may be more prevalent among shiftworkers, and night workers in particular, because these workers have limited access to healthy food (i.e., many restaurants and stores are closed from midnight to 6 A.M.). Shiftworkers' irregular hours also encourage sporadic dietary habits. However, research has largely discredited this theory, showing few differences between the nutritional intake of day workers and shiftworkers (e.g., Lennernaes, Hambraeus, & Åkerstedt, 1994). Although this issue has not been fully resolved, the biological effects of circadian disruption or sleep deficit may be contributing factors (Vener, Szabo, & Moore, 1989).

Cardiovascular disorders are another class of serious medical conditions that appear to differentially affect shiftworkers. In a 15-year longitudinal study, Knutsson, Åkerstedt, Jonsson, and Orth-Gomer (1986) found increased risk factors of cardiovascular disease (e.g., smoking) in shiftworkers as well as increased morbidity from cardiovascular disease as their years in shiftwork increased. Occupations with a high percentage of shiftworkers also are at greater risk for heart disease (Costa et al., 2000). In their meta-analyses of epidemiological studies on shiftwork and heart disease, Boggild and Knutsson (1999) recently demonstrated that shiftworkers are at a 40% greater risk for cardiovascular mortality or morbidity than are comparable day workers.

The etiology of cardiovascular disorders, like gastrointestinal disorders, is unknown (Åkerstedt & Knutsson, 1997). However, many cardiovascular risk factors, such as gastrointestinal symptoms, sleep dysfunction, and smoking, are common in shiftworkers. Shiftwork may also function as a stressor, increasing the stress response over time; elevated blood pressure, heart rate, cholesterol levels, and changes in glucose and lipid metabolism are the unfortunate consequences (Costa, 1996).

The impact of shiftwork on women's reproductive functions has garnered interest as women have increasingly entered previously male-dominated occupations, such as industrial jobs. Because shiftwork disrupts cyclic processes in the body, such as sleep and digestion, its negative influence on the female menstrual cycle is predictable. For example, compared to their day-working counterparts, female shiftworkers suffer irregularities in cycle length or pattern (Hatch, Figa-Talamanca, & Salerno, 1999), lower rates of pregnancies and deliveries (Nurminen, 1989), and premature delivery and low birth-weight infants (Nurminen, 1989).

Safety

Safety has long been an issue for shiftworkers because shiftwork is prevalent in jobs that require handling toxic or dangerous industrial processes (e.g., the steel and chemical industries) or jobs that involve protecting public safety and health (e.g., police, medical professions). Due to the physical and psychological dysfunction that usually accompany shiftwork, researchers have warned that serious errors and injury may occur, especially on the night shift (Costa, 1996). Folkard and his colleagues (Folkard, Åkerstedt, Macdonald, Tucker, & Spencer, 2000; Smith, Folkard, & Poole, 1994), by carefully controlling for a priori risk (i.e., work conditions) across shifts, have conclusively shown that more frequent accidents and injuries occur on the night shift. In addition, they found that the accidents and injuries occurring on the night shift are more serious (i.e., requiring medical attention rather than first aid; Smith et al., 1994). Relative to day workers, workers on the night shift are also more often involved in automotive accidents while driving home after work (Monk, Folkard, & Wedderburn, 1996). The circadian malaise, sleep loss, and fatigue common to shiftworkers are undoubtedly to blame.

Studies such as Folkard et al. (2000) and Smith et al. (1994) are quite valuable because few studies have been able to control for a priori accident risk across shifts. In fact, different shifts are often quite diverse in terms of the amount of supervision, experience of the workers, degree of workload, and type of work, all of which can influence accident risk. For example, supervision is frequently reduced at night, whereas workers on the night shift tend to be less experienced (at least in the United States). The day shift typically has the heaviest workload (with the largest workforce), and maintenance and repair activities are usually reserved for the night shift. These differences in workload and type of work activities across shifts can obscure true on-shift accident rates (Costa et al., 2000; Smith, Silverman, et al., 1997).

Improving Shiftwork Effectiveness

In this section we introduce some of the most commonly used methods to increase shiftwork effectiveness. The interested reader should consult more comprehensive treatments of this topic (e.g., Knauth, 1993; Smith et al., in press).

The optimal design of shift systems to promote circadian harmony and health is probably the most frequently cited method to increase shiftwork effectiveness. Although shiftwork researchers agree that no best shift system exists, they acknowledge that some systems are preferable to others.

From a health perspective, fixed shifts are preferable to rotating shifts because shiftworkers are not required to change their hours of work and sleep. However, permanent night workers rarely retain adaptation to their schedules because they typically revert to a diurnal (day) orientation on their rest days to engage in family or social activities. In this case night workers create their own rotating shift schedule, thus requiring readaptation to night work after every rest break.

Rotating shifts can assume any shift speed, from very rapid (two to three days per shift) to very slow (three to four weeks per shift). One of the most common rotating schedules in the United States is weekly rotating, in which shiftworkers change their schedule every seven days. Unfortunately, this schedule is one of the worst from a circadian (health) perspective because just as the shiftworker's circadian rhythms are beginning to adapt to the new schedule, adaptation must begin again. (Adaptation to an eight-hour shift change typically takes from 10 to 14 days, if it occurs at all.) Because total adaptation so rarely occurs, even on very slowly rotating shift schedules, shiftwork experts frequently recommend very rapidly rotating shift schedules. Such schedules allow shiftworkers to retain their diurnal orientation, while minimizing blocks of night work (i.e., only two to three days) (Folkard, 1992; Knauth, 1993). Any schedule that minimizes night work should protect the worker's health.

One of the most popular trends in shift scheduling is to combine shiftwork with the compressed workweek (CWW), a topic we revisit later in this chapter. CWW typically results in shift length being extended from eight to 12 hours. This type of arrangement is very popular with shiftworkers because it allows longer blocks of free time. Unfortunately, the effects of fatigue may increase with increased hours of work, especially if night shifts are involved. Surprisingly, empirical investigations of the impact of 12-hour shift systems on health and sleep have been mostly positive; compared to workers on eight-hour shifts, workers on 12-hour shifts have not generally shown greater decrements in sleep, health, or well-being (Smith, Folkard, Tucker, & Macdonald, 1998). However, these studies have not considered some factors, such as the hardships 12-hour shifts may exert on older or female shiftworkers. Older workers, who may experience fatigue or medical problems more frequently than their younger counterparts, could be at a disadvantage when working extended hours. Due to their home and child-care responsibilities, female shiftworkers under CWW could face increased conflicts between home and work schedules. Future research is needed to resolve these issues.

On a smaller scale, organizations may add bright lights to shiftworkers' work areas to aid adaptation to shift changes. Several years ago, researchers found that exposure to very bright light (2,500 lux; indoor illumination is about 500 lux) suppresses the nocturnal secretion of the brain hormone melatonin and therefore delays sleep. These effects have been documented in work settings with shiftworkers (e.g., Stewart, Hayes, & Eastman, 1995). However, bright lights have generally not been popular with organizations or their shiftworkers as a method to enhance shiftwork effectiveness. Installing and maintaining such lighting systems can be expensive for organizations. Also, shiftworkers must adhere to a strict schedule of exposure to bright light over time, which may prove to be impractical and difficult to maintain (Smith et al., in press).

Workers have long used sleep aids to enhance sleep and therefore increase on-shift effectiveness. However, most sleep aids require a doctor's prescription and are plagued with serious side effects (Walsh, 1990). However, one of the newest pharmacological sleep aids, melatonin, is an over-the-counter medication that has few, if any, short-term side effects. Melatonin is a naturally occurring brain hormone that initiates sleep. In pill form this natural hormone has proven to be effective and safe in promoting sleep and in aiding adaptation to schedule changes. However, melatonin is not a controlled drug, so its purity cannot be ensured, and its interactions with other drugs are unknown. Because the chronic effects of melatonin are also unknown, shiftworkers should take melatonin for extended periods only under medical supervision (Arendt & Deacon, 1997). Additional research must be conducted to determine melatonin's full impact on the human body.

Summary

Shiftwork is any arrangement of work hours that differs from the typical daylight hours. This type of arrangement developed from the need to manage modern industrial technologies and globalization. Researchers have discovered that alternating sleep and work hours disrupts the body's biological rhythms and sleep, which predisposes shiftworkers to develop serious medical conditions, such as psychological, gastrointestinal, cardiovascular, and reproductive disorders. Work-related safety is also an issue because many shiftworkers' jobs involve either exposure to dangerous industrial chemicals and processes or the protection of public safety and health. The evidence suggests that night work, not shiftwork per se, is the most critical factor in the development of disease and work-related accidents and injuries.

Regardless of its negative impact on health and safety, shiftwork is considered to be a necessary component of many modern work organizations. Therefore, scholars have proposed various methods to decrease shiftwork's negative health effects and increase its effectiveness. The optimal design of shift systems to promote circadian function is probably the most frequently cited method. Although shiftwork experts agree that no best shift system exists, some systems are definitely better than others. CWW has recently been combined with shiftwork schedules. This combination is very popular with shiftworkers because it allows them longer blocks of free time. Surprisingly, studies suggest that shiftwork and CWW are not generally deleterious to health. On a smaller scale, companies have used bright lighting (2,500 lux or greater) to illuminate work environments during the night shift, thus delaying the onset of sleep. Melatonin, one of the newest pharmacological sleep aids, has proven to be effective in initiating sleep. Organizations have successfully used bright lights and melatonin to help shiftworkers adjust to schedule changes, although both have disadvantages.

Telework and Other Work Arrangements

Since 1980, profound changes in the way work is performed and structured have occurred. One trend is an increase in telework. Although a variety of definitions exist, *telework* is generally understood to be any work performed outside of the traditional office environment or workplace that employs communication technology to keep coworkers and other relevant constituents connected (e.g., clients, customers) (Gray, Hodson, & Gordon, 1993). These employees may work at home or at a satellite office or location. Teleworkers may also be on the road, connected to their organization via a computer and modem (Olson & Primps, 1984). Because teleworkers often work from home and do not have to commute to work, telework is sometimes referred to as telecommuting (Nilles, 1994).

Teleworkers potentially employ a variety of technologies in their everyday work, including telephones, faxes, and e-mail, as well as other more sophisticated technologies (e.g., video conferencing) (Finholt, 1997). Technology becomes central to the teleworker as a mechanism for maintaining contact with coworkers and other relevant constituents. Technological advancements in telecommunications and information technologies have made it easier for workers to work out of remote locations or even their own homes (Hone, Kerrin, & Cox, 1998).

Organizations have become increasingly interested in this alternative to the traditional office environment for several reasons. One reason is financial; companies can significantly reduce overhead costs, including the costs associated with office space (Venkatesh, 2000). Satellite offices, for example, are typically much less expensive to operate than a central office (Olson & Primps, 1984), and home offices reduce organizational costs even further. Also, office space has increased in cost over the years, whereas costs associated with information technologies have decreased (Hill, Miller, Weiner, & Colihan, 1998).

A second reason for the increasing popularity of telework stems from the difficulties faced by dual-career families—especially those with children—in balancing work and home responsibilities (e.g., Kelloway & Gottlieb, 1998). Teleworking from home allows one of the parents to be at home, thus obviating (or lessening) the need for child care and other logistical difficulties arising from balancing home and work responsibilities.

Third, telework is sometimes perceived as a method of attracting and retaining top-level employees, because this work arrangement offers prospective and current employees both scheduling flexibility and work autonomy (Ellison, 1999). The fact that telework reduces or eliminates commuting time and related costs is a further incentive for choosing this type of work arrangement (Claes, 2000).

Given the potential advantages associated with telework, it is not surprising that the number of North American telework employees has surged upward. For example, Hamilton (1987) noted that an estimated 200 U.S. companies experimented with a variant of telework. In terms of individual workers, Wells (1997) reported U.S. prevalence estimates ranging from 9 million to 42 million.

Defining Telework

Unfortunately, no conceptual or operational consensus exists among telework researchers concerning the definition of telework, which has contributed greatly to the variable prevalence estimates (Hone et al., 1998). For example, some definitions require the use of information technologies, whereas others do not. Moreover, some definitions stipulate the home as the primary workplace, whereas others define telework more broadly to include any work performed outside the traditional office (Ellison, 1999; Hone at al., 1998). Further complicating matters, some researchers conceptually distinguish telework from telecommuting (Ellison, 1999). Without a clear consensus on the definition, consolidating the empirical research examining the relations among telework and various health-related criteria is difficult. As the reader will soon discover, the research presents a somewhat mixed picture regarding the potential effects of telework, and this may be at least in part an unfortunate consequence of definitional imprecision.

Perhaps one of the most insidious problems in defining telework is the fact that many workers often spend a portion of their week engaged in telework. Is there a cutoff, such that individuals must spend a fixed proportion of their time outside the main office before researchers consider them to be teleworkers? Typically, telework research has proceeded in a quasi-experimental manner, categorizing workers as either teleworkers or nonteleworkers (based upon some arbitrary cutoff) (Hone et al., 1998) and comparing the two groups on various outcome variables (e.g., stress, productivity). However, this practice is tantamount to categorizing what might be best considered a continuous variable.

Hone et al. (1998) suggested a reconceptualization of telework, whereby telework is not considered a categorical variable. Specifically, they proposed that telework be conceptualized according to two core dimensions: (1) the proportion of time the employee spends outside the traditional workplace and (2) the extent to which the employee uses information technologies while teleworking. For example, relations between these continuous variables and selected outcomes may be examined, without categorizing individuals as teleworkers based upon arbitrary cutoffs (e.g., how many days the worker spends at home). Beyond cleaning up the construct, the proposed reformulation permits the development of theoretically guided links between specific telework dimensions and work outcomes. For instance, proportion of time spent teleworking might predict employees' feelings of isolation, and extent of technology use might predict eye strain (Hone et al., 1998). In sum, future research may greatly benefit from reconceptualizing the telework construct: a conceptual definition recognizing the simple fact that telework is multidimensional and not an all-ornone proposition.

Selected Research Findings: Stress and Health

Telework research presents a mixed array of findings. Again, conceptual and methodological issues likely contribute to the varied results across studies. However, little empirical research has investigated telework (Hill et al., 1998). Thus, definitional issues notwithstanding, too few studies may exist to allow researchers to draw conclusions.

In terms of stress and health, a commonly mentioned variable is social isolation: the idea that teleworkers may feel isolated, leading to stress and health-related strains (Claes, 2000). This finding has been commonly cited as the most serious antecedent of teleworker stress (Venkatesh, 2000). However, the limited empirical research does not suggest that teleworkers necessarily suffer from isolation. For example, Trent, Smith, and Wood (1994) found that regular trips to the office may decrease teleworking employees' feelings of isolation. These results underscore the problems arising when telework is transformed into a categorical variable. That is, the way telework is currently treated or defined ignores the possibility that individuals may spend a considerable amount of time at the central office as well as in remote locations. Indeed, Ramsower (1985) found (in a quasi-experimental study) that part-time teleworkers did not report significant decreases in face-to-face communications with colleagues, whereas full-time teleworkers reported significant decreases in this regard. Interestingly, Ramsower also found that even full-time teleworkers reported that the use of technology (e.g., e-mail) somewhat mitigated against feelings of isolation.

Kurland and Bailey (1999) suggested that where telework occurs is an important factor when researchers consider social isolation. Teleworkers who work from satellite offices have opportunities for social contact unavailable to those working entirely from home. Mobile teleworkers may have considerable social contact, depending upon the extent to which their work requires face-to-face interactions.

One area of increasing concern is the implication of telework for work-home balance. Kelloway and Gottlieb (1998) found that women reported significantly less stress when they perceived they had scheduling flexibility, a common feature of telework. The researchers indicated that the flexible scheduling that teleworking affords can assist women in balancing the dual roles of worker and mother, thereby reducing stress from role overload. Kelloway and Gottlieb noted that this result is consistent with Karasek's demand-control stress model (1979), insofar as telework increases the amount of employees' perceived control over their work (see also

Hone at al., 1998). Likewise, Hill, Hawkins, and Miller (1996) reported that the majority of teleworkers (i.e., 75% of 249 respondents to an electronic survey) indicated positive spillover as a result of their working arrangements. That is, most teleworkers reported that telework allowed them to better balance work and home responsibilities.

Other research and theory, however, suggest that telework may have negative implications for increasing role stress when telework occurs at home. For instance, Olson and Primps (1984) discovered that the lack of separation between work and home created significant stress due to the employee's continuous juggling of work and home pressures throughout the workday. Olson and Primps suggested that the implications of telework for work-home balance may ultimately depend upon whether children are at home during the workday. Standen, Daniels, and Lamond (1999) discussed a number of ways in which telework may lead to negative spillover between the work and home. For example, they proposed that telework invites role conflict to the extent that mutual intrusions arise between work and home responsibilities. Moreover, given the lack of physical separation between work and home, less psychological separation exists between the employee's two roles. This lack of psychological separation may make it more difficult for the individual to dismiss work stressors while engaging in family-related activities and also more difficult to filter out home-related stressors while working.

Perhaps the most informative research involving telework has examined both teleworker and prospective teleworker perceptions of stress associated with telework. For example, Hill et al.'s (1998) analyses of self-report survey data suggested that teleworkers believe their work arrangements do not promote stress, with the possible exception of stress emanating from negative work-home spillover. Olson and Primps's (1984) study revealed that some teleworkers perceive that telework may have negative effects on promotion opportunities. Claes (2000) collected questionnaire data from prospective teleworkers and found that respondents indicated two potential disadvantages associated with telework: social isolation and role ambiguity. Stress resulting from role ambiguity may be a serious problem if information technologies are insufficient for clarifying teleworkers' roles and responsibilities and if teleworkers have little physical contact with colleagues (cf. Shamir & Salomon, 1985).

Introducing Telework: Stress-Related Considerations

When an organization decides to introduce telework as a type of work arrangement, a number of issues and concerns inevitably arise. In particular, the introduction of telework may inadvertently introduce unwanted work stress, leading to decrements in both productivity and employee health.

One concern involves managers' readiness and willingness to adopt telework. Managers do not have the same level of everyday contact with teleworkers, which could make tracking and controlling performance relatively difficult (Olson & Primps, 1984). This situation could increase the stress levels of managers, affecting their willingness to support new telework initiatives. Indeed, lack of management support due to a loss of control over workers has been cited as one of the major roadblocks to the adoption of telework in some organizations (Ellison, 1999). Motivating and effectively leading employees who are not physically present also introduces new challenges for managers and other organizational leaders.

The introduction of telework also has implications for teleworkers' performance evaluations. Specifically, teleworkers may be evaluated more on the results (i.e., outcomes) of their work and less on the process of work (Kurland & Bailey, 1999). Given the teleworker's reliance on computers, the computer may be set up to monitor performance and count what managers deem to be objective indices of performance. This type of monitoring is commonly referred to as electronic performance monitoring (EPM). Beyond concerns that focus on countable outcomes rewards quantity as opposed to quality of performance, close scrutiny of individual behaviour through EPM has generally been assumed to be stressful (e.g., see Aiello & Kolb, 1995).

Another potential concern is the stress associated with the use of complex technologies. Without proper orientation and training, teleworkers may have difficulty effectively communicating with colleagues or coworkers, and this may increase both role ambiguity and feelings of isolation. Recent research has begun to explore how best to train teleworkers on new and emerging technologies (e.g., Venkatesh, 2000). Any organization contemplating the inclusion of telework must consider the training and development component to ensure maximal job satisfaction and productivity, while minimizing the effects of stress.

Employees need training in how to cope effectively with any perceived stress resulting from telework. Effective telework coping strategies have been shown to predict increases in self-reports of psychological health (Norman, Collins, Conner, Martin, & Rance, 1995). Managers of teleworkers also require special training (Caudron, 1992). Managers must learn how to supervise employees who are not physically present, including how to motivate and provide corrective feedback.

Another important issue is the selection of employees for telework. Little is known about the special skills, abilities, and other characteristics necessary for being an effective teleworker. Research has suggested that the teleworker must be self-motivated, with excellent time-management skills, and must be able to work alone with little direct supervision (Armstrong-Stassen, 1998). Selecting individuals who are poorly equipped to handle the unique demands of this work arrangement will almost certainly lead to increases in worker stress and decreases in employee health and productivity.

A further concern is ergonomic: What safeguards need to be put in place to ensure that a home or remote office is set up ergonomically (see Gottlieb, Kelloway, & Barham, 1998)? When the employee is working at home, who decides upon the correct lighting and placement of key work components (e.g., reading lights, computer monitor)? If organizations are not vigilant with respect to the physical workspace of teleworkers, stress and health strains are the likely outcomes. Relatedly, organizations must be aware of any relevant legislation pertaining to telework (cf. Claes, 2000), especially if a poorly designed workspace contributes to an accident or injury.

Finally, organizations should consider the kinds of jobs suitable for telework. For instance, a university course can be taught exclusively through electronic means; whether this is a good idea, however, is another matter. The majority of current telework can be classified as either professional (e.g., computer programmer) or clerical (e.g., data entry). Beyond this distinction, however, a framework is needed to determine what jobs are most suitable for telework and whether specific jobs may be modified to make telework possible (Armstrong-Stassen, 1998; Claes, 2000). A recipe for increasing worker stress is to choose jobs that are not suitable for telework or to redesign jobs (so that they may be performed at remote sites) in ways that make the jobs less challenging and interesting.

Other Work Arrangements

One relatively new work arrangement is the compressed workweek (CWW). Armstrong-Stassen (1998) reviewed current research on this scheduling variant (the four-day, 40-hour week is the most common) and reported mixed results for the relations between CWW and employee fatigue. Fatigue effects may be more pronounced in older workers, and women may be more likely to experience time-related stress as a result of CWW. Overall, little is known about how to choose an optimal compressed schedule. Optimal CWW schedules likely depend upon the nature of the job as well as characteristics of the workforce (e.g., predominantly older) (Dunham, Pierce, & Castaneda, 1987). Regardless of the particular schedule adopted, however, employee acceptance of compressed schedules is critical if the arrangement is ultimately going to be effective (cf. Latack & Foster, 1985). As we discussed previously, these issues become even more salient when CWW is combined with shiftwork.

Another popular alternative work arrangement is flextime. Flextime permits employees to control the time they start and leave work, although they typically must work a certain fixed number of hours (and they usually must be present during certain critical periods of the workday) (Armstrong-Stassen, 1998). Gottlieb et al. (1998) reported that 73 percent of U.S. employers include flextime as a scheduling option. Although earlier reviews were generally quite supportive (Golembiewski & Proehl, 1978), Armstrong-Stassen's (1998) review highlighted more recent research revealing that flextime does not always lead to positive outcomes.

In terms of stress and health-related outcomes, Kelloway and Gottlieb (1998) found that flextime was associated with reductions in self-reported stress. Also, Barling and Barenburg (1984), in a sample of working mothers, discovered that flextime was associated with reductions in role conflict and self-reported depression.

Future flextime research should examine the implications of various flextime parameters (e.g., how much freedom employees have to set their schedules) for organizationally and personally relevant strains (e.g., employee health). In addition, researchers should examine the implications of flextime for the performance of teams (Armstrong-Stassen, 1998). For example, the unavailability of certain team members at specific times may lead to increased stress levels for those team members who are at work. Moreover, does flextime increase managerial stress? Employees may be present at work when managers are not (e.g., early or late), and employees may not be there when managers require their assistance. The implications of flextime for managerial stress await future research.

Summary

Telework represents an alternative work arrangement that may have substantial benefits for both employees and organizations. However, our review points to a number of stress-related issues associated with this type of work. Given the lack of consensus regarding the meaning and measurement of telework, drawing definitive conclusions about the links between telework and employee stress and health is difficult. However, the available evidence does portray telework as a promising work arrangement with plenty of what we might call stress-related land mines. It is vital that organizations appreciate the implications of telework for a variety of personnel and organizational issues, including staffing, training, performance appraisal, motivation, and job design or redesign. To implement telework and ignore such issues is to invite the strong possibility that the human costs will far outweigh any financial savings.

Similar to telework, the other work arrangements discussed in this chapter, CWW and flextime, do not have a sufficient research literature on worker stress and health for us to draw definitive conclusions. CWW may be problematic, especially for older and women workers, both of whom probably have difficulty dealing with extended work hours. Flextime, a popular work arrangement, may reduce stress and strains for some types of workers who require flexibility in their daily schedules.

General Discussion

Our brief overview of the research literatures on shiftwork, telework, CWW, and flextime reveals that these work arrangements can have an impact on worker health and safety. Decades of shiftwork research have conclusively demonstrated that the experience of shiftwork, especially night work, is associated with the development of health strains, accidents, and injuries. The research literatures on telework and the other work arrangements are not as extensive, but they suggest that these arrangement are related to perceptions of stress and health strains.

Implications for Personal and Organizational Health

In the beginning of this chapter, we stated that beyond being different ways to structure or schedule the workday, shiftwork and telework appear to differ in terms of their goals, historical roots, and research literatures. Shiftwork grew out of organizational and societal goals to provide products and services to a global (24-hour) population. In this case, personal (workers') goals are definitely subordinate to organizational and societal goals. However, telework and similar work arrangements grew out of both organizational and personal goals. For example, the organization's goals are to reduce costs and to attract and retain valued employees. Personal goals are to better manage dual-career and alternative lifestyles. Because of the serious health implications of altering the body's cyclic functions, shiftwork researchers adopted a medical model (Wood, 1986) to guide their studies. This type of model focuses on the presence or absence of disease; specifically, disease develops when body functions are disrupted. Telework and similar arrangements concentrate on the psychosocial indicators of disease (e.g., self-reported stress, stressors, and strains). This "softer" medical model's emphasis is on the presence or absence of negative psychosocial reactions to work. As a consequence of their somewhat differing focus, shiftwork's research literature emphasizes physical health and safety, whereas telework's literature emphasizes psychological health.

But are these two arrangements really that different? Both are variants of the medical model, so both emphasize the presence or absence of disease, physiological or psychosocial. Both acknowledge health as simply the absence of disease, yet the absence of disease does not necessarily indicate health (e.g., "flourishing and a sense of well-being," see Hoffman & Tetrick, Chapter One of this volume). The techniques to increase shiftwork effectiveness and the caveats about introducing telework discussed earlier in this chapter represent attempts to reduce disease. These attempts have met with limited or mixed success, and some are merely speculative. Yet how can scientists hope to design and implement work arrangements to enhance health when disease has not been adequately reduced? Both workers and organizations must address the manifestations of disease in themselves (e.g., weight control) and the workplace (e.g., hazard-free workstations), respectively, before healthy organizations can become a reality. This issue is undoubtedly one of the greatest challenges now facing behavioral scientists.

Directions for Future Research

Throughout this chapter we have mentioned topics that future research should address. We continue that theme. However, the topics we discuss here are, in our opinion, of paramount importance in work arrangements research.

Historically, well-established theory has not guided shiftwork research. Although scholars have offered several models of the proposed relations between shiftwork and strains, most served merely as heuristic frameworks (e.g., Monk, 1988). Few researchers attempted to test empirically all or parts of their model (cf., Smith et al., 1999). The relation between the experience of shiftwork and health strains is undisputed. However, if scientists hope to circumvent the development of health strains in shiftworkers, they must understand the mediating processes (e.g., disruption of sleep or family life, heavy workload, coping efforts) between the experience of shiftwork and the development of illness. For example, if shiftworkers cope unsuccessfully with the disruption of family life caused by working night shifts, and if unsuccessful coping is directly or indirectly related to illness onset (e.g., gastrointestinal symptom), scholars should develop interventions designed to improve coping. This goal, however, can only be achieved through the systematic development and test of theory.

Telework research can also benefit from the development of theory. According to Hone et al. (1998), two dimensions are crucial in defining telework: the amount of time the employee spends outside the traditional workplace and the extent that he or she uses information technology. These dimensions can be used to develop a process model of telework and health strains. For example, the amount of time the employee spends in telework affects feelings of social isolation and stress, which in turn influence perceptions of psychological health (e.g., depression). The extent of technology use, however, may cause eye and back strain, and physical stress that in turn influences physical health (e.g., the development of musculoskeletal disorders). Again, optimal methods to introduce and manage telework cannot be determined without the systematic development and test of theory.

Finally, scientists know relatively little about the manner in which these different work arrangements interact or the prevalence of multiple arrangements in industry. One exception is the existing data on combining shiftwork with CWW, which are surprisingly positive. However, other combinations may not fare as well. For example, consider a night worker who is also a teleworker on a compressed schedule. The social isolation and stress (compounded by working at night) might create a work environment that encourages fatigue and human error, especially at the end of the shift cycle. Future research must address such issues if the goals of attaining healthy workers and organizations are ever to be achieved.

References

Aiello, J. R., & Kolb, K. J. (1995). Electronic performance monitoring: A risk factor for workplace stress. In S. L. Sauter & L. R. Murphy (Eds.), *Organizational risk factors for job stress* (pp. 163–179). Washington, DC: American Psychological Association.

Åkerstedt, T., & Knutsson, A. (1997). Cardiovascular disease and shift work. *Scandinavian Journal of Work, Environment, and Health, 23,* 241–242.

Angersbach, D., Knauth, P., Loskant, H., Karvonen, M. J., Undeutsch, K., & Rutenfranz, J. (1980). A retrospective cohort study comparing complaints and diseases in day and shift workers. *International Archives of Occupational and Environmental Health, 45,* 127–140.

Arendt, J., & Deacon, S. (1997). Treatment of circadian rhythm disorders— melatonin. *Chronobiology International, 14,* 185–204.

Armstrong-Stassen, M. (1998). Alternative work arrangements: Meeting the challenges. *Canadian Psychology, 39,* 108–123.

Barling, J., & Barenbrug, A. (1984). Some personal consequences of "flexitime" work schedules. *Journal of Social Psychology, 123,* 137–138.

Boggild, H., & Knutsson, A. (1999). Shift work, risk factors and cardiovascular disease. *Scandinavian Journal of Work, Environment, and Health, 25,* 85–99.

Caudron, S. (1992, July). Working at home pays off. *Personnel Journal, 73,* 52–55.

Claes, R. (2000). Meaning of atypical work: The case of potential telehomeworkers. *European Review of Applied Psychology, 50,* 27–36.

Costa, G. (1996). The impact of shift and night work on health. *Applied Ergonomics, 27,* 9–16.

Costa, G., Folkard, S., & Harrington, J. M. (2000). In P. Baxter, P. H. Adams, T. C. Aw, A. Cockcroft, & J. M. Harrington (Eds.), *Hunter's diseases of occupations* (9th ed.). London: Arnold.

Dunham, R. B., Pierce, J. L., & Castaneda, M. B. (1987). Alternative work schedules: Two field quasi-experiments. *Personnel Psychology, 40,* 215–242.

Ellison, N. B. (1999). Social impacts: New perspectives on telework. *Social Science Computer Review, 17,* 338–356.

Finholt, T. A. (1997). The electronic office. In C. L. Cooper & D. M. Rousseau (Eds.), *Trends in organizational behavior* (Vol. 4, pp. 29–42). New York: Wiley.

Folkard, S. (1992). Is there a "best compromise" shift system? *Ergonomics, 35,* 1453–1463.

Folkard, S., Åkerstedt, T., Macdonald, I., Tucker, P., & Spencer, M. (2000). Refinement of the three-process model of alertness to account for trends in accident risk. In S. Hornberger, P. Knauth, G. Costa, & S. Folkard (Eds.), *Shiftwork in the twenty-first century: Challenge for research and practice.* Frankfurt am Main: Peter Lang.

Golembiewski, R. T., & Proehl, C. W. (1978). A survey of the empirical literature on flexible work hours: Character and consequences of a major innovation. *Academy of Management Review, 3,* 837–853.

Gottlieb, B. H., Kelloway, E. K., & Barham, E. (1998). *Flexible work arrangements.* New York: Wiley.

Gray, M., Hodson, N., & Gordon, G. (1993). *Teleworking explained.* New York: Wiley.

Hamilton, C. A. (1987). Telecommuting. *Personnel Journal, 166*(4), 91–101.

Hatch, M. C., Figa-Talamanca, I., & Salerno, S. (1999). Work stress and menstrual patterns among American and Italian nurses. *Scandinavian Journal of Work, Environment, and Health, 25,* 144–150.

Hill, E. J., Hawkins, A. J., & Miller, B. C. (1996). Work and family in the virtual office: Perceived influences of mobile telework. *Family Relations, 45,* 293–301.

Hill, E. J., Miller, B. C., Weiner, S. P., & Colihan, J. (1998). Influences of the virtual office on aspects of work and work/life balance. *Personnel Psychology, 51,* 667–683.

Hone, K. S., Kerrin, M., & Cox, T. (1998). CORDit: A multidimensional model for evaluating the psychological impact of teleworking. *European Psychologist, 3,* 227–237.

Karasek, R. (1979). Job demands, job decision latitude, and mental strain: Implications for job redesign. *Administrative Science Quarterly, 24,* 285–308.

Kelloway, E. K., & Gottlieb, B. H. (1998). The effect of alternative work arrangements on women's well-being: A demand-control model. *Women's Health: Research on Gender, Behavior, and Policy, 4,* 1–18.

Knauth, P. (1993). The design of shift systems. *Ergonomics, 36,* 15–28.

Knutsson, A., Åkerstedt, T., Jonsson, B. G., & Orth-Gomer, K. (1986). Increased risk of ischemic heart disease in shift workers. *Lancet, 2,* 89–92.

Kurland, N. B., & Bailey, D. E. (1999). Telework: Advantages and challenges of working here, there, anywhere, anytime. *Organizational Dynamics, 28*(2), 53–67.

Latack, J. C., & Foster, L. W. (1985). Implementation of compressed work schedules: Participation and job redesign as critical factors for employee acceptance. *Personnel Psychology, 38,* 75–92.

Lennernaes, M., Hambraeus, L., & Åkerstedt, T. (1994). Nocturnal eating and serum cholesterol of three-shift workers. *Scandinavian Journal of Work, Environment, and Health, 20,* 401–406.

Monk, T. H. (1988). Coping with the stress of shiftwork. *Work and Stress, 2,* 169–172.

Monk, T. H., Folkard, S., & Wedderburn, A. I. (1996). Maintaining safety and high performance on shiftwork. *Applied Ergonomics, 27,* 17–23.

Nilles, J. M. (1994). *Making telecommuting happen: A guide for telemanagers and telecommuters.* New York: Van Nostrand Reinhold.

Norman, P., Collins, S, Conner, M., Martin, R., & Rance, J. (1995). Attributions, cognitions, and coping styles: Teleworkers' reactions to work-related problems. *Journal of Applied Social Psychology, 25,* 117–128.

Nurminen, T. (1989). Shift work, fetal development, and course of pregnancy. *Scandinavian Journal of Work, Environment, and Health, 15,* 395–403.

Olson, M., & Primps, S. (1984). Working at home with computers: Work and non-work issues. *Journal of Social Issues, 40,* 97–112.

Ramsower, R. M. (1985). *Telecommuting: The organizational and behavioral effects of working at home.* Ann Arbor, MI: UMI Research Press.

Shamir, B., & Salomon, I. (1985). Work-at-home and the quality of working life. *Academy of Management Review, 10,* 455–464.

Smith, C. S., Folkard, S., & Fuller, J. A. (in press). Shiftwork and working hours. In J. Quick & L. Tetrick (Eds.), *Handbook of occupational health psychology.* Washington, DC: American Psychological Association.

Smith, C. S., Robie, C., Folkard, S., Barton, J., Macdonald, I., Smith, L., Spelten, E., Totterdell, P., & Costa, G. (1999). A process model of shiftwork and health, *Journal of Occupational Health Psychology, 4,* 207–218.

Smith, C. S., Silverman, G. S., Heckert, T. M., Brodke, M. H., Hayes, B. E., Silverman, M. K., & Mattimore, L. K. (1997). Shift-related differences in industrial injuries: Application of a new research method in fixed-shift and rotating-shift systems. *International Journal of Occupational and Environmental Health, 3,* 46–52.

Smith, L., Folkard, S., & Poole, C. J. (1994). Increased injuries on night shift. *Lancet, 244,* 1137–1139.

Smith, L., Folkard, S., Tucker, P., & Macdonald, I. (1998). Work shift duration: A review comparing eight hour and twelve hour shift systems. *Occupational and Environmental Medicine, 55,* 217–229.

Standen, P., Daniels, K., & Lamond, D. (1999). The home as a workplace: Work-family interaction and psychological well-being in telework. *Journal of Occupational Health Psychology, 4,* 368–381.

Stewart, K. T., Hayes, B. C., Eastman, C. I. (1995). Light treatment for NASA shiftworkers. *Chronobiology International, 12,* 141–151.

Trent, J. T., Smith, A. L., & Wood, D. L. (1994). Telecommuting: Stress and social support. *Psychological Reports, 74,* 1312–1314.

U.S. Congress, Office of Technology Assessment. (1991). Biological rhythms: Implications for the worker (OTA-BA-463). Washington, DC: U.S. Government Printing Office.

Vener, K. J., Szabo, S., & Moore, J. G. (1989). The effect of shift work on gastrointestinal (GI) function: A review. *Chronobiologia, 16,* 421–439.

Venkatesh, V. (2000). Creating an effective training environment for enhancing telework. *International Journal of Human-Computer Studies, 52,* 991–1005.

Walsh, J. K. (1990). Using pharmacological aids to improve waking function and sleep while working at night. *Work and Stress, 4,* 237–243.

Wells, S. (1997, August 17). For stay-home workers, speed bumps on the telecommute. *New York Times,* p. C1.

Wood, P.H.N. (1986). Health and disease and its importance for models relevant to health research. In B. Z. Nizetic, H. G. Pauli, & P. G. Svensson (Eds.), *Scientific approaches to health and health care* (pp. 57–70). Copenhagen: World Health Organization.

CHAPTER 10

The Work and Family Interface

Conflict, Family-Friendly Policies, and Employee Well-Being

Pamela L. Perrewé
Darren C. Treadway
Angela T. Hall

Research examining the interface between work and family has increased dramatically, partially due to the changing and evolving nature of balancing work and family responsibilities (Edwards & Rothbard, 2000; Kinnunen & Mauno, 1998). Since 1980, work environments have seen an increase of dual-income couples (Edwards & Rothbard, 2000; Zedeck, 1992), single-parent families (Edwards & Rothbard, 2000; Zedeck, 1992), and individuals involved in the care of elder parents (Friedman & Galinsky, 1992). This chapter examines the current status of work-family conflict research as well as the impact of organizational family-friendly policies, highlighting the Family and Medical Leave Act (FMLA) (1993).

Work-Family Conflict Defined

Work-family conflict is a form of interrole conflict in which engaging in one role interferes with engaging in the other (Greenhaus & Beutell, 1985). Work-family conflict has been conceptualized as a two-dimensional construct (e.g., Frone, 2000a; Frone, Yardley, & Markel, 1997; Netemeyer, Boles, & McMurrian, 1996), where work interferes with family (work-to-family conflict) and family interferes with work (family-to-work conflict).

The rationale for the effects of both work-to-family and family-to-work conflict on work and family outcomes is twofold. First, both types of work-family conflict have been found to affect one another indirectly through role overload and distress (Frone et al., 1997). Therefore, we would expect each to affect both work and family outcomes, either directly or indirectly through the other. Second, we would expect that the domain that is the source of the conflict might also be the domain affected by that conflict, via coping strategies. If the employee perceives work to be the source of conflict in work-to-family conflict, the individual may adopt such coping strategies as reducing effort or time on the job (i.e., withdrawing from work) in the hope of reducing the interference of work with family. In support of this, recent research has demonstrated relationships between work-to-family conflict and intentions to leave one's job (Grandey & Cropanzano, 1999; Kirchmeyer & Cohen, 1999) as well as relationships between work-to-family conflict and continuance of organizational commitment (Casper, Martin, Buffardi, & Erdwins, 2002).

Researchers have generally offered three perspectives to explain the work-family interface: compensation, segmentation, and spillover. The compensation hypothesis states that the roles an individual enacts within the home and work domains act in equilibrium (e.g., Greenglass & Burke, 1988). That is, individuals will attempt to compensate for deficiencies in one domain of their life by excelling in the other. The segmentation perspective, on the other hand, posits that individuals can compartmentalize the work and family domains, thus, experiencing strain only within the stressful domain. The current state of the field, however, favors the spillover perspective of work-family conflict. Researchers argue that individuals carry the attitudes and behaviors of one domain over into the other (e.g., Leiter & Durup, 1996). Although early research in this area treated work-family conflict as a global, bidirectional construct, more recent work has specified the direction of the interference between work and family roles (Carlson & Perrewé, 1999; Kossek & Ozeki, 1998). Simply stated, the line of demarcation separating work and family domains is blurred by the fact that situations that occur at work spill over into the family and vice versa.

The Context of Conflict

Identifying the exact conditions under which work-family conflict will occur is difficult. Individuals within the same organization and position may experience or perceive environmental stimuli differently. Although some classes of job condition variables can be considered objective stressors, the literature contains many studies of perceived stressors but few studies examining the same stressors from an objective and perceived perspective (Spector & Goh, 2001). Regardless, research has established several organizational and individual factors that affect the likelihood that an organizational member will experience work-family conflict. Among the most researched and salient antecedents to work-family conflict are work stressors, social support, gender, and care-giving roles. We will examine these general antecedents to work and family conflict.

Work Stressors

A stressor can be conceptualized as any perceived aspect of the environment that harms, threatens, or challenges a worker (Fox, Dwyer, & Ganster, 1993). Research has consistently demonstrated a relationship between the existence of work stressors and increased work-family conflict. Early work on work-family conflict focused on the number of hours one worked (Burke, Weir, & DuWors 1980) and the variability and flexibility of work schedules (Pleck, Staines, & Lang, 1980) as significant job stressors. Often, as an alternative to addressing each stressor individually, authors have acknowledged work-related stressors as a collection of variables that produce strain on the employee. For example, Grandey and Cropanzano (1999) used composite measures of overall job stressors and found that this overall scale was directly related to work-family conflict. Whether treated as individual or composite indicators, these stressors have a negative impact on the work and home environment. Most of the research on the work and family interface has focused on job demands, control, and role stressors.

Job Demands

An examination of the stress literature suggests a relationship between the task demands of a job and mental and physical health of an employee (Ganster & Schaubroeck, 1991). High task demands are characterized by employee perceptions of a large amount of work and the need to work hard to meet deadlines (Fox et al., 1993). The construct is broad and could as easily be operationalized as work pressure, the "frequency with which individuals perceive high job-related demands resulting from heavy workloads and responsibilities" (Frone, Russell, & Cooper, 1995, p. 3), or the number of hours worked (Wallace, 1997). The degree to which these demands exist affects the strain the worker experiences in the work-family relationship.

The demands placed on an individual at the workplace would obviously have a detrimental impact on that individual's ability to balance work and family roles. The greater the strain imposed by the work domain, the more difficult it is to achieve the desired balance between work and family. Simply increasing the amount of time or tasks necessary in a job will not in itself increase the amount of work-family conflict. For example, Wallace (1997) found that increasing the number of hours worked did not consistently generate interference in the work and family domains of lawyers. This suggests that the relationship between work-family conflict and job demands may be affected by individual and contextual variables.

Research also suggests that the amount of demands the family places on the individual may be a product of the national culture of which he or she is a member. For example, individuals in collectivist cultures may experience less work-family conflict because sacrificing one's family time for the greater good closely resembles their national values (Yang, Chen, Choi, & Zou, 2000). Although we are not suggesting that work-family conflict is a parochial concept, we believe scholars have much to gain from evaluating the impact that national and even professional values have on the work-family relationship.

Control

New human resources policies and working arrangements seem to provide researchers with a logical point of fusion for applying the control perspective of strain in the work-family interface. Karasek's model (1979) presents work strain as a concept comprising two domains: role demands and control. Because previous sections have discussed the influence of role demands in the work-family relationship, we will briefly focus our attention on the issue of defining the nature and impact of control in the work-family interface.

Control, which Ganster (1989) argues is the ability to exert some influence over one's personal environment so that the environment becomes more rewarding or less threatening, is often viewed as moderating the stressor-strain relationship. The application of Karasek's model (1979) presents strain as occurring only when an individual is unable to exercise enough discretion to overcome the demands of the job. Control research has, however, also focused on the direct and beneficial effects of control (Terry & Jimmieson, 1999). Thus, one might envision organizational policies that provide individuals with greater control over their work—such as flextime, job sharing, or participation in managerial decisions that affect their tasks—should reduce the overall strain caused by family-work or work-family interference. Yet Thomas and Ganster (1995) noted the lack of research linking organizational policies and procedures to employee perceptions of control.

Control has been found to lower levels of distress and increase levels of calmness in both the work and family domains (Williams & Alliger, 1994). Although little research has addressed the impact of specific organizational interventions, several studies have sought to evaluate the effects of increased perceptions of control on the work-family relationship. Thomas and Ganster (1995) found that those work arrangements that gave employees greater control over work and family matters helped employees manage the often conflicting demands of work and family. For example, lower control over scheduling, personal time off, and overtime work increases the likelihood that an individual will experience work-family conflict (Pleck et al., 1980). Further, research has established that the positive effects of control on work-family conflict exist regardless of family structure or gender (Duxbury, Higgins, & Lee, 1994).

Closely related to control is the concept of autonomy, the degree of discretion one has in the work environment. Studies have generally shown that job autonomy is associated with low levels of reported work-family conflict (e.g., Parasuraman, Purohit, Godshalk, & Beutell, 1996). To the extent that autonomy enables a person to control the timing and even the work location, a person would be able to better balance work and family demands. Even if greater demands accompany job autonomy, the employee's ability both to structure and control the work schedule (Pleck et al., 1980) and decide on how to perform a job may be beneficial in balancing both work and family roles.

A large body of research provides evidence for the negative relationship between control and numerous indicants of psychological well-being including burnout, depression, irritability, and health symptoms such as cardiovascular disease. Terry and Jimmieson (1999) provide a comprehensive review regarding the reduction of the negative indicants of health associated with control.

Role Stressors

Scholarship about the work-family conflict is predicated on the belief that the role stressors from work and family are inherently conflicting. Typical role stressors include role conflict, role ambiguity, and role overload, all of which have been found to directly and positively relate to work-family conflict (Frone et al., 1997).

Role conflict exists when two or more sets of role pressures exist in an individual's workspace and when complying with any one of these pressures impedes the accomplishment of another (Kahn, Wolfe, Quinn, & Snoek, 1964). Bacharach, Bamberger, and Conley (1991) found that role conflict in the workplace contributed to an increase in the worker's experienced work-family conflict. Further, this study found that role conflict contributes to increased role overload, which in turn increases work-family conflict.

Role overload is the feeling of having too much to do in too little time. As the pressures from the work and family domains exceed the individual's ability to effectively cope with those demands, the inability to effectively accomplish the tasks of either role leads to work-family conflict. For example, Frone et al. (1995) found that role overload in the work and family domains were directly related to work-to-family conflict and family-to-work conflict, respectively. Furthermore, the interference of family in work directly affects the degree of work overload that the individual experiences (Frone et al., 1997). The reverse is also true: the interference of work-to-family directly affects the degree of role overload experienced in the family domain (Wallace, 1997). These cross-domain relationships again demonstrate the nature of spillover in the work and home domains.

Envisioning situations in which two roles conflict or perhaps even contribute to overload is easy, but what if the roles are not defined? Role ambiguity exists when the individual is unclear as to his or her job tasks or manager's expectations. Although research often evaluates role ambiguity as part of a composite measure of role stressors, its distinct effect on complicating the compliance of work and family role demands is evident. Juggling work and family roles is more difficult when their demands are unexpected (Williams & Alliger, 1994). Thus, when the individual is unable to accurately define the boundaries of the roles, that person is more likely to experience lower fulfillment of those roles and higher work-family conflict.

Social Support

The amount of emotional and instrumental assistance, or social support (House, 1981), one receives has consistently been found to temper the extent of work-family conflict the individual experiences. In general, two types of social support exist: instrumental and emotional. Instrumental social support is tangible in that one receives it directly from others. Emotional support is the individual's perception of the availability of thoughtful, caring individuals with whom he or she can share thoughts and feelings. This second type of support is generally considered to be the more important of the two (House, 1981).

Research has consistently established the stress-reducing benefits of the existence of one or both types of social support (Perrewé & Carlson, 2002). The existence, or absence, of social support appears to serve as a primary predictor of work-family conflict. Consistent with this relationship, social support has been found to directly reduce perceived role stressors, time demands, and work-family conflict (Carlson & Perrewé, 1999).

Social support exists in both the family and work domains. An individual may experience support from work or from home. Yet researchers have found social support's ability to reduce work-family conflict to be primarily domain specific; that is, social support from the family has its greatest stress-reducing effects on family strain, and social support from work has its greatest stress-reducing effects on work strain (e.g., Frone et al., 1997).

Social Support at Work

Research has demonstrated that employees who receive social support from their coworkers and supervisors experience less work-family conflict than do those who do not receive this support (Frone et al., 1997; Thomas & Ganster, 1995).

Supportive supervisors accommodate individuals who are struggling to balance the work and family domains. These supervisors may allow employees to implement their own work schedules or may simply offer a sympathetic ear. When this type of support is present, employees experience lower work-family conflict than in situations in which this support is not present (Goff, Mount, & Jamison, 1990).

Coworkers provide much the same type of support as does the supervisor. They may provide a sounding board for a fellow employee's problems or may "cover" an employee's work in times of family crisis. Research notes the positive associations between coworker and supervisor social support and employee satisfaction (e.g., Carlson & Perrewé, 1999; Thomas & Ganster, 1995). From an organizational perspective, the benefits of encouraging supportive supervision are not limited to the reduction of work-family conflict. Social support has been found to increase satisfaction and contribute to the well-being of dual-career couples (Parasuraman & Greenhaus, 1994). Research further indicates that social support may influence mortality through changes in the cardiovascular, endocrine, and immune systems (Uchino, Cacioppo, & Kiecolt-Glaser, 1996). Improvement in overall well-being of employees inherently translates to lower absence and turnover in the organizational setting.

Social Support from the Family

Support from the family can take both instrumental and emotional forms. Schedule flexibility, a comforting talk, and assistance with household duties are just a few common ways that family members try to help each other in times of stress. Empirical investigations have found that family social support leads to an overall reduction in work-family conflict (e.g., Ayree, 1992).

It is easy to envision the positive effects that family social support has on family-to-work and work-to-family conflict. Research has demonstrated that the employee who receives social support from spouse and family is less likely to experience family interference with work (Frone et al., 1997). Furthermore, both instrumental and emotional types of social support from the family have been found to reduce family interference with work conflict (Adams, King, & King, 1996; Parasuraman et al., 1996).

Family social support has far-reaching benefits for the employee. Research supports the notion that supportive family environments contribute to a greater sense of life (Adams, King, & King, 1996) and family (Carver & Jones, 1992) satisfaction for the employee. Taken together with the work-related benefits of family social support, this type of support appears most beneficial to employees' well-being and satisfaction.

Gender

Perhaps one of the most salient issues in regard to work-family conflict is that of gender. Gender roles appear to affect significantly the amount and type of responsibility individuals experience in the home (Gutek, Searle, & Klepa, 1991). Early research found that women spend more time working in both the household (Denmark, Shaw, & Ciali, 1985) and in the workplace and household combined (Pleck, 1985) than do their male counterparts. These increased demands would seem to present women with a greater challenge in balancing work and family roles. Although research has presented some conflicting results about the gender-work-family relationship, gender continues to be an important consideration for future research in work-family conflict.

Pleck (1977) offered a seminal conceptualization of this phenomenon, arguing that men and women would differ in their interrole conflict domain. Traditional gender roles cast men as the financial providers; thus, men's work life would interfere with the family domain more than would women's work life. In contrast, females traditionally have responsibilities for the home and therefore are likely to experience more interference of family life in the work domain.

Although Pleck's conceptualization (1977) may be intuitively appealing, research does not necessarily bear it out. One general finding that seems to challenge Pleck's assertions is that women and men often perceive the same amount of work-family conflict (e.g., Duxbury & Higgins, 1991; Kinnunen & Mauno, 1998). Further, Frone, Russell, and Barnes (1996) argued that social identity theory would predict that family-to-work conflict would be more detrimental for men, whose self-image is more closely tied to their work, and work-to-family conflict would be more detrimental to women, whose self-image is more closely tied to home and family. However, Frone et al. (1996) found very little evidence for gender differences and concluded that work-family conflict, regardless of the direction, may be detrimental to the health of both men and women. Recent research by Frone (2000b) supports his earlier conclusion that growing evidence shows that work-family conflict is equally deleterious to the health of male and female employees. This does not, however, mean that gender differences in the work and family interface do not exist.

Nelson and Burke (2002) recently edited a book of essays dealing with gender, occupational stress, and well-being. They concluded that, in regard to gender, (1) there is a complex interplay between work and home dynamics, gender, and health, and (2) social support seems to be a critical area of study because women both seek and benefit from social support differently than men.

Perrewé and Carlson (2002) examined the interactive effects of social support (one form of coping) and gender on conflict and satisfaction in both the work and family contexts. Their results suggested that social support played a greater role for women than men in reducing family interference with work but was insignificant for reducing work interference with family. Further, women were able to better use social support than men were to obtain greater work and family satisfaction. Although findings suggest that women may benefit more than men do from social support, particularly social support stemming from the family, research is still needed to examine the specific role of gender in the work and family interface.

Role of the Caregiver

Women's increased presence in the workplace, coupled with the aging of baby boomers' parents, has made many women caregivers of both children and aging parents. Over the past decade, research has focused on evaluating the extent to which this caregiving role affects the work and family interface. Not surprisingly, research has found that family-related absence from work is more common among women, the traditional family caregivers, than among men (Boise & Neal, 1996).

Over the next twenty years, the elderly population will significantly increase due to the aging of the baby-boomer generation. The aging of this segment of the population will ensure that a larger number of employees will continue to balance work demands and roles as caregiver to an elder parent. Research has begun to focus attention on this phenomenon and its relationship to work withdrawal behaviors (e.g., Barling, MacEwen, Kelloway, & Higginbottom, 1994). This work provides evidence toward establishing a direct relationship between work and elder-care conflict and work withdrawal (i.e., lateness, leaving work early, time on the telephone).

Given these demographic changes in the workforce and in the population as a whole, the impact of caregiver roles on work-family conflict is a more salient topic than ever. These changes will create a large group of employees who are "sandwiched" between caregiving roles, simultaneously caring for both children and elder parents. Therefore, many organizational researchers argue that the study of the strain from the family domain should include the role of caregiver for both children and aging family members (Buffardi, Smith, O'Brien, & Erdwins, 1999).

The Effects of Conflict

Work-family conflict has been associated with a number of dysfunctional outcomes including general work withdrawal behaviors (Gignac, Kelloway, & Gottlieb, 1996; Kirchmeyer & Cohen, 1999; Thomas & Ganster, 1995), decreased family and occupational well-being (Kinnunen & Mauno, 1998), psychological costs and physical complaints (Frone, 2000a), and job and life dissatisfaction (Netemeyer et al., 1996; Kossek & Ozeki, 1998; Perrewé, Hochwarter, & Kiewitz, 1999). The following sections will examine the work-related outcomes resulting from conflict within the work and family interface.

Work Withdrawal Behaviors

Work withdrawal involves minimizing time in one's specific work role and includes being absent, arriving late, daydreaming, and making personal telephone calls at work. Blau (1998) noted problems in the vague definition of withdrawal behaviors and specifically suggested that researchers use the term *physical work withdrawal* to refer to behaviors such as absenteeism and lateness. Although physical work withdrawal behaviors can be predicted by attitudes such as job satisfaction (Hanisch & Hulin, 1991), less is known about the effects of perceived work-family conflict on these behaviors.

A limited amount of research has focused specifically on work-family conflict and physical work withdrawal behaviors. The literature shows conflicting results in regard to absenteeism. These seem more a result of the method of measurement than of a significant change in the phenomenon of interest. For example, Goff et al. (1990) found that work-family conflict predicted absenteeism, but Thomas and Ganster (1995) did not find a relationship between the two. The issue in these conflicting findings may be in the use of a unidirectional, composite measure of work-family conflict. In evaluating work-family conflict bidirectionally, Gignac et al. (1996) and Kirchmeyer and Cohen (1999) found a significant relationship between family-to-work conflict and absenteeism but not between work-to-family conflict and absenteeism. Gignac et al. (1996), however, also found a significant relationship between work-to-family conflict and a number of behavioral work outcomes, such as missed meetings. Although limited, this research suggests positive relationships between work-family conflict and work withdrawal behaviors.

Well-Being

Employee well-being is a broad concept and, although definitions vary greatly, Danna and Griffin (1999) suggest that the study of well-being should take into account the "whole person." Further, they suggest its measurement should include life experiences (e.g., life satisfaction, happiness), generalized job-related experiences (e.g., job satisfaction, job attachment), as well as more facet-specific dimensions (e.g., satisfaction with pay or coworkers).

Researchers in the work-family context have used well-being to address a number of dysfunctional outcomes, including burnout (Bacharach et al., 1991; Burke, 1988), unpleasant moods (Frone, 2000b; Williams & Alliger, 1994), depression (Thomas & Ganster, 1995), decreased family and occupational well-being (Kinnunen & Mauno, 1998), and psychological costs and physiological concerns (Frone, 2000b).

In a recent and comprehensive national study, Frone (2000b) presented data that indicated both types of work-family conflict were positively related to mood disorders, anxiety, and substance dependence. Specifically, Frone found that individuals experiencing work-to-family conflict were 3.13 times more likely to have a mood disorder, 2.46 times more likely to have an anxiety disorder, and 1.99 times more likely to experience a substance disorder than were individuals who were not experiencing this type of conflict. Individuals experiencing family-to-work conflict were 29.66 times more likely to have a mood disorder, 9.49 times more likely to have an anxiety disorder, and 11.36 times more likely to have a substance dependence than were individuals not experiencing this type of stress. These results suggest a far-reaching impact of the work and family interface on employee well-being. Not only does work-family conflict have an impact on general psychological distress (e.g., Frone et al., 1996), but work-family conflict may be associated with more severe detriments to the health of employees (Frone, 2000b).

Performance

As a number of researchers have duly noted, the work-family conflict literature contains little research examining work-related performance outcomes (Frone et al., 1997; Kossek & Ozeki, 1998). The few studies examining this relationship have, not surprisingly, reported a negative association between work-family conflict and work performance. The magnitude of the relationship, however, may depend upon the type of work-family conflict measure used (i.e., bidirectional or unidirectional), as well as the operationalization of performance (i.e., objective versus subjective).

As we have previously discussed, work-family conflict is a multidimensional concept in which work may interfere with family or family with work. Implicitly, these dimensions suggest that the value of one domain is taking precedence over the other, and we should therefore expect suboptimal performance of the role in the inferior domain. To that point, Netemeyer, Boles, and McMurrian (1996) found family-to-work conflict was more strongly related to poor sales performance than was work-to-family conflict. Simply, work-related performance suffers when family demands interfere with the accomplishment of task demands of the job.

Job and Life Satisfaction

As individuals experience conflict between their family and work roles, satisfaction in both job and life domains decreases. A recent meta-analysis reviewing the relationship between work-family conflict and job and life satisfaction found a consistent negative relationship to exist among all types of work-family conflict and job and life satisfaction (Kossek & Ozeki, 1998). Again, it appears that the manner in which researchers measure work-family conflict has an impact on the outcomes related to that conflict. For example, Kossek and Ozeki (1998) found that a unidimensional measure of family-to-work conflict was less strongly related to job and life satisfaction than was a bidirectional measure of work-family conflict. Clearly, the nature of the work-family conflict (i.e., examining the differential impact of work-to-family and family-to-work conflict) would add much to its predictive validity.

New Directions

The paradigm in work-family research is gradually changing from an evaluation of the consequences and antecedents of conflict to the following: (1) an increased level of attention to crossover stress, specifically, individuals' reaction to the job stress of those with whom they interact regularly; (2) an understanding of the role of work and family involvement, work and family values, and value attainment in the work and family interface; (3) an examination of the effectiveness of organizational family-friendly work policies; and (4) an examination of how the FMLA has affected the work and family interface.

Crossover Stress

As we discussed earlier, researchers have generally offered three perspectives to explain the work-family interface: compensation, segmentation, and spillover. A fourth process that links the family and work domains is called crossover (Bolger, DeLongis, Kessler, & Wethington, 1989).

Bolger et al. (1989) and Wethington (2000) distinguish between spillover and crossover. Spillover is a contagion of demands and subsequent strain from one domain of a person's life to another. In crossover, demands and their subsequent strains cross over between closely linked individuals. Thus, stress that the individual experiences in the workplace leads to the individual's spouse at home experiencing strain. Whereas spillover is an intra-individual transmission of stress or strain, crossover is a dyadic, interindividual transmission of stress or strain.

Examining the crossover effects of stress and strain using the dyad as the unit of analysis contributes to our understanding of the complexities of multiple roles in different domains. This framework enables the investigation of the ripple effect of stress as it starts at the workplace and crosses over to the spouse, who in turn conveys the consequences to the coworkers, who transmit it to their spouses at home, and so on (Westman, 2002). This expansion also shows how stress, generated either at home or at the workplace, moves from the micro-level (couple, family) to meso-level social systems (workplace-management, team, or group), thereby potentially affecting the entire organization. This involves the multiple levels of analysis that researchers must use when investigating the work-family interface: individuals, partners, families, teams, and organizations. The clear advantage of crossover research is that it is based on observation of two partners. Collecting data from both partners reduces confounding and enables researchers to control for each partner's stress.

Finally, examining positive as well as negative crossover is important. Empathy could just as easily involve the sharing of another's positive affect, moods, and emotions. One possible reason for the neglect of the investigation of positive crossover is that stress research relies heavily on medical perspectives that emphasize negative effects (see Chapter One of this volume; Westman, 2002). Given the recent thinking that health is more than the absence of illness (e.g., Danna & Griffin, 1999; Chapter One), conceptualizing and examining positive crossover might help to push work and family research into a new realm of physical, mental, and social well-being—not merely the absence of illness.

Work and Family Involvement, Values, and Value Attainment

An encouraging line of research evaluates the importance of work and family involvement as well as values and value attainment in regard to the work and family interface. This line of research demonstrates how organizations might benefit from assisting employees in attaining those values that they have deemed important in their lives.

Work Involvement

Work involvement represents the psychological involvement and salience of the work role for an individual (Beutell & Wittig-Berman, 1999). Although seemingly a positive organizational outcome, work involvement has consistently been found to predict work-family conflict (e.g., Beutell & Wittig-Berman, 1999). Further, work involvement has been found to exacerbate the use of alcohol and the deterioration of health in employees experiencing work pressure and role ambiguity (Frone et al., 1995). Although work involvement has been found to be related to a number of dysfunctional outcomes, it also has a positive relationship with job satisfaction (Adams et al., 1996). Thus, work involvement is associated with both positive (e.g., job satisfaction) and negative (e.g., work-family conflict) outcomes.

Family Involvement

Family involvement can be conceptualized as the psychological participation in and centrality of the family in a person's life (Beutell & Wittig-Berman, 1999). Research findings have offered a more complex picture of the relationship between family involvement and work-family conflict. Family involvement leads to a greater propensity for family to interfere with work (Adams et al., 1996). Yet a greater degree of family involvement is also related to social support from the family, which in turn reduces work-family conflict. These relationships illustrate the delicate balancing act at the core of the work-family interface.

Work and Family Values

Although a great deal of research has examined how values that individuals hold affect their behavior, rarely have scholars related values research to the area of work-family conflict. The incorporation of values into the work-family conflict literature is important because life-role values are central to organizing meaning and action for working people. Values motivate action and are the basis from which individuals define their roles (England & Harpaz, 1983). Thus, value expression represents the individual's identity or self-concept (Katz & Kahn, 1978).

Life-role values consist of the individual's system of values regarding work and family domains. They are based on what the individual believes to be important or central to or makes a priority in his or her life (Carlson and Kacmar, 2000). From this perspective two employees who are experiencing the same work stressors may experience work-family conflict to varying degrees, based upon the value each individual places on family or work roles. For example, in a recent empirical investigation, Adams et al. (1996) concluded that although work-family conflict was related to job-life satisfaction, the level of involvement or importance an employee assigns to work and family roles is also an important predictor of these relationships.

Carlson and Kacmar (2000) examined the effect of different life-role values in the work-family conflict process. In general, when family values were high, the antecedents from the work domain were more salient and had a greater impact on conflict and satisfaction. Similarly, when work values were highly central to an individual, antecedents from the family domain had a greater impact. Interestingly, the relationship between the directions of work-family conflict did not differ based on values. These results suggest that the values one places on the work or family domain are important considerations in understanding the nature of work-family conflict.

Value Attainment

The value attainment hypothesis suggests that the degree to which an individual's work or family situation allows her or him to attain those life values will determine the degree of satisfaction that person obtains from job or family (George & Jones, 1996; Perrewé et al., 1999). Locke (1976) reasoned that satisfaction was an emotional response based upon two factors: (1) an estimate between what is valued and what is obtained and (2) an estimate of the importance of the value to the individual. Thus, the concept of value attainment offers an important explanation for why individuals who are experiencing the same work-family conflict experience different degrees of job and life satisfaction.

One argument is that an individual experiences conflict between work and family demands because of incongruent values (Perrewé et al., 1999). Specifically, value incongruence can occur between the target individual and a pivotal family member (i.e., lack of value similarity) or between the target individual and the organization (i.e., lack of value congruence). Perrewé et al. conclude that work-family conflict leads to job and life dissatisfaction for individuals because this conflict frustrates the attainment of important work and family values. Because individuals are sometimes unable to attain important values from their work or home environment due to work-family conflict, perhaps family-supportive organizational policies can help to provide some assistance.

Family-Supportive Work Environments

Family-supportive policies represent an increasing number of organizational interventions that are designed to provide assistance to employees coping with competing demands of work and home. In general, family-friendly work programs have been designed to reduce the negative indicants of employee health and well-being due to perceived work-family conflict. Day care, flextime, and telecommuting are just a few of the formal programs that organizations implement to assist their employees (Thomas & Ganster, 1995). For the individuals who use these programs, they represent instrumental support. More importantly, they have the potential to create an environment in which employees perceive that significant emotional support is available. We will discuss the two most popular family-friendly organizational initiatives, schedule flexi-bility and dependent care benefits (Friedman & Johnson, 1997). Further, because no formal program can help employees unless the organizational culture supports it, we will discuss the support-ive workplace.

Schedule Flexibility

Organizational research has demonstrated that organizational pro-grams that provide employees with alternatives to the traditional 8 A.M. to 5 P.M., five-day per week schedule can be beneficial. Flexi-ble work schedules have been found to increase productivity and job satisfaction (Baltes, Briggs, Huff, Wright, & Neuman, 1999). Further, Thomas and Ganster (1995) found that schedule flexi-bility regarding work arrangements gave employees greater con-trol over work and family matters, thereby helping employees manage the often conflicting demands of the two domains. Stud-ies that have shown that job autonomy is associated with low levels of reported work-family conflict (e.g., Parasuraman et al., 1996) offer additional support for the benefits of flexible work schedules.

Interestingly, simply offering flexible work schedules may not convince employees to take advantage of scheduling options, par-ticularly if employees do not believe management supports these types of programs. In other words, employees may be concerned that their supervisor may not view favorably their use of a flexible work schedule. For example, Kossek, Barber, and Winters (1999) found that when managers provide an example and make visible to others that flexible work arrangements are viable options for employees, employees are more likely to use such schedules themselves.

Benefits for Dependent Care

Organizations often provide dependent care benefits as a means to reduce the conflict in the work and family interface. Benefits for dependent care include programs that range from helping em-ployees arrange for child and elder care to paying for leaves of ab-sence in order to care for a dependent. Little systematic research, however, has been done on these types of programs. Interestingly, Kossek and Ozeki (1998) found that organizational policies initi-ated to help employees meet family responsibilities have not gen-erally had the desired impact of reducing levels of work-family conflict. Clearly, additional research into the costs and benefits of family-supportive work policies is needed. One promising area of research that could increase the effectiveness of family-supportive policies is the organizational support for these policies.

The Supportive Workplace

When organizations formally adopt family-supportive programs (e.g., schedule flexibility), some managers may not be supportive. Interestingly, many organizations have adopted family-friendly policies, but they implement these programs case by case, giving discretion to the managers. Although many alternative work arrangements may result from employees negotiating individually with their managers as needed (Scandura & Lankau, 1997), not all managers may support these programs.

Only a few studies have examined the role of the informal workplace and the supportive manager when adopting family-friendly work policies. Powell and Mainiero (1999) concluded that permission to have an alternative work arrangement is often based on the manager's previous experiences and beliefs. Thomas and Ganster (1995) found that perceptions of managerial support were linked to lower levels of work-family conflict and psychological and physiological strain. Finally, Carlson and Perrewé (1999) concluded that the quality of the relationship employees have with their supervisors, coworkers, and subordinates reduces the perceived stressors that lead to work-family conflict. Additional research is needed on the role of a supportive work environment as a mechanism to increasing the effectiveness of family-friendly work policies.

The following section examines a specific type of family-friendly work policy, the 1993 FMLA. We discuss FMLA eligibility and benefits, legal rights, employee concerns, and possible consequences, as well as its future.

The 1993 FMLA

Recognizing the tension between work and family demands that many individuals experience, the U.S. Congress enacted the FMLA of 1993. Table 10.1 gives a summary of its major provisions. To date, the FMLA is the most comprehensive and far-reaching U.S. federal legislation that addresses work-family issues.

Table 10.1. Summary of Major FMLA Provisions.

Purpose	Congress enacted the FMLA in order to "(1) balance the demands of the workplace with the needs of families, (2) promote the stability and economic security of families, and (3) promote national interests in preserving family integrity" (29 U.S.C.A. 2601).
Covered employers	FMLA applies to employers that employ 50 or more employees during at least 20 weeks of the year (29 U.S.C.A. 2611).
Eligible employees	Employees who work at least 1,250 hours during the preceding year are eligible. This includes those who work for public agencies (state, local, and federal) and those who work for private sector employers. An employee must have worked for at least 12 months and at a location where at least 50 employees were employed within a 75-mile radius (29 U.S.C.A. 2611).
Reinstatement to former position	Except for certain highly compensated employees, an employee returning from FMLA leave must be restored to the same or similar position (29 U.S.C.A. 2614).
Form of leave	An eligible employee is entitled to up to 12 workweeks of leave during any 12-month period. The employee may take the leave all at once, use it intermittently, or take it via a reduced work schedule (29 U.S.C.A. 2612).
Entitlement to leave	An eligible employee is entitled to leave for (1) the birth or adoption of a child; (2) the placement of a child with the employee for foster care; (3) caregiving of a spouse, child, or parent who has a serious health condition; or (4) the employee's own serious health condition (29 U.S.C.A. 2612).
Benefits protection	Employers are prohibited from revoking any benefit accrued by an employee prior to taking FMLA leave. Employers must maintain health insurance coverage for employees who take FMLA leave, provided that the employee was covered by a group health plan prior to the leave (29 U.S.C.A. 2614).
Penalties for FMLA violations	Employers who violate the provisions of the FMLA may be ordered to pay damages, such as back wages and attorneys' fees and costs, in addition to punitive damages (29 U.S.C.A. 2617; Askew, 2000; Ripple, 1999).

Eligibility and Benefits Under the FMLA

Parlo (2000) estimated that the FMLA covers approximately 70% of the U.S. workforce. Covered employees include those who work for public agencies (state, local, and federal) and those who work for private sector employers, provided that the organization employed 50 or more employees during at least 20 weeks of the previous year (Askew, 2000). To be eligible for FMLA leave benefits, an employee must have worked for at least 12 months for the employer and at a location where at least 50 employees were employed within a seventy-five mile radius. The FMLA requires employers to grant employees leave for up to 12 weeks due to serious personal illness, serious family illness, the birth of a newborn child, or the adoption of a child. The FMLA also provides for intermittent leave or reduced work schedules in order for an employee to heal from a serious medical condition or to care for a spouse, parent, or child (biological, foster, adopted, stepchild, or legal ward) who has a serious medical condition. Employers must maintain health insurance for employees who exercise FMLA leave benefits. Moreover, upon their return to work, employees are generally entitled to hold the same position or one comparable to the position they held before taking leave.

Legal Rights Under the FMLA

The FMLA contains provisions that prohibit employers from retaliating or discriminating against employees who use the leave benefits under the act. For example, noncompliant employers may be required to pay the prevailing employee-plaintiff compensatory damages, such as back wages, and attorneys' fees and costs (Askew, 2000).

Employee Concerns and Antecedents to Usage

Although many employees fall under the purview of the FMLA and would benefit from the use of this initiative, most do not take advantage of the program (Twomey & Jones, 1999). A major reason that Twomey and Jones cite for the failure to invoke FMLA benefits is that the FMLA does not require an employer to pay an employee for the employee's time on FMLA leave. Moreover, the economic reality is that lower income workers can rarely afford to take unpaid leave. Another factor cited for employees' reluctance to use the FMLA for reasons other than illness is the negative effect that taking such leave might have on employees' careers. This is so even in spite of the FMLA's gender-neutral language forbidding discrimination and retaliation (Ripple, 1999). Further, the limited preliminary research conducted on potential family leave usage indicates that men may be considerably more reluctant to use family leave benefits because they fear that it will negatively affect their careers (Twomey & Jones, 1999).

Consequences and Future of the FMLA

Although few empirical studies have investigated the effect the FMLA has on important organizational outcomes, neither organizations nor researchers should discount the potential benefits of the FMLA. Grover and Crooker (1995) found that employees in organizations that offered family-supportive benefits evidenced greater organizational commitment and lower intention to quit their jobs. Moreover, previous research suggests that these positive organizational outcomes of work-family initiatives may outweigh potential costs, especially in light of the presumed effect such policies have in decreasing work-family conflict (Thomas & Ganster, 1995). Because the FMLA is a family-supportive policy, this research suggests that organizations may realize positive outcomes when employees exercise their leave benefits. Clearly, further research is needed to investigate organizational and employee outcomes resulting from use of the FMLA.

Conclusion

This chapter examined organizational and individual antecedents to the conflict that employees experience in the work and family interface. Research has determined that typical antecedents leading to work and family conflict include job demands, a lack of control over one's work, a lack of social support, and multiple caregiving roles. We also addressed gender as a possible contributor to work and family conflict. Although a handful of studies have demonstrated gender differences in regard to perceptions of conflict, ability to cope, and experienced strain, research findings have not been consistent. In fact, Frone (2000b) recently concluded that little evidence supports gender differences in the relationship between work-family conflict and well-being. Perhaps gender differences play a role in the way individuals cope (e.g., Greenhaus & Beutell, 1985; Perrewé & Carlson, 2002) rather than the degree to which they differentially experience conflict.

We also examined the negative effects of work-family conflict for the organization and the employee. Research generally supports the findings that work-family conflict (regardless of the direction of conflict) can lead to work withdrawal behaviors, decreased well-being and satisfaction, and increased psychological and physical complaints. Interestingly, Frone (2000b) found that family-to-work conflict was more strongly related to psychiatric disorders than was work-to-family conflict. Frone suggested that perhaps the attributions individuals make regarding the source of the conflict might explain these findings. Specifically, Frone argued that if individuals hold their organizations responsible for work-to-family conflict, they make external attributions. However, if individuals hold themselves responsible for family-to-work conflict, they may feel an inability to manage their own family lives that in turn may lead to more detrimental disorders.

Further, this chapter examined some organizational family-friendly work policies and the FMLA, looking at their impact on reducing the negative indicants of health and well-being for employees. Although research examining the impact of family-friendly work policies on work-family conflict and employee well-being is limited, it does appear that top management support is needed before employees will trust enough to feel comfortable taking advantage of these benefit opportunities. Although having access to family-supportive programs such as flextime has been found to be important in reducing family-to-work conflict (Frone & Yardley, 1996), research on meso-level factors (e.g., organizational culture, leadership) that may influence this relationship has been very limited. Additional research is needed to determine the types of family-friendly policies that can reduce work-family conflict for employees and subsequently reduce the negative indicants of health for employees. Additionally, future research should consider the process through which individuals decide whether or not to make an accommodation request. Research is needed to determine both the antecedents as well as the consequences of requesting accommodations (Baldridge & Veiga, 2001). Clearly, simply developing strategies and family-friendly work policies will not reduce work-family conflict unless the corporate culture allows employees to feel comfortable taking advantage of these policies.

Finally, this chapter has proposed that much can be gained from analyzing the work-family interface at multiple levels. The strategic human resource management literature has largely ignored the role of family-friendly policies in the workplace. Delery and Doty (1996) noted that researchers have consistently identified seven strategic human resource practices, none of which addresses the work-family interface. Thus, we suggest that future research seek to examine two central questions at the strategic level: (1) What configuration of family-friendly policies yields the greatest benefit for organizational outcomes? (2) When evaluated with other strategic human resource practices, to what degree are family-friendly policies capable of generating a sustainable competitive advantage for organizations? These questions are important for research but also for the success of strategic human resource practices.

In closing, we note that the research on the work and family interface is far from exhausted. Perhaps examining some of the new suggested research streams emphasizing family-friendly work programs, the FMLA, meso-level factors, work and family involvement, values, and value attainment may help to unravel some of the mixed findings in the literature on work and family conflict. Although work-family conflict can clearly have serious debilitating effects on employees, perhaps the more interesting approach to examining the work and family interface is the focus on individual and organizational health. The new positive psychology movement may offer an alternative to the tendency of organizational scientists to focus on negative outcomes. Studying employees is much more than trying to fix what is wrong with them. It is about identifying and nurturing their strongest qualities and helping them find niches in which they can best live out these strengths (Seligman and Csikszentmihalyi, 2000).

References
Adams, G., King, L., & King, D. (1996). Relationships of job and family involvement, family social support, and work-family conflict with job and life satisfaction. *Journal of Applied Psychology, 81,* 411–420.

Askew, K. J. (2000). The Family and Medical Leave Act of 1993: 2000 Update. *Employer and labor relations law for the corporate counsel and the general practitioner, 52,* 483–497.

Ayree, S. (1992). Antecedents and outcomes of work-family conflict among married professional women: Evidence from Singapore. *Human Relations, 45,* 813–837.

Bacharach, S. B., Bamberger, P., & Conley, S. (1991). Work-home conflict among nurses and engineers: Mediating the impact of role stress on burnout and satisfaction at work. *Journal of Organizational Behavior, 12,* 39–53.

Baldridge, D. C., & Veiga, J. F. (2001). Toward a greater understanding of the willingness to request an accommodation: Can requesters' beliefs disable the Americans with Disabilities Act? *Academy of Management Review, 26,* 85–99.

Baltes, B. B., Briggs, T. E., Huff, J. W., Wright, J. A., & Neuman, G. A. (1999). Flexible and compressed workweek schedules: A meta-analysis of their effects on work-related criteria. *Journal of Applied Psychology, 84,* 496–513.

Barling, J., MacEwen, K. E., Kelloway, E. K., & Higginbottom, S. F. (1994). Predictors and outcomes of elder-care-based interrole conflict. *Psychology and Aging, 9,* 391–397.

Beutell, N. J., & Wittig-Berman, U. (1999). Predictors of work-family conflict and satisfaction with family, job, career, and life. *Psychological Reports, 83,* 893–903.

Blau, G. (1998). On the aggregation of individual withdrawal behaviors into larger multi-item constructs. *Journal of Organizational Behavior, 19*(5), 437–451.

Boise, L., & Neal, M. B. (1996). Family responsibilities and absenteeism: Employees caring for parents versus employees caring for children. *Journal of Managerial Issues, 8*(2), 218–238.

Bolger, N., DeLongis, A., Kessler, R., & Wethington, E. (1989). The contagion of stress across multiple roles. *Journal of Marriage and the Family, 51,* 175–183.

Buffardi, L. C., Smith, J. L., O'Brien, A. S., & Erdwins, C. J. (1999). The impact of dependent-care responsibility and gender on work attitudes. *Journal of Occupational Health Psychology, 4,* 356–367.

Burke, R. J. (1988). Some antecedents and consequences of work-family conflict. *Journal of Social Behavior and Personality, 3*(4), 287–302.

Burke, R. J., Weir, T., & DuWors, R. E. (1980). Work demands on administrators and spouse well-being. *Human Relations, 33,* 253–278.

Carlson, D. S., & Kacmar, K. M. (2000). Work-family conflict in the organization: Do life role values make a difference? *Journal of Management, 26,* 1031–1054.

Carlson, D. S., & Perrewé, P. L. (1999). The role of social support in the stressor-strain relationship: An examination of work-family conflict. *Journal of Management, 25,* 513–540.

Carver, M. C., & Jones, W. H. (1992). The family satisfaction scale. *Social Behavior and Personality, 20,* 71–84.

Casper, W. J., Martin, J. A., Buffardi, L. C., & Erdwins, C. J. (2002). Work-family conflict, perceived organizational support, and organizational commitment among employed mothers. *Journal of Occupational Health Psychology, 7,* 99–108.

Danna, K., & Griffin, R. W. (1999). Health and well-being in the workplace: A review and synthesis of the literature. *Journal of Management, 25,* 357–384.

Delery, J. E., & Doty, D. H. (1996). Modes of theorizing in strategic human resource management: Test of universalistic, contingency, and configurational performance predictions. *Academy of Management Journal, 39,* 802–835.

Denmark, F. L., Shaw, J. S., & Ciali, S. D. (1985). The relationship among sex roles, living arrangements, and the division of household responsibilities. *Sex Roles, 12,* 617–625.

Duxbury, L. E., & Higgins, C. A. (1991). Gender differences in work-family conflict. *Journal of Applied Psychology, 76,* 60–74.

Duxbury, L., Higgins, C., & Lee, C. (1994). Work-family conflict: A comparison by gender, family type, and perceived control. *Journal of Family Issues, 15,* 449–466.

Edwards, J., & Rothbard, N. (2000). Mechanisms linking work and family: Clarifying the relationship between work and family constructs. *Academy of Management Review, 25,* 178–199.

England, G. E., & Harpaz, I. (1983). Some methodological and analytic considerations in cross-national comparative research. *Journal of International Business Studies, 14,* 49–59.

Family and Medical Leave Act of 1993, 29 U.S.C.A. § 2601 *et seq.* (West, 2002).

Fox, M. L., Dwyer, D. J., & Ganster, D. C. (1993). Effects of stressful job demands and control on physiological and attitudinal outcomes in a hospital setting. *Academy of Management Journal, 36,* 289–318.

Friedman, D. E., & Galinsky, E. (1992). Work-family issues: A legitimate business concern. In S. Zedeck (Ed.), *Work, families, and organizations* (pp. 168–207). San Francisco: Jossey-Bass.

Friedman, D. E., & Johnson, A. A. (1997). Moving from programs to culture change: The next stage for the corporate work-family agenda. In S. Parasuraman & J. H. Greenhaus (Eds.), *Integrating work and family: Challenges and choices for a changing world* (pp. 192–208). Westport, CT: Quorum Books.

Frone, M. R. (2000a). Interpersonal conflict at work and psychological outcomes: Testing a model among young workers. *Journal of Occupational Health Psychology, 5,* 246–255.

Frone, M. R. (2000b). Work-family conflict and employee psychiatric disorders: The national comorbidity survey. *Journal of Applied Psychology, 85,* 888–895.

Frone, M. R., Russell, M., & Barnes, G. M. (1996). Work-family conflict, gender, and health-related outcomes: A study of employed parents in two community samples. *Journal of Occupational Health Psychology, 1,* 57–69.

Frone, M. R., Russell, M., & Cooper, M. L. (1995). Job stressors, job involvement, and employee health: A test of identity theory. *Journal of Occupational and Organizational Psychology, 68,* 1–11.

Frone, M. R., & Yardley, J. K. (1996). Workplace family-supportive programmes: Predictors of employed parents' importance ratings. *Journal of Occupational and Organizational Psychology, 69,* 351–366.

Frone, M. R., Yardley, J. K., & Markel, K. S. (1997). Developing and testing an integrative model of the work-family interface. *Journal of Vocational Behavior, 50,* 145–167.

Ganster, D. C., & Schaubroeck, J. (1991). Work stress and employee health. *Journal of Management, 17,* 235–271.

George, J., & Jones, G. (1996). The experience of work and turnover intentions: Interactive effects of value attainment, job satisfaction, and positive mood. *Journal of Applied Psychology, 81,* 318–325.

Gignac, M.A.M., Kelloway, E. K., & Gottlieb, B. H. (1996). The impact of caregiving on employment: A mediational model of work-family conflict. *Canadian Journal on Aging, 15,* 525–542.

Goff, S. J., Mount, M. K., & Jamison, R. L. (1990). Employer supported child care, work-family conflict, and absenteeism: A field study. *Personnel Psychology, 43,* 793–809.

Grandey, A., & Cropanzano, R. (1999). The conservation of resources model and work-family conflict and strain. *Journal of Vocational Behavior, 54,* 350–370.

Greenglass,, E. R., & Burke, R. J. (1988). Work and family precursors of burnout in teachers: Sex difference. *Sex Roles, 18,* 215–229.

Greenhaus, J. H., & Beutell, N. J. (1985). Sources of conflict between work and family roles. *Academy of Management Review, 10,* 76–88.

Grover, S. L., and Crooker, K. J. (1995). Who appreciates family-responsive human resource policies: The impact of family-friendly policies on organizational attachment of parents and non-parents. *Personnel Psychology, 48,* 271–288.

Gutek, B., Searle, S., & Klepa, L. (1991). Rational versus gender-role explanations for work-family conflict. *Journal of Applied Psychology, 76,* 560–568.

Hanisch, K. A., & Hulin, C. L. (1991). General attitudes and organizational withdrawal: An evaluation of a causal model. *Journal of Vocational Behavior, 39,* 110–128.

House, J. S. (1981). *Work stress and social support.* Reading, MA: Addison-Wesley.

Kahn, R., Wolfe, D., Quinn, R., & Snoek, D. (1964). *Organizational stress: Studies in role conflict and ambiguity.* New York: Wiley.

Karasek, R. A. J. (1979). Job demands, job decision latitude, and mental strain: Implications for job redesign. *Administrative Science Quarterly, 24,* 285–309.

Katz, D., & Kahn, R. (1978). *The social psychology of organizations* (2nd ed.). New York: Wiley.

Kinnunen, U., & Mauno, S. (1998). Antecedents and outcomes of work-family conflict among employed women and men in Finland. *Human Relations, 51,* 157–177.

Kirchmeyer, C., & Cohen, A. (1999). Different strategies for managing the work/non-work interface: A test for unique pathways to work outcomes. *Work and Stress, 13,* 59–73.

Kossek, E. E., Barber, A. E., & Winters, D. (1999). Using flexible schedules in the managerial world: The power of peers. *Human Resource Management, 38,* 33–46.

Kossek, E. E., & Ozeki, C. (1998). Work-family conflict, policies, and the job-life satisfaction relationship: A review and directions for organizational behavior–human resources research. *Journal of Applied Psychology, 83,* 139–149.

Leiter, M. P., & Durup, M. J. (1996). Work, home, and in-between: A longitudinal study of spillover. *Journal of Applied Behavioral Science, 32,* 29–47.

Locke, E. (1976). The nature and causes of job satisfaction. In M. D. Dunnette (Ed.), *Handbook of industrial and organizational psychology* (pp. 1297–1349). Skokie, IL: Rand McNally.

Nelson, D. L., & Burke, R. J. (Eds.). (2002). *Gender, work stress, and health.* Washington, DC: American Psychological Association.

Netemeyer, R. G., Boles, J. S., & McMurrian, R. (1996). Work and family variables, entrepreneurial career success, and psychological well-being. *Journal of Applied Psychology, 81,* 400–410.

Parasuraman, S., & Greenhaus, J. H. (1994). Determinants of support provided and received by partners in two-career relationships. In L. A. Helsop (Ed.), *The ties that bind* (pp. 121–145). Calgary, Alberta: Canadian Consortium of Management Schools.

Parasuraman, S., Purohit, Y. S., Godshalk, V. M., & Beutell, N. J. (1996). Work and family variables, entrepreneurial career success, and psychological well-being. *Journal of Vocational Behavior, 48,* 275–300.

Parlo, C. A. (2000). Recent developments under the Family and Medical Leave Act. *Proceedings of the 29th Annual Institute on Employment Law, 2,* 599–662.

Perrewé, P. L., & Carlson, D. (2002). Do men and women benefit from social support equally? A field examination within the work and family context. In D. L. Nelson & R. J. Burke, (Eds.), *Gender, work stress, and health* (pp. 101–114). Washington, DC: American Psychological Association.

Perrewé, P. L., Hochwarter, W. A., & Kiewitz, C. (1999). Value attainment: An explanation for the negative effects of work-family conflict on job and life satisfaction. *Journal of Occupational Health Psychology, 4,* 318–326.

Pleck, J. H. (1977). The work-family role system. *Social Forces, 24,* 417–427.

Pleck, J. H. (1985). *Working wives/working husbands.* Thousand Oaks, CA: Sage.

Pleck, J., Staines, G., & Lang, L. (1980). Conflicts between work and family life. *Monthly Labor Review, 103*(3), 29–32.

Powell, G. N., & Mainiero, L. A. (1999). Managerial decision making regarding alternative work arrangements. *Journal of Occupational and Organizational Psychology, 72,* 41–56.

Ripple, M. (1999). Supervisors beware: The Family and Medical Leave Act may be hazardous to your health. *Journal of Contemporary Health Law and Policy, 16,* 273–303.

Scandura, T. A., & Lankau, M. J. (1997). Relationships of gender, family responsibility, and flexible work hours to organizational commitment and job satisfaction. *Journal of Organizational Behavior, 18,* 377–391.

Seligman M., & Csikszentmihalyi, M. (2000). Positive psychology. *American Psychologist, 55,* 5–14.

Spector, P. E., & Goh, A. (2001). The role of emotions in the occupational stress process. In P. L. Perrewé & D. C. Ganster (Eds.), *Research in occupational stress and well-being: Vol. 1. Exploring theoretical mechanisms and perspectives* (pp. 195–232). New York: Elsevier.

Terry D., & Jimmieson, N. (1999). Work control and employee well-being: A decade review. In C. L. Cooper & I. T. Robertson (Eds.), *International review of industrial and organizational psychology* (Vol. 14, pp. 95–148). New York: Wiley.

Thomas, L., & Ganster, D. (1995). Impact of family-supportive work variables on work-family conflict and strain: A control perspective. *Journal of Applied Psychology, 80,* 6–15.

Twomey, R., & Jones, G. (1999). The Family and Medical Leave Act of 1993: A longitudinal study of male and female perceptions. *Employee Rights and Employment Policy Journal, 3,* 229–249.

Uchino, B. N., Cacioppo, J. T., & Kiecolt-Glaser, K. G. (1996). The relationships between social support and physiological processes: A review with emphasis on underlying mechanisms and implications for health. *Psychological Bulletin, 119,* 488–531.

Wallace, J. E. (1997). It's about time: A study of hours worked and work spillover among law firm lawyers. *Journal of Vocational Behavior, 50,* 227–248.

Westman, M. (2002). Crossover of stress and strain in the family and in the workplace. In P. L. Perrewé & D. C. Ganster (Eds.), *Research in occupational stress and well-being: Vol. 2. Historical and current perspectives on stress and health* (pp. 143–181). New York: Elsevier.

Wethington, E. (2000). Contagion of stress. *Advances in Group Processes, 17,* 229–253.

Williams, J., & Alliger, M. (1994). Role stressors, mood spillover, and perceptions of work-family conflict in employed parents. *Academy of Management, 37,* 837–868.

Yang, N., Chen, C. C., Choi, J., & Zou, Y. (2000). Sources of work-family conflict: A Sino-U.S. comparison of the effects of work and family demands. *Academy of Management Journal, 43,* 113–123.

Zedeck, S. (1992). Introduction: Exploring the domain of work and family concerns. In S. Zedeck (Ed.), *Work, families, and organizations: Frontiers of industrial and organizational psychology* (Vol. 5, pp. 1–32). San Francisco: Jossey-Bass.

CHAPTER 11

Workplace Health Promotion

Amanda Griffiths
Fehmidah Munir

This chapter provides a brief review of the development of workplace health promotion (WHP). It focuses on the possible effectiveness of WHP programs for both individual and organizational health. It considers a definition of WHP and the nature of its objectives, then gives a brief account of WHP's history and an outline of the range of programs represented in the scientific literature. The chapter explores the effectiveness of such programs, and the broad challenges involved in their evaluation, and follows this with an in-depth account of one type of program—targeted at encouraging smoking cessation. The chapter concludes with a discussion of future research needs and an exploration of the benefits that would be obtained by adopting a more theoretical, integrated, and comprehensive approach than has hitherto been common.

Definition and Objectives of WHP

Early definitions of health promotion describe it as a process of enabling people to increase control over, and to improve, their health (World Health Organization [WHO], 1984). Specifically, WHO (1998, p. 21) suggests that "informing, motivating and supporting individuals, groups, and societies to lead a more healthy life can best be done by focusing systematically on where people live, work, and play." Although the workplace is widely recognized as a useful place in which to undertake such activities, the reasons for this belief vary (Cox, 1997). From a public health point of view, the workplace has often been thought of as a convenient access point to the majority of an adult population. Workplaces provide easy access to a relatively stable population: people spend a large amount of their adult life at work, and public health agencies can make use of established administrative and communication channels. In addition, the workplace has readily available social support (known to be an important influence on health-related behavior), and arranging access to expert health staff is straightforward. Thus, many see the workplace as a convenient stage on which to tackle major risks to public health. From a human resource management (HRM) perspective, at least in countries where employers have financial responsibility for individual employees' ill health, WHP is often seen as a potential means to reduce health-related costs. An occupational health specialist might take another perspective, viewing WHP as a means to improve the individual employee's health, fitness, and availability for work. And again, because researchers widely recognize that the workplace can itself represent a source of danger and ill health (Chu, Driscoll, & Dwyer, 1997), safety managers, occupational hygienists, and more recently occupational health psychologists are also concerned more generally to protect and promote both individual and organizational health.

The literature cites a number of reasons why employers implement WHP programs. They include the politics of being seen to invest in staff; enhancing the organization's external image; recruiting and retaining high performers; improving organizational commitment; enhancing physical and psychological health; decreasing absenteeism; improving work performance and morale; facilitating resistance to stress; and for some, an altruistic concern for employee welfare (Cox, Gotts, Boot, & Kerr, 1988; Warner, Wickizer, Wolfe, Schildroth, & Samuelson, 1988). Although these reasons may have validity and appear self-justifying, they have not necessarily always been based on scientific evidence about the effectiveness of WHP (Griffiths, 1996). However, because many programs have not been accompanied by any formal goals or objectives (Fielding & Piserchia, 1989), evaluating their effectiveness is problematic—an issue we cover later in this chapter.

What exactly is WHP promoting? As Hofmann and Tetrick discussed in Chapter One of this volume, the early medical concept of health is more readily defined than the concept of wellness. Health has traditionally referred to the absence of disease. Wellness or well-being, on the other hand, is more value-laden and subjective, and it may even be spurious in that it might derive from factors that are ultimately detrimental to an individual's long-term health. Narcotic-induced euphoria, for example, might well enhance feelings of well-being but would not stand up to much scrutiny as reflecting true well-being or a state of positive health (Downie, Tannahill, & Tannahill, 1996). True well-being implies some kind of reference to a conception of an agreed good life, as determined by cultural or social mores. However, even physical ill health has a subjective component in terms of nausea or discomfort and objective components such as disease or injury. The relationship between disease, fitness, and well-being is complex. It is possible to feel well when one is not particularly fit or is even in the presence of a serious disease. It is also possible to be free of illness yet not feel well. Even objective states are relative: joint stiffness might be normal for a 60-year-old but a sign of illness in a 20-year-old (Downie et al., 1996). Thus, ill health can be said to be abnormal, unwanted, or incapacitating. These higher order considerations are absent from the majority of the WHP literature and yet are crucial in considering WHP's actual and potential impact.

U.S. and European surveys (Fielding, 1991; Malzon & Lindsay, 1992; Stokols, Pelletier, & Fielding, 1996) suggest that the most common WHP activities carried out at the workplace concern smoking cessation, health risk appraisals (screening), back care, fitness or physical activity and exercise, stress management, and diet. But WHP can also involve individual stress management, relaxation training, hypertension control, cardiovascular health, sleeping patterns, advice on alcohol consumption, time management skills training, psychotherapy and counseling, career planning, work-life balance, retirement planning, and financial or legal advice. Health promotion programs are usually implemented to help people better their health rather than necessarily to help them attain a specific level of health. Health is, after all, not an absolute concept. Thus, the objectives of health promotion might be usefully described as "the balanced enhancement of physical, mental, and social facets of positive health, coupled with the prevention of physical, mental, and social ill health" (Downie et al., 1996, p. 26). Nonetheless, WHP has traditionally focused on the prevention of ill health rather than on the promotion of positive well-being, and on the prevention of physical ill health rather than mental or social ill health.

In the field of general health promotion, the prevention of ill health has been described as taking four forms: (1) prevention of onset of disease via risk reduction (e.g., taking measures to lower blood pressure and serum cholesterol in order to prevent cardiovascular disease), (2) prevention of disease progression via early detection (e.g., screening for cancers), (3) prevention of avoidable complications (e.g., pressure sores), and (4) prevention of the recurrence of an illness (Downie et al., 1996). In the specific case of health promotion in the workplace, various other more organizationally specific goals may be added to this list of individually focused concerns, such as reducing absence, turnover, and health-care costs and increasing job satisfaction or productivity.

Shephard (2002) has pointed out that attempting to change the behavior of employees raises ethical and professional issues. In particular, he notes that individuals maintain a right to remain unhealthy should they so wish and that health workers should consider the multifactorial influences on health before blaming individuals entirely for lifestyle-related disease. A strong culture of blaming individuals for unhealthy behavior has distracted attention from the more challenging question of why people behave the way they do. Further ethical issues can arise in conflicts of interest between the employer's financial concerns, the individual's privacy, and physicians' concerns about the well-being of their patients (Stokols et al., 1996). These more philosophical considerations and current views as to the nature and purpose of WHP have only a short history but one that has been influenced by a background of broader political and economic concerns.

History and Development of WHP

The United States, where health care is largely funded by employers and the insurance industry, has a longer history of concern for employee health than do countries such as the United Kingdom, which has a history of health care being provided by the state. In many countries the first generation of WHP-related activities was established for a variety of reasons, not all of which were ostensibly concerned with health (Chu, Driscoll, & Dwyer, 1997; Wilson, Holman, & Hammock, 1996). For example, some organizations introduced nonsmoking policies solely for reasons of hygiene or safety. In the United States many early programs related to employers' concerns about the effects of employee alcoholism. A second generation of WHP programs typically concerned a single health-related risk factor, targeted at particular populations. For example, fitness programs emerged in the 1960s in the United States (Shephard, 1986) but usually were offered only to white-collar workers or senior management. In the United Kingdom, an interest in fitness grew from the sporting activities of work-related social clubs in the 1970s (largely among blue-collar workers) as well as from corporate clubs offering benefits to senior managers. The vast majority of studies on WHP concerns these individual, single-issue programs focusing on the modification of individual behavior. They have usually ignored the broader and multifactorial influences of employee health such as those of the organization, family, and community or matters of economy. However, the workplace cannot be treated as an isolated and independent component of the system in which it exists; it is one of many influential factors in a wider environment (Baker, Israel, & Schurman, 1996; Cox & Griffiths, 1998). For example, in order to facilitate sustainable dietary change, a national multisectoral strategy is required, involving economic, agricultural, and retail policies that increase the availability of healthy foods (WHO, 1998), not simply changes in employees' knowledge, attitudes, or behavior and an overview of organizational catering facilities. Partly as a result of health workers' coming to understand this, a third generation of WHP programs that offers a more comprehensive range of interventions to all employees is now emerging. We will examine these in the discussion section at the end of this chapter.

Nowadays, WHP programs are more commonly offered to all sections of the workforce, but there is still some inequality in provision. A European survey revealed that WHP was more likely to be offered to permanent employees in large organizations (Stokols et al., 1996) and in state-owned organizations from affluent countries whose governments supported those activities (Malzon & Lindsay, 1992). In such organizations the workforce is predominantly male. In contrast, small and privately owned organizations in less affluent countries with predominantly female workforces are less likely to provide such programs. Scholars have also suggested that organizations with authoritarian management styles are less likely to implement WHP programs than are those whose style is democratic (Witte, 1993). The availability of WHP is also limited among highly mobile workers; those employed by small businesses; workers who reside in rural areas; racially and ethnically diverse populations; and individuals working in the mining, agriculture, construction, and manufacturing industries (Stokols et al., 1996).

Who are the dominant actors in health promotion? Researchers have expressed concerns recently about the overmedicalization of health, where health becomes a commodity that can be bought by insurance or paid for by the state, where a so-called expert (usually medical) is on hand to solve all problems, and where ordinary people no longer feel empowered to take responsibility for their own health (Downie et al., 1996). It is increasingly clear that health promotion is a multidisciplinary activity, involving medical practitioners, sociologists, educationalists, management scientists, psychologists, lawyers, and economists; more recently, health promotion has been accompanied by a widening definition of health and a recognition of the importance of perceptual, cognitive, and behavioral issues.

Evaluating the Effectiveness of WHP Programs

Many of the early studies on WHP were characterized by poor methodology. Since the 1990s, their quality has improved considerably—a matter that we will discuss in this section.

The current consensus on the effectiveness of WHP is one of "cautious optimism" (Heaney & Goetzel, 1997, p. 306). Published reviews state that the evidence is suggestive for exercise; suggestive or indicative for diet and cholesterol; indicative for weight control (Wilson et al., 1996); indicative for stress management; and suggestive and weak, respectively, for alcohol and HIV/AIDS, largely due to the lack of experimental studies (Wilson, 1996). Among the major programs that have been more carefully evaluated than many is the Johnson and Johnson program (e.g., Breslow, Fielding, Herrmann, & Wilbur, 1990; Holzbach et al., 1990; Jones, Bly, & Richardson, 1990). For example, Holzbach et al. (1990) reported on a two-year evaluation of a comprehensive health program covering exercise, weight control, nutrition, smoking cessation, stress management, high blood pressure, and alcohol. Findings indicated that the program had a positive effect on exercise and smoking cessation, coupled with a reduction of both costs of corporate health benefits and absenteeism among employees. Holzbach et al. also found significant improvements in organizational commitment. Not all WHP programs report positive effects, particularly when researchers evaluate them carefully. However, note that the failure of some studies to demonstrate statistically significant change does not necessarily mean that no such change is taking place; it may mean instead that change remains to be demonstrated by studies using rigorous research designs (Dishman, Oldenburg, O'Neal, & Shephard, 1998). These latter authors also point out that even a statistically very small positive effect such as increasing the fitness of six people per 100 may be equivalent to reducing the population risk for cardiovascular morbidity and overall mortality to a meaningful degree.

Stokols et al. (1996) have suggested that the following strategies show promise for increasing WHP effectiveness:

- Intensive marketing at sign-up and continuation stages
- Targeting all employees but providing higher intensity support for those with demonstrated high risk factors
- Facilitating employee participation
- Implementing comprehensive programs that target multiple risks
- Providing telephone counseling and support

Research on the effectiveness of WHP programs appears to be increasing. Stokols et al. (1996) point out that although increasing evidence indicates the health and cost benefits of certain WHP interventions (they cite examples such as weight loss, smoking cessation, hypertension control, health risk appraisals, and behavioral change programs), many studies have not been rigorously designed. Typical methodological flaws have been lack of attention to sampling theory (notably the issue of the self-selection of participants or "preaching to the converted"), an absence of baseline measures, an absence of control or comparison groups, an overreliance on self-report data, the use of all-or-none measures, short-term follow-ups, introduction of bias by program advocates conducting evaluations, failure to state formal objectives or to examine drop-out rate (adherence), and failure to assess organizational outcomes or cost effectiveness (Griffiths, 1996). Nonetheless, Fielding (1991) describes how the designs used in published studies have undergone considerable improvement with time. Whereas initial studies rarely employed control groups, later studies used nonrandomized control groups; several of the most recent studies have employed random assignment of participants to groups. At the same time, researchers have moved away from the use and evaluation of single-component health promotion programs toward the consideration of multicomponent programs.

Evaluation of the effectiveness of a WHP program could take place on at least four different levels (Cox & Griffiths, 1998). Level one represents the impact of health promotion on individuals and its effects on their perceptions, beliefs and values, and health-related behavior and health status. Level two represents the impact and aggregated effects of health promotion at the level of the community (public health) or the working group (occupational health) and particularly on those diseases and disorders that are made manifest in or particular to those environments. Level three represents the impact of health promotion on the function and productivity or performance of organizations, and level four represents its impact on the wider socioeconomic environment. In the early days of WHP, the focus was largely on level one, but since 1990 an increasing number of publications have taken the broader view. In addition to the issue of what researchers are measuring, one may evaluate programs in terms of how they measure it. Researchers can ask questions about the psychometric robustness of these measures, including concerns about reliability, validity, sensitivity, and lack of bias (Cox & Griffiths, 1998). In addition, it is important to know when something is measured; most studies have only short-term follow-ups. Future research needs to involve large-scale and long-term evaluations. For example, researchers suggest that exercise is particularly beneficial for the overall health and cognitive function of older people. Recent models of the neurobiological bases of such findings suggest that exercise enhances brain health and plasticity, increasing levels of brain-derived neurotrophic factor and other growth factors, stimulating neurogenesis, increasing resistance to brain insult and improving learning and mental performance (Cotman & Berchtold, 2002). The impact in organizational terms of any such cognitive benefits (e.g., creativity and innovation) may become evident only after a prolonged period (Stokols et al., 1996). Paradoxically, very long-term evaluations may show that successful interventions are associated with increased longevity and therefore increased costs such as those related to pensions (Shephard, 1992).

Not many studies stand up to rigorous scrutiny against the principles we outlined earlier. However, putting into effect methodologically sound interventions in working organizations is extremely challenging. There is a danger that a counsel of perfection will put off attempts to make real progress. Exhortations from the traditional positivist scientific establishment to achieve such methodological perfection may discourage progress and throw the baby out with the bathwater. Instead, researchers should acknowledge any weaknesses of their designs against experimental principles and attempt to solve the issues raised creatively, developing new techniques (Griffiths, 1999). Further, interventions are complex, unpredictable social processes, and yet researchers fondly believe that what they intended as the intervention is what actually happened. This is not usually the case (Nytrø, Saksvik, Mikkelsen, Bohle, & Quinlan, 2000). It is not usually a question of employees receiving or not receiving an intervention; it is not a simple dichotomy. Many subtle levels of variation occur. One way forward might be to identify and use these natural individual variations as part of an evaluation (Randall, Griffiths, & Cox, 2001). Yin (1994) has described these as partial comparisons and attempts to exploit the uncontrolled nature of work settings. Note that much of the published WHP evaluation research is exclusively quantitative and outcome-related. Examination of qualitative and process issues—very important in explaining significant and nonsignificant changes after interventions—have been largely ignored (Griffiths, 1999).

Having outlined the general issues involved in evaluating WHP programs, we now turn to an account of the most common type of program: smoking cessation.

Workplace Smoking Cessation Programs

The scientific literature on workplace smoking cessation programs is impressive both in its quantity and quality. It highlights the wide variety of psychological, behavioral, social, organizational, economic, legislative, and ethical issues presented by WHP. For these reasons we provide a relatively in-depth coverage of this issue as an example.

Researchers estimate that about 120,000 adults in the United Kingdom and 430,000 in the United States die from smoking each year (Callum, 1998; U.S. Department of Health and Human Services, 1993), figures that place smoking as the largest preventable cause of premature mortality in both countries. Further, studies associate passive smoking, or exposure to environmental tobacco smoke, with lung cancer as well as respiratory and cardiovascular disease; researchers have suggested it as the third largest preventable cause of premature mortality (Eisner, Smith, & Blanc, 1998; Glantz & Parmley, 1991). According to WHO (1999), smoking is responsible for over 22,000 deaths per year in the European Union. Although some have argued that, with a lower life expectancy, smokers will consume fewer health resources in later life, smokers' lifetime health-care expenditures in fact usually outweigh those of nonsmokers (Parrott, Godfrey, & Raw, 2000). Smoking is fostered not only via the addictive properties of nicotine but also by the tobacco industry's continued and aggressive marketing and its alleged efforts to engineer increased addictiveness (Quinn, Sengupta, & Cleary, 2001). Public health authorities in many countries now recognize that tobacco control is important for the health of current and future generations. Workplaces are a common site for action.

In addition to the direct medical and, more recently, medico-legal costs of smoking, employers have cause for concern. Smokers have been reported to use more health resources than do their nonsmoking colleagues (Penner & Penner, 1990). They lose time in smoking breaks and are absent from work through sickness significantly more often, even demonstrating a dose-response relationship between number of cigarettes smoked per day and days absent (Hawker & Holtby, 1988). This equates to a significant loss of productivity. Prospective cohort studies have shown that workplace productivity increases and absenteeism decreases among employees after they have given up smoking, again with a demonstrated dose-response relationship between years of cessation and absenteeism (Halpern, Shikiar, Rentz, & Khan, 2001). Cost estimates of the effects of smoking on productivity have been produced largely from North America but also from Canada, Australia, and the United Kingdom (Parrott et al., 2000); in all cases the losses are considerable. In addition, insurance premiums may be higher as a result of claims for fire damage, and tobacco smoke can damage plant infrastructure and machinery. Studies have also suggested that smokers have substantially more occupational injuries and accidents than do nonsmoking employees (Halpern et al., 2001). Programs to assist smokers to give up smoking thus clearly offer significant potential for cost savings in the workplace, as well as health benefits for smokers and nonsmokers alike. It is not clear whether accident and injury rates would improve among former smokers, because these may be associated with other more persistent social and behavioral characteristics. This literature does not frequently explore the multifactorial nature of workplace health.

Employers have become aware of the need to protect employees (both smokers and nonsmokers) from the health dangers of smoking, and themselves from incurred costs, and many have introduced a range of initiatives designed to promote smoking cessation. As with many WHP programs, these vary from more general systems-level initiatives such as smoking policies to very specific, individually focused smoking cessation programs. Evaluating the effectiveness of these activities brings an interesting array of methodological challenges, such as determining how to measure cessation, deciding how often and when to follow up with employees, establishing meaningful comparison groups, or exploring data creatively when predetermined comparison groups are not possible (as is often the case).

Smoking policies attempt to create an environment where either smoking is completely prohibited or nonsmoking is the accepted norm, with smoking restricted to separate areas. Complete prohibition at work can be difficult to police and may encourage illicit behaviors or unwanted effects such as smokers congregating at the entrance to workplaces, which may lead to accumulated smoking debris and a poor public image (Parry, Platt, & Thomson, 2000). Nonetheless, evidence is accumulating from longitudinal studies that employees in workplaces with smoking bans have higher and faster rates of smoking cessation than do employees in workplaces that permit smoking (Longo, Johnson, Kruse, Brownson, & Hewett, 2001; Moskowitz, Zihua, & Hudes, 2000). Workplace policies that ban smoking are twice as effective at reducing consumption and prevalence as are policies that permit smoking in designated areas (Fichtenberg & Glantz, 2002). Researchers and policymakers are suggesting that the widespread adoption of total smoking bans at the workplace could have major beneficial health effects both for smokers and for nonsmokers exposed to environmental tobacco use (Eriksen & Gottlieb, 1998; Task Force on Community Preventive Services, 2001). A recent review of the effects of smoke-free policies at work concluded that the combined effects of reduced prevalence of smokers together with a lower consumption per continuing smoker reduced cigarette consumption by 29% (Fichtenberg & Glantz, 2002). These authors calculate that if all workplaces that are currently not smoke-free in the United States and the United Kingdom were to become smoke-free, cigarette consumption for the entire adult population would drop by 4.5 and 7.6%, respectively, costing the tobacco industry $1.7 billion annually in lost sales. They estimate that achieving the same result via increases to the tax on tobacco would require a 47% increase in the United States and a 24% increase in the United Kingdom.

In addition to the positive effects of a smoke-free workplace for smokers' health, studies have documented the benefits for nonsmokers' health. A study of bartenders in San Francisco, California, demonstrated convincingly that the reduced exposure to environmental tobacco smoke that followed the prohibition of tobacco in California bars and taverns in January 1998 was associated with the fast-improved respiratory health of those employees (Eisner et al., 1998). Another study suggests that as much as half of the economic benefits of workplace smoking cessation programs benefit the broader community (Warner, Smith, Smith, & Fries, 1996).

Smoking cessation programs vary enormously in their scope and methods. They may be tailored to age (e.g., adolescents or adults), to reasons for smoking (e.g., nicotine dependence, affect control, boosting self-esteem, coping with stress), or to motivation and readiness for change (Singleton & Pope, 2000). They may focus on self-help or on counselor assistance (the latter generally proves more effective) and on individual or group programs. Although many people prefer to stop smoking on their own, most seem to benefit from assistance (Singleton & Pope, 2000). In studies comparing behavioral strategies with cognitive strategies, researchers observe that people generally find the latter easier to adopt (Glasgow, Klesges, Mizes, & Pechacek, 1985); another study strongly recommends telephone advice and support, as well as mass media campaigns (Task Force on Community Preventive Services, 2001). In general, multicomponent, intensive programs yield the best results (Eriksen & Gottlieb, 1998).

An important point about the effectiveness of workplace smoking cessation programs is that they do not focus on smokers who are highly motivated and ready to quit; they target smokers at all stages of readiness and create a climate supportive of the desired behaviors. Because of this, although broad-based workplace programs often result in lower cessation rates than do initiatives that target only smokers who are ready to quit, these programs may have a greater impact on workforce health in the long term by persuading smokers who had not otherwise considered quitting to do so (Sorensen, 2001).

According to Sorensen (2001), other important factors for success or failure include the following:

- Senior management commitment (visible in mission statements, budgetary allocations, and nomination of responsible persons)
- Supervisor and coworker support
- An effective initial analysis of employees' perceptions of their own needs and priorities
- The involvement of unions
- Clear policies on taking breaks (such that smoking is not the only valid reason for stopping work)
- Awareness of work stress
- Allowing workers to attend programs on company time
- Programs structured around the schedules of line workers
- Free access, particularly for workers on low incomes, to nicotine replacement therapy, which has been shown to double quit rates

Low participation rates present a major challenge for all WHP programs. Smoking cessation initiatives are no exception. Although many smokers are aware of the health risks and many would like to give up smoking, very few take advantage of available cessation programs, even though these are reasonably effective. Greenwood (2001) suggests that low participation rates of smoking cessation interventions may be partly due to four causes: (1) too narrowly cast recruitment messages, (2) inadequate exposure to campaign messages, (3) lack of immediate and accessible enrollment mechanisms, and (4) the use of passive communication channels (e.g., mass media campaigns or bulletin board flyers) as opposed to active channels (live telephone recruitment or interpersonal communication).

Research has shown that providing incentives (money, competitions, prizes, and so on) increases participation rates in workplace smoking cessation programs (Koffman, Lee, Hopp, & Emont, 1998), sometimes to twice the participation rates of worksites that offer no incentives (Hennrikus et al., 2002). However, it is not yet clear whether incentives also translate into increased cessation rates (in addition to participation rates) among a group of smokers. Perhaps with large enough incentives, and as part of a comprehensive program, incentives might boost success (cessation) rates.

In summary, studies have carefully demonstrated that workplace smoking cessation programs have clear benefits for smoking and nonsmoking employees and for their employing organizations. Research has also indicated where and how organizations may yet achieve further increases in smoking cessation rates.

WHP Programs: Discussion

Although we have reason to be cautiously optimistic about the effectiveness of WHP programs, the current literature points to three useful steps forward: (1) a recognition of the value of a more theoretical approach, (2) better integration among key stakeholders, and (3) more comprehensive programs. We will now explore these in more detail.

First, researchers suggest that greater use should be made of theories and frameworks; most of the literature has been and remains atheoretical (Griffiths, 1996; Sonstroem, 1988). In this respect, the stages of change model (Prochaska and DiClemente, 1984) has proved to be a useful development from the traditional all-or-none evaluation strategies in some, but not all, populations. The model defines behavior and intentions in five stages:

1. Precontemplation: no intention to change, denying the need to change
2. Contemplation: intention to change
3. Preparation: making small changes
4. Action: regular engagement in change
5. Maintenance: continuation of change efforts

The use of this framework facilitates a more sensitive evaluation (Jaffee, Lutter, Rex, Hawkes, & Bucaccio, 1999; Titze, Martin, Seiler, Stronegger, & Marti, 2001; Wilson et al., 1996). Similarly, using motivation theory, one study measured employees' stage of readiness for exercise and found that those who received motivationally tailored exercise programs were significantly more likely to show increases in motivation and to increase the amount of time they spent exercising than did those who received a standard program (Marcus et al., 1998). Research has demonstrated the dangers of an inadequate understanding of people's perceptions of health risks, as the following example demonstrates. Smokers in general tend to demonstrate "optimistic bias" in that they minimize their personal health risks (Weinstein, 1998). Studies have used imagination tasks (known to be helpful in reducing optimistic bias) in which participants imagine a severe smoking-related disease happening to them as a result of their smoking and to describe the consequences (Myers & Frost, 2002). Results of these studies show that for people who are relatively optimistic (unrealistic) about smoking-related disease, such interventions engage them in thinking about the severe consequences of their behavior, succeed in changing their risk perceptions to more realistic levels, and may be successful in motivating them to quit. However, for those smokers who are already motivated to quit (and are therefore realistic about their chances of contracting smoking-related diseases) such interventions have the opposite effect to that intended. Perhaps these individuals find the intervention scenario so threatening that they react defensively, resulting in comparative optimism. Several large interventions concerned with reducing the risk of cancer have been based on a combination of theoretical and program-planning models such as cognitive theory (Bandura, 1986), the theory of reasoned action (Ajzen & Fishbein, 1980) as well as the transtheoretical model and its stages of change construct (Prochaska & DiClemente, 1984). These theoretically based studies also support the value of measuring moderating or mediating factors (e.g., perceived benefits or social support) as part of intervention evaluations. And finally, as well as increasing success among participants, Linnan, Sorenson, Colditz, Klar, and Emmons (2001) suggest that more sociobehavioral theoretical approaches, which acknowledge the many levels of influences on participation (intrapersonal, interpersonal, and organizational) in WHP programs, offer a promising way forward for increasing participation.

Second, researchers increasingly call for adoption of a more integrated approach to WHP and protection, one that moves away from regarding the workplace simply as a stage for attempting to change individuals. The current split between health promotion, occupational health, safety and job design, for example—both in terms of the actors responsible, activities undertaken, and disciplines involved—exists largely for historical reasons that may no longer be justifiable in today's workplace. Stokols et al. (1996, p. 137) suggest the emergence of a paradigm shift away from individually focused corporate programs to a concern for the broader "ecology of work and health." WHP may be best integrated with other related systems (such as occupational health, health and safety, human resource management) and more generally into organizational systems culture, as has traditionally been the case in Scandinavia and Finland. There are calls for increasing collaboration among the various stakeholders within organizations (Cox, 1997) as well as suggestions that it may be more useful to view WHP activities in terms of roles (decision maker, advocate, expert, change facilitator, deliverer, participant) rather than occupational titles (Wynne, 1994). Many of the sections on future areas for research in the conclusions of recent papers call for more investigation into the basic processes that facilitate the implementation of WHP programs and boost individual participation; these are not clinical but rather psychological, social, and organizational issues.

It may be that interfunction rivalry at the organizational level presents a barrier to an integrated approach. In addition to pleas for a broader organizational approach to WHP, Stokols et al. (1996) point out that although isolated organization-specific wellness interventions may have made sense when the nature, security, and location of work was stable, current trends mean that such stability no longer applies to many in today's workforce. For these reasons the researchers recommend that future research should adopt a broader view that sees the workplace as part of a larger community system in which organizational, educational, medical, technological, and regulatory strategies combine to affect the health of employees and their families.

Third, many leading authorities on WHP are recommending the potential benefits of more comprehensive programs rather than one-off short-duration programs that target only one health risk. In Germany and the Nordic countries, for example, WHP has traditionally been a much wider concept, more akin to human resource development and a systems-level issue, typically embracing many activities that could be seen as providing healthful effects. Other chapters in this volume discuss some of these activities. A comprehensive approach is more theoretically informed and takes account of the literature that demonstrates the multiple factors influencing worker health and the associations between health risk behaviors. Scholars know, for example, that smokers are more likely to demonstrate other unhealthy behaviors such as alcoholism (Ryan, Zwerling, & Endel, 1992) and that the presence of additional health risk factors strongly affects the prognosis for smoking cessation (Sherwood, Hennrikus, Jeffery, Lando, & Murray, 2000). Some have suggested, for example, that young men may respond to an underdemanding work environment by increasing their alcohol consumption, as do employees under stress (Emmons, Linnan, Shadel, Marcus, & Abrams, 1999; Hemmingsson & Lundberg, 1998; Ragland, Greiner, Yen, & Fisher, 2000). Evidence suggests that exercise may serve as a gateway to change in other health behaviors (Emmons et al., 1999). These interrelationships between risk factors have important implications for the design and delivery of interventions. Comprehensive programs address several risk factors (both individual and organizational) simultaneously, allowing for the possibility that different factors may appeal to different sections of the workforce and enhancing opportunities for sustainable change (Price, Mackay, & Swinburn, 2000). Examples of evaluated comprehensive programs are rare, but we will provide two.

In the Netherlands a comprehensive WHP program was designed for Brabantia, a manufacturer of household goods. Interventions concerned both individual lifestyle changes (nutrition, alcohol and drug consumption, stress, smoking behavior, headaches, back pain, and physical exercise) together with changes in working conditions (management training and reorganization of the production line). Overall, the program resulted in favorable changes in working conditions, fewer health risks, and reduced absenteeism (Maes, Veroenven, France, & Scholten, 1998). In Canada the Royal Bank Financial Group implemented a program designed to enhance the work-life balance of its 60,000 employees. Its various elements, used by up to half of its employees at any given time, include a family care information service, formal flextime policies, and a considerable amount of flexibility on the part of supervisors to respond to the individual needs of employees. In the late 1980s, 81 percent of employees reported being able to manage their work-family responsibilities more easily; 70% reported lower levels of stress; 52% reported increased participation in exercise, education, and community activities; and 65% perceived themselves to have more energy. There were no detrimental effects on retention, employee commitment, nor on customer satisfaction (Tombari & Spinks, 1999). Such programs recognize that conflict between the demands of home and working lives can be stressful and that broad-ranging and flexible solutions are available. For example, when a union dispute closed many schools in Toronto, the same company installed emergency schooling at its offices for its employees' children within 48 hours of the announcement (Tombari & Spinks, 1999).

As these programs have demonstrated, a comprehensive approach to WHP can include attention to both individual and systemwide factors. Some programs have examined organizationally relevant variables such as staff morale and attitudes toward the organization. Some evidence suggests that the implementation of WHP programs is associated with improved attitudes toward the organization, even among those employees who have not actually used the services provided (e.g., Holzbach et al., 1990). However, research thus far has found less impressive evidence for favorable organizational outcomes compared with positive outcomes for individuals. For example, research reveals that physical activity programs usually result in an increase in individual exercise behavior (e.g., Emmons et al., 1999), but a review of eight studies suggested that their effect on organizational outcomes was limited for absenteeism; inconclusive for job satisfaction, job stress; and employee turnover; and nonexistent for productivity (Proper, Staal, Hildebrandt, van der Beek, & van Mechelsen, 2002). The authors of this review point out that the methodological limitations of these studies preclude firm conclusions. Although challenging for researchers, a wider range of productivity and organizational effectiveness measures needs to be incorporated in future studies (Stokols et al., 1996; Wilson et al., 1996).

Many now think that offering comprehensive programs, particularly for high risks groups, may be the most effective way forward (Heaney & Goetzel, 1997; Sorensen, 2001; Wilson et al., 1996). The example of smoking cessation demonstrates the various levels at which WHP can operate: changing legislation, changing organizational policies, reducing exposure to workplace hazards, modifying job design, encouraging social support at work, involving key players outside work, providing links to relevant community initiatives, and promoting health enhancement by targeting individual workers' behavior. Traditionally, many of these activities have been the territory of policymakers, safety specialists (health protection), human resource management, and organizational psychology, whereas only individual behavior change has usually been conceptualized as health promotion.

A comprehensive approach to WHP offers promise, targeting one of the major challenges to public health today—health inequalities. Blue-collar workers are known to be at higher risk than are white-collar workers. They have high exposure to work-related hazards; are more likely to be smokers, particularly if they are shift workers (and smoking may act synergistically with toxic agents in the workplace); and have fewer opportunities to participate in WHP programs. Furthermore, they are less likely to participate in those programs, may demonstrate various other unhealthy behaviors, and may be less successful in changing those behaviors

(Sorensen, 2001). These multiple risks for this and other vulnerable groups could be addressed in comprehensive workplace programs. Sorensen et al. (1996) have noted examples of increased success rates, especially for skilled and unskilled workers, in programs that integrate health promotion with health protection. (In their study an integrated program yielded a smoking cessation rate of 17.3%, compared with 12.7% from single interventions.) In an integrated system, all activities can work in synergy to promote employee health. This represents a step-change from conceptualizations of WHP as aiming to encourage individuals to take more responsibility for their own health. The new approach locates responsibility for promoting individual health with employees, with their managers, and with the wider community.

References

Ajzen, I., & Fishbein, M. (1980). *Understanding attitudes and predicting behavior.* Englewood Cliffs, NJ: Prentice Hall.

Baker, E., Israel, B., & Schurman S. (1996). The integrated model: Implications for worksite health promotion and occupational health and safety practice. *Health Education Quarterly, 23,* 175–190.

Bandura, A. (1986). *Social foundations of thought and action: A social cognitive theory.* Englewood Cliffs, NJ: Prentice Hall.

Breslow, L., Fielding, J., Herrmann, A., & Wilbur, C. (1990). Worksite health promotion: Its evolution and the Johnson and Johnson experience. *Preventative Medicine, 19,* 13–21.

Callum, C. (1998). *The UK smoking epidemic: Deaths in 1995.* London: Health Education Authority.

Chu, C., Driscoll, T., & Dwyer, S. (1997). The health-promoting workplace: An integrative perspective. *Australian and New Zealand Journal of Public Health, 21,* 377–386.

Cotman, C. W., & Berchtold, N. C. (2002). Exercise: A behavioral intervention to enhance brain health and plasticity. *Trends in Neurosciences, 25,* 295–301.

Cox, T. (1997). Workplace health promotion. *Work and Stress, 11,* 1–5.

Cox, T., Gotts, G., Boot, N., & Kerr, J. H. (1998). Physical exercise, employee fitness, and the management of health at work. *Work and Stress, 2,* 71–77.

Cox, T., & Griffiths, A. J. (1998). Rethinking the nature of workplace health promotion: A European model. In J. Mossink & F. Licher (Eds.), *Costs and benefits of occupational safety and health* (pp. 307–313). The Hague: NIA-TNO.

Dishman, R., Oldenburg, B., O'Neal, H., & Shephard, R. (1998). Worksite physical activity interventions. *American Journal of Preventive Medicine, 15,* 344–361.

Downie, R. S., Tannahill, C., & Tannahill, A. (1996). *Health promotion: Models and values.* Oxford: Oxford University Press.

Eisner, M. D., Smith, A. K., & Blanc, P. D. (1998). Bartenders' respiratory health after establishment of smoke-free bars and taverns. *Journal of the American Medical Association, 280,* 1909–1914.

Emmons, K. M., Linnan, L. A., Shadel, W. G., Marcus, B., & Abrams, D. B. (1999). The Working Healthy Project: A worksite health promotion trial targeting physical activity, diet, and smoking. *Journal of Occupational and Environmental Medicine, 41,* 545–555.

Eriksen, M. P., & Gottlieb, N. H. (1998). A review of the health impact of smoking control at the workplace. *American Journal of Health Promotion, 13,* 83–104.

Fichtenberg, C. M., & Glantz, S. A. (2002). Effects of smoke-free workplaces on smoking behaviour: Systematic review. *British Medical Journal, 325,* 188–194.

Fielding, J. E. (1991). Health promotion at the worksite. In G. M. Green & F. Baker (Eds.), *Work, health, and productivity* (pp. 256–276). Oxford: Oxford University Press.

Fielding, J. E., & Piserchia, P. (1989). Frequency of worksite health promotion activities. *American Journal of Health Promotion, 79,* 16–20.

Glantz, S. A., & Parmley, W. W. (1991). Passive smoking and heart disease. *Circulation, 83,* 1–11.

Glasgow, R. E., Klesges, R. C., Mizes, J. S., & Pechacek, T. F. (1985). Quitting smoking: Strategies used and variables associated with success in a stop-smoking contest. *Journal of Clinical and Consulting Psychology, 53,* 905–912.

Greenwood, T. (2001). Comparing smoking cessation interventions for work-site disease management. *Disease Management and Health Outcomes, 9,* 565–576.

Griffiths, A. (1996). The benefits of employee exercise programmes: A review. *Work and Stress, 10,* 5–23.

Griffiths, A. (1999). Organizational interventions: Facing the limits of the natural science paradigm. *Scandinavian Journal of Work, Environment, and Health, 25,* 589–596.

Halpern, M. T., Shikiar, R., Rentz, A. M., & Khan, Z. M. (2001). Impact of smoking status on workplace absenteeism and productivity. *Tobacco Control, 10,* 233–238.

Hawker, R., & Holtby, I. (1988). Smoking and absence from work in a population of student nurses. *Public Health, 102,* 161–167.

Heaney, C., & Goetzel, R. (1997). A review of health-related outcomes of multi-component worksite health promotion programs. *American Journal of Health Promotion, 11,* 290–308.

Hemmingsson, T., & Lundberg, I. (1998). Work control, work demands, and work social support in relation to alcohol consumption among young men. *Alcoholism in Clinical and Experiential Research, 22,* 921–927.

Hennrikus, D. J., Jeffery, R. W., Lando, H. A., Murray, D. M., Brelje, K., Davidann, B., Baxter, J. S., Thai, D., Vessey, J., & Liu, J. (2002). The SUCCESS Project: The effect of the program format and incentives on participation and cessation in worksite smoking cessation programs. *American Journal of Public Health, 92,* 274–279.

Holzbach, R., Pierchia, P., McFadden, D., Hartwell, T., Herrmann, A., & Fielding, J. E. (1990). Effect of a comprehensive health promotion program on employee attitude. *Journal of Occupational Medicine, 32,* 973–978.

Jaffee, L., Lutter, J. M., Rex, J., Hawkes, C., & Bucaccio, P. (1999). Incentives and barriers to physical activity for working women. *American Journal of Health Promotion, 13,* 215–218.

Jones, R., Bly, J., & Richardson, J. (1990). A study of a work-site health promotion program and absenteeism. *Journal of Occupational Medicine, 32,* 95–99.

Koffman, D. M., Lee, J. W., Hopp, J. W., & Emont, S. L. (1998). The impact of including incentives and competition in a workplace smoking cessation program on quit rates. *American Journal of Health Promotion, 13,* 105–111.

Linnan, L., Sorenson, G., Colditz, G., Klar, N., & Emmons, K. M. (2001). Using theory to understand the multiple determinants of low participation in worksite health promotion programs. *Health Education and Behavior, 28,* 591–607.

Longo, D. R., Johnson, J. C., Kruse, R. L., Brownson, R. C., & Hewett, J. E. (2001). A prospective investigation of the impact of smoking bans on tobacco cessation and relapse. *Tobacco Control, 10,* 267–272.

Maes, S., Veroenven, C., France, K., & Scholten, H. (1998). Effects of a Dutch work-site wellness-health program: The Brabantia project. *American Journal of Public Health, 88,* 1037–1041.

Malzon, R. A., & Lindsay, G. B. (1992). *Health promotion at the worksite: A brief survey of large organizations in Europe.* Copenhagen: World Health Organization Regional Office for Europe.

Marcus, B. H., Emmons, K. M., Simkin-Silverman, L. R., Linnan, L. A., Taylor, E. R., Bock, B. C., Roberts, M. B., Rossi, J. S., & Abrams, D. B. (1998). Evaluation of motivationally tailored vs. standard self-help physical activity interventions at the workplace. *American Journal of Health Promotion, 12,* 246–253.

Moskowitz, J. M., Zihua, L., & Hudes, E. S. (2000). The impact of workplace smoking ordinances in California on smoking cessation. *American Journal of Public Health, 90,* 757–761.

Myers, L. B., & Frost, S. (2002). Smoking and smoking cessation: Modifying perceptions of risk. In D. Rutter & L. Quine (Eds.), *Changing health behaviour.* Buckingham: Open University Press.

Nytrø, K., Saksvik, P. O., Mikkelsen, A., Bohle, P., & Quinlan, M. (2000). An appraisal of key factors in the implementation of occupational stress interventions. *Work and Stress, 14,* 213–225.

Parrott, S., Godfrey, C., & Raw, M. (2000). Costs of employee smoking in the workplace in Scotland. *Tobacco Control, 9,* 187–192.

Parry, O., Platt, S., & Thomson, C. (2000). Out of sight, out of mind: Workplace smoking bans and the relocation of smoking at work. *Health Promotion International, 15,* 125–133.

Penner, M., & Penner, S. (1990). Excess insured health care costs from tobacco-using employees in a large group plan. *Journal of Occupational Medicine, 32,* 521–523.

Price, G., Mackay, S., & Swinburn, B. (2000). The Heartbeat Challenge programme: Promoting healthy changes in New Zealand workplaces. *Health Promotion International, 15,* 49–55.

Prochaska, R., & DiClemente, C. (1984). *The transtheoretical approach: Crossing traditional boundaries of change.* Irwin, IL: Homewood.

Proper, K. I., Staal, B. J., Hildebrandt, V. H., van der Beek, A. J., & van Mechelsen, W. (2002). Effectiveness of physical activity programs at worksites with respect to work-related outcomes. *Scandinavian Journal of Work, Environment, and Health, 28,* 75–84.

Quinn, J., Sengupta, S., & Cleary, H. (2001). The challenge of effectively addressing tobacco control within a health-promoting NHS Trust. *Patient Education and Counselling, 45,* 255–259.

Ragland, D. R., Greiner, B. A., Yen, I. H., & Fisher, J. M. (2000). Occupational stress factors and alcohol-related behavior in urban transit operators. *Alcoholism—Clinical and Experimental Research, 24,* 1011–1019.

Randall, R., Griffiths, A., & Cox, T. (2001). Using the uncontrolled work setting to shape the evaluation of work stress interventions. In C. E. Weikert & E. Torkelsen, (Eds.), *European Academy of Occupational Health Psychology conference proceedings series: Occupational health psychology: Europe 2001* (pp. 145–149). Nottingham: University of Nottingham, Institute of Work, Health, and Organisations.

Ryan, J., Zwerling, C., & Endel, J. O. (1992). Occupational risks associated with cigarette smoking: A prospective study. *American Journal of Public Health, 82,* 29–32.

Shephard, R. (1986). *Fitness and health in industry.* Basel: Karger.

Shephard, R. (1992). A critical analysis of work-site fitness programs and their postulated economic benefits. *Medicine and Science in Sports and Exercise, 24,* 354–370.

Shephard, R. (2002). Issues in worksite health promotion: A personal viewpoint. *Quest, 54,* 67–82.

Sherwood, N. E., Hennrikus, D. J., Jeffery, R. W., Lando, H. A., & Murray, D. M. (2000). Smokers with multiple behavioral risk factors: How are they different? *Preventive Medicine, 31,* 299–307.

Singleton, M. G., & Pope, M. (2000). A comparison of successful smoking cessation interventions for adults and adolescents. *Journal of Counselling and Development, 78,* 448–453.

Sonstroem, R. J. (1988). Psychological models. In R. K. Dishman (Ed.), *Exercise adherence: Its impact on public health* (pp. 125–153). Champaign, IL: Human Kinetics Books.

Sorensen, G. (2001). Worksite tobacco control programs: The role of occupational health. *Respiration Physiology, 128,* 89–102.

Sorensen, G., Thompson, B., Glanz, K., Feng, Z., Kinne, S., DiClemente, C., Emmons, K., Heimendinger, J., Probart, C., & Lichtenstein, E. (1996). Work site-based cancer prevention: Primary results from the Working Well Trial. *American Journal of Public Health, 86,* 939–947.

Stokols, D., Pelletier, K., & Fielding, J. (1996). The ecology of work and health: Research and policy directions for the promotion of employee health. *Health Education Quarterly, 23,* 137–158.

Task Force on Community Preventive Services. (2001). Recommendations regarding interventions to reduce tobacco use and exposure to environmental tobacco smoke. *American Journal of Preventive Medicine, 20*(2S), 10–15.

Titze, S., Martin, B. W., Seiler, R., Stronegger, W., & Marti, B. (2001). Effects of a lifestyle physical activity intervention on stages of change and energy expenditure in sedentary employees. *Psychology of Sport and Exercise, 2,* 103–116.

Tombari, N., & Spinks, N. (1999). The work/family interface at Royal Bank Financial Group: Successful solutions—a retrospective look at lessons learned. *Women in Management Review, 14,* 186–193.

U.S. Department of Health and Human Services. (1993). Cigarette smoking-attributable mortality and years of potential life lost: United States 1990. *Morbidity and Mortality Weekly Reports, 42,* 645–649.

Warner, K. E., Smith, R. J., Smith, D. G., & Fries, B. E. (1996). Health and economic implications of a work-site smoking-cessation program: A simulation analysis. *Journal of Occupational and Environmental Medicine, 38,* 981–992.

Warner, K. E., Wickizer, T. M., Wolfe, R. A., Schildroth, J. E., & Samuelson, M. H. (1988). Economic implications of workplace health promotion programs: Review of the literature. *Journal of Occupational Medicine, 30,* 106–112.

Weinstein, N. D. (1998). Accuracy of smokers' risk perceptions. *Annals of Behavioral Medicine, 20,* 135–140.

Wilson, M. (1996). A comprehensive review of the effects of worksite health promotion on health-related outcomes: An update. *American Journal of Health Promotion, 11,* 107–108.

Wilson, M., Holman, P., & Hammock, A. (1996). A comprehensive review of the effects of worksite health promotion on health-related outcomes. *American Journal of Health Promotion, 10,* 429–435.

Witte, K. (1993). Managerial style and health promotion programs. *Social Science and Medicine, 36,* 227–235.

World Health Organization (WHO). (1984). *The concept and principles of health promotion.* Copenhagen: WHO Regional Office for Europe.

World Health Organization (WHO). (1998). *Health for all in the twenty-first century.* Copenhagen: WHO Regional Office for Europe.

Wynne, R. (1994). *Workplace health promotion: A specification for training.* Dublin: European Foundation for the Improvement of Living and Working Conditions.

Yin, R. K. (1994). *Case study research: Design and methods* (2nd ed.). Thousand Oaks, CA: Sage.

CHAPTER 12

Using Workers' Compensation to Promote a Healthy Workplace

Karen Roberts

Work-related disability is costly to U.S. employers. Estimates of the exact cost vary depending on what exactly is included in the estimate, but regardless of who is making the estimate, the costs are always high. The National Safety Council (1998) estimated total costs of occupational injuries and illnesses for 1997 to be $127.7 billion. Among the more readily measured components of that cost is that associated with workers' compensation, although even those costs are not easy to measure accurately (Thomason, Schmidle, & Burton, 2001). According to the National Academy of Social Insurance (2001), employers spent $53.3 billion on workers' compensation in 1999. One would expect that, with such a high price tag, workers' compensation would be high on the list of human resource (HR) priorities; that understanding the structure, implementation, and intricacies of workers' compensation programs would be among some of the most valued HR competencies; that workers' compensation would be a prominent research topic; and that the major business school curricula would include at least one course on workers' compensation.

But this is not the case. For example, a search of the table of contents for the 1999–2000 issues of *HRFocus,* a newsletter for HR professionals, yielded no articles on workers' compensation. A search over the same time period of *HRMagazine,* a bimonthly publication of the Society for Human Resource Managers, a leading organization for HR professionals, found that one article out of the 450 published over the past two years was on workers' compensation. A review of two of the leading HR management academic journals, *Academy of Management Journal* and *Personnel Psychology,* found a total of four articles in workers' compensation published over the period 1995–2000.

The framework presented in Chapter One of this volume is useful to analyzing why workers' compensation in the United States, although long a part of the employment relationship, remains underdeveloped as an HR tool in promoting health at work. The framework arrays organizational goals along two axes: intrinsic to extrinsic on one axis and short-term to long-term on the other. Organizations are seen as having to balance competing goals so as to satisfy the short- and long-term intrinsic and extrinsic goals. Using the terminology of the framework, workers' compensation has played a marginal role in HR strategies to promote a healthy workplace because more external stakeholders are integral to the design, history, and operation of workers' compensation than is common in other HR arenas. As a result, employers have found it difficult to develop policies for and to control and manage workers' compensation within their organizations. This externalization of stakeholders is manifest in two ways.

First, from a managerial perspective, workers' compensation is the domain of several different disciplines—insurance, medical, and employment relations—none of which completely dominates the program or its implementation. For example, as I will discuss in detail in this chapter, workers' compensation in the United States is as much an insurance phenomenon as an HR function. Insurance carriers are external stakeholders with substantially different goals from those of the organization. Workers' compensation policies and practices, more than most other aspects of workplace safety and health, have been designed to meet the short-term goals of extrinsic stakeholders—insurance carriers—and are thus suboptimal for the organization. Further, although other HR functions are governmentally mandated or otherwise regulated, workers' compensation is the only major employment program that is state-based and differs substantially across states. As a result, generalizing about the program's problems and potential solutions is difficult, as is deciding how to adapt it to meet larger organizational objectives. In addition, those managing workers' compensation within organizations must always be mindful of governmentally imposed constraints.

Second, an environment of mutual distrust surrounds the implementation of workers' compensation. As will be discussed later, filing a workers' compensation claim often converts an employee who had been previously viewed as valuable into someone who may be trying to defraud the organization. Again in terms of the paradigm that structures this volume, distrust converts injured employees from intrinsic to extrinsic stakeholders whose goals may diverge from those of the organization.

The next section of this chapter describes what I will refer to as the historic deal, when the various workers' compensation statutes were passed and the underlying assumptions that color the relationships of the various actors in the system, most importantly employees and employers, were set. The next section of the chapter provides a brief overview of the basic design of workers' compensation through an examination of the broad statutory provisions that underlie the program and introduces the concept of insurance carriers as third parties to the workers' compensation relationship. The following section discusses the primary safety incentives within workers' compensation: its pricing and statutorily required safety programs. A discussion of how workers' compensation could be adapted to contribute to and reinforce a broad workplace strategy to create and maintain a healthy workplace then follows, and the final section suggests directions for future research.

Workers' Compensation's Historic Deal and the Environment of Distrust

Workers' compensation is one of the oldest employee benefit programs in the United States. The first constitutional workers' compensation state statute went into effect in Wisconsin in 1911 (Plumb & Cowell, 1998). By 1920, all but six states had statutes. Historians still debate why these statutes were passed. Some argue that it was to improve the social welfare of workers who became hurt in the line of duty but lacked resources to sue for restitution. Others argue that workers were beginning to win these suits and employers wanted protection from this type of liability and more predictable injury-related costs (Berkowitz & Berkowitz, 1985; Spieler & Burton, 1998).

The Historic Deal

Regardless of the true motivation behind these initial legislative efforts, all of them represented a deal between workers and employers. The old tort system where workers had had to sue their employers for compensation for a work-related injury was replaced by a no-fault insurance system. The implementation of workers' compensation required all employers to purchase insurance covering any injuries that occurred at the workplace (Berkowitz & Berkowitz, 1985).[1] Workers were now entitled to income replacement and medical benefits regardless of fault, relieving them of the burden of having to prove any sort of negligence on the part of the employer or to run the risk of losing in court.

In exchange for these benefits, however, employers gained what is referred to as the protection of exclusive remedy (Larson, 1988; Worrall & Appel, 1985). Exclusive remedy refers to the fact that in exchange for their employers accepting full liability for work-related injuries, employees accept that their benefits are limited to those provided under the workers' compensation statute and that they may not sue their employers for further benefits or damages. Some see a social benefit to this arrangement: it stabilizes the employment relationship by reducing uncertainty for both employees and employers when injury does occur (Hyatt & Law, 2000; Spieler & Burton, 1998).

Like most deals, workers' compensation represented a compromise in which each partner gained some rights at the expense of others. Workers get a guarantee of medical care and benefit payments that partially replace income without having to resort to tort, but those benefits are capped regardless of fault or negligence, and statute makes no allowance for pain and suffering or punitive damages, tools that are common in other personal injury tort claims. Employers get limited liability but are obligated to pay benefits regardless of whether the employee is partially or wholly at fault.

The people who actually struck the deal are not those who now live with it, and many tensions and unintended consequences plague program implementation today. For example, employers worry that the exclusive remedy protection is being eroded, although the evidence on that is inconclusive (Larson, 1988; Atkinson, 2000). And employees fear that the scope of what is considered a compensable injury is narrowing, which the evidence suggests is occurring (Spieler & Burton, 1998), and that employers are looking for ways to introduce contributory negligence provisions into the law, which has happened in several states (Ellenberger, 1998; Shainblum, Sullivan, & Frank, 2000).

Further, it is not uncommon that one or both parties fail to understand how they are bound by the deal. Employers often do not fully understand that for the most part they take the employee as they find him, meaning that poor employee health or safety habits that may have contributed to a claim do not relieve the employer from paying benefits. And employees often do not realize that workers' compensation is their only recourse even if the employer fails to maintain a safe workplace or is obstructive if a workers' compensation claim is filed (Roberts & Gleason, 1994).

Environment of Distrust

An important legacy from this historic arrangement is a basic assumption about the nature of work that was made when these programs were designed. That assumption, which comes out of economics, is that people do not want to work (Ehrenberg & Smith, 1996). This assumption is fundamental to labor economics and guided the framers of both workers' compensation and social security disability insurance in the United States. Most societies expect people to work; however, certain classes of people are exempt: the very young, the very old, and the very sick (Stone, 1984). To the extent possible, society will try to provide economically for these groups or otherwise protect them from the obligation of work. The definitional boundaries for these groups shift across countries, time, political climates, and economic conditions. Of these three groups, the most controversial is the sick, primarily because membership in that group is hard to define. Inability to work is not simply a medical condition but also depends on economic, psychological, and educational factors (Leonard, 1986). Also, within a certain range, individuals can define themselves as being too sick to work (Kreider, 1999). One of the challenges of providing benefits to those too sick to work, and one that takes an inordinate amount of programmatic resources within workers' compensation, is accurately determining who fits into that category (Stone, 1984).

The primary mechanism in workers' compensation to deal with this human condition where individuals do not want to work was to set benefits equal to less than 100% of lost income (Worrall & Appel, 1985). The underlying logic was that if individuals were fully reimbursed for all of their lost income, they would avoid work. Although few argue that individuals deliberately injure themselves in order to collect benefits, many argue that individuals do respond to increases in benefit levels. Specifically, as benefits rise, individuals may become more careless at work, be more likely to file a workers' compensation claim, or be more likely to extend the duration of work absences (Hyatt, 1996; Worrall & Appel, 1985).

The empirical results with regard to the effect of benefit generosity on duration of disability unambiguously support a duration effect. Although there is a range of estimates, the research supports the hypothesis that increases in benefit level, whether in the form of temporary total or permanent partial benefits, lead to longer absences from work (Butler & Worrall, 1985; Hyatt, 1996; Meyer, Viscusi, & Durbin, 1995). The empirical work examining the effect of benefit generosity on claiming behavior is less clear. The reason for the ambiguity is that uncoupling claims rates from underlying injury rates is difficult (Durbin & Butler, 1998). Those studies that are unable to disentangle claims and injury rates show some support for a response to benefits (Hirsch, McPherson, & DuMond, 1997; Krueger, 1990). However, one study was able to separate the effects of claiming from injury by examining a sample whose respondents had all been identified as having a work-related injury by physicians but only some of whom had filed a workers' compensation claim. When it is possible to separate out the effect of injury in this way, the economic incentive effect of benefit replacement rates on filing a workers' compensation claim depends on model specification and thus should be characterized as unstable (Biddle & Roberts, in press).

This assumption about individual motivation to work is fundamental to understanding how workers' compensation operates in workplaces today. Because workers' compensation provides income for not working, scholars commonly describe it as creating a moral hazard (Butler & Worrall, 1991; Dionne, 1982; Krueger, 1990). In general terms, *moral hazard* can be defined as "the tendency of insurance protection to alter an individual's motive to prevent loss" (Shavell, 1992, p. 280). In the context of workers' compensation, the availability of income replacement benefits gives employees an incentive to exaggerate the severity of an existing injury, to miss more work than necessary, or to inaccurately attribute an injury to work.[2] This type of moral hazard arises out of information asymmetries, in which the worker has more complete knowledge of the cause and severity of his or her injury than does the employer.[3] This asymmetry is the basis for substantial distrust between employers and employees. Employers have reason to distrust that an employee is actually injured, that the injury occurred at work, and that the person is legitimately absent from work.

One practical manifestation of this sort of ambiguity is employer and insurance company emphasis on reducing workers' compensation fraud (Ellenberger, 1998). One of the basic guidelines for employer response to injury is to investigate the injury as soon as possible. Employers are counseled to control costs by looking for inconsistencies in employee accounts and for witnesses who will contradict the injured employee's version of the accident (McGavin, 2001).

The response on the part of employees is predictably negative. Individuals with work-related injuries often report a shift in the employer's attitude: the employer who used to see a valued employee now sees a problem employee (Roberts & Gleason, 1994). One study found that 16.5% of workers who had been injured on the job failed to file a workers' compensation claim for fear of retribution, indicating that a substantial number of employees, one out of six, do not trust their employers to view their injuries as legitimate (Rosenman et al., 2000). Another study of workers' compensation claimants in Florida and Minnesota found that between 21.4 and 33.4% of workers feared that filing a claim would either result in their being fired or negatively affect future promotions

(Markiewicz & Roberts, 2001). Other research has found that injured workers report feeling stigmatized once they file a claim, isolated from coworkers, and viewed with distrust by their supervisors (Ray, 1986; Roberts & Gleason, 1991, 1994; Sum, 1996). In summary, what is meant to be a straightforward no-fault insurance system often operates in an environment of distrust that has negative implications for the employment relationship.

Overview of Workers' Compensation Laws and Programs

What we refer to as workers' compensation in the United States is actually a collection of 54 separate programs, one in each state and the District of Columbia, one for federal workers, and two for interstate commerce workers (one for maritime and harbor workers, another for railroad workers). Although each of these programs is unique, certain features are common across jurisdictions. All have the same purpose: to provide income replacement and medical benefits to individuals injured in the course of employment until their health is restored.

Benefits for Temporary Wage Loss

There is a range of income replacement benefits. All states provide for short-term disability benefits called temporary total benefits. The norm replacement rate is 66⅔% of lost average weekly wages. A few states have higher statutory replacement rates, and a few use a formula based on after-tax weekly earnings (U.S. Department of Labor, 2001). All states have a benefit maximum, typically 100% of the state's average weekly wage, and a few states have a minimum benefit. These benefits are a statutory entitlement for injuries and illnesses that "arise in the course of employment" (Nackley, 1984, p. 13).

Operational definitions of what injuries and illnesses arise from work vary from state to state both because of differences in statute as well as divergent paths of workers' compensation case law. Philosophically, states can be thought of as taking either an increased risk or a positional risk approach. The increased risk approach requires that the nature of the job increases the likelihood that an injury or illness occurred for workers' compensation to cover the illness or injury. That is, something about the design or require- ments of the job must put an employee at risk, not just the fact that the injury occurred on the job. The positional risk approach re- quires only that, for a claim to be covered, the person be exposed to some hazard that resulted in the illness or injury while on the job (Larson, 1953). For example, a firefighter who suffered burns would be covered under both approaches. However, an employee who was burned because the office building caught fire would only be covered according to the positional approach: being at work put that employee in a position to be injured. The second ap- proach tends to define coverage more broadly than the first; how- ever, there are gray areas regardless of approach. Whether an injury is covered is a source of conflict in workers' compensation (Thomason, Hyatt, & Roberts, 1998). This ambiguity is one of the features that makes workers' compensation especially complex for organizations that operate in multiple states.

Assuming that the insurance carrier accepts an injury as work- related, most states pay temporary total benefits until a worker reaches what is referred to as maximum medical improvement: ei- ther the injured employee is fully recovered or it is apparent that full recovery is not likely. The condition has stabilized, and the worker's health can be assessed. In the vast majority of cases, work- ers fully recover from injuries and return to work at their pre- injury job and earnings level. At this point, employees who are not fully recovered can be thought of as either having a permanent total or a permanent partial disability. A permanent total disability designation is reserved for the most serious injuries, but states do vary according to whether they take economic and educational fac- tors into account in making the determination.

Benefits for Permanent Disability

The more difficult permanent disability cases are those that entail partial disability. State programs can be thought of as arrayed on a benefit philosophical continuum, with states where benefits are based on degree of impairment at one end and states where ben- efits are based on wage loss at the other (Berkowitz & Burton, 1987). Wage-loss states pay benefits only for income lost due to the work-related disability. Thus, workers who are able to return to work at or above their pre-injury wage will receive no additional income benefits regardless of any lingering impairment. A worker who returns to the workforce but with reduced earnings will receive benefits equal to the replacement rate times the difference between earnings before and after the injury. Advocates of this approach argue that it is both simpler to administer and fully in the original spirit of workers' compensation to compensate workers for income lost during their recovery.

At the opposite end of the continuum, impairment ratings systems allow for the determination of the degree of remaining impairment and compensate the worker accordingly. In most cases the benefits are expressed in terms of number of weeks of lost time (and so again a percent of pre-injury earnings). Impairment usually is expressed in terms of percent of body function lost; although many states have impairment schedules that attach precise impairment ratings to the industrial loss of specific body parts (Chamber of Commerce, 2000). Critics argue that the impairment ratings are almost always a source of dispute and encourage gamesmanship that increases costs. Advocates of impairment systems argue the opposite, that they are less contentious and improve the motivation to return to work (Spieler & Burton, 1998).

Financing Workers' Compensation

In addition to differences in benefits, states vary considerably in how they finance workers' compensation. Broadly, three types of financing systems exist: open competition, administered pricing, and exclusive state funds. In the first two systems, employers purchase workers' compensation insurance from private insurance carriers in what is called the voluntary market. The primary distinction between open competition and administered pricing is that in states with open competition, insurance carriers are free to set their own base rates. In contrast, in states with administered pricing, the state insurance commissioner on the advice of the state rating organization sets the base insurance rates administratively.

In both systems private insurance carriers carry out the administration of claims, that is, the handling of benefit payments and other claim decisions. All but three states allow employers to self-insure, so that the employer does not purchase insurance but rather pays benefits directly. This option is typically available only to the largest of employers that are sufficiently capitalized to gain state approval to serve as their own insurer (U.S. Department of Labor, 2001). About 22% of all workers' compensation benefits are paid by self-insured employers (Mont, Burton, Reno, & Thompson, 2001). Even a self-insured employer will commonly use a third-party administrator to actually handle claims. Often but not always, this administrator is an insurance carrier that contracts to do claims management. Only the largest employers have in-house claims managers.

In states with exclusive state funds, the state agency prices, offers, and administers workers' compensation.[4] The states with exclusive state funds collect premiums from employers to fund benefits and claims management; however, because state funds are not profit-making organizations, the rates they charge have traditionally been less cyclically sensitive than those carriers charge (Plumb & Cowell, 1998; Williams, 1986). Through the 1970s most states used the administered pricing system. However, beginning in the 1980s, states began to allow for open competition in an effort to lower workers' compensation premiums. Currently, there are 38 open-competition states and five exclusive state funds, with the remainder using administered pricing (Thomason, Schmidle, & Burton, 2001).

Regardless of whether the workers' compensation program is financed and administered through private insurance carriers or a state fund, workers' compensation brings a third party into the employment relationship, a characteristic that differentiates workers' compensation from most other HR programs and practice areas. As such, insurance carriers are external stakeholders whose interests are usually quite different from the employees and employers who are thought to be central to the workers' compensation transaction. Carriers typically play an active role in that transaction.

The basic incentives for carriers are to collect as much premium (revenue) as possible, then minimize the cost of the claim. There are various ways to minimize claims costs. Although it has become axiomatic in the disability management literature that a rapid return to work benefits all parties, carriers can minimize costs in other ways, including restricting access to medical care, cutting off benefits, or denying claims altogether. Protecting the employment relationship is not intrinsic to the carrier's cost-minimization goals, although it usually is for the employer. Employers are counseled to use carriers that share their philosophies toward claims management (McGavin, 2001), but this does not always occur.

In terms of the framework that guides this volume, disability management requires balancing the intrinsic goals of the organization, which sees the employee as part of any strategy to promote health and minimize effects of injury and illness, and the extrinsic goals of the carrier, which sees the employee as a claimant and thus a cost. More enlightened carriers may see a benefit to selling insurance to an employer that promotes health and develops an integrated approach toward that objective, but the benefits of a healthier workplace do not necessarily accrue to insurance carriers or meet their goals.

Prevailing Models

Most observers would agree that since the 1900s, the U.S. workplace has become safer, although the trend toward improved safety appears to have flattened through the 1980s and 1990s (Burton & Chelius, 1997; Durbin & Butler, 1998). Many reasons might explain this: changing technologies, passage and implementation of the Occupational Safety and Health Act of 1970, employers' recognition that a healthy workplace is more productive, and various incentives contained within the workers' compensation system. Because undoubtedly all of these factors have contributed to improved workplace safety, it is difficult to disentangle their independent effects. Workers' compensation statutes address safety in two forms. The traditional form is through pricing. More recently, some states have revised their laws to include incentives or requirements for safety committees or loss control services.

Pricing to Promote Safety: Experience Rating

The primary mechanism through which workers' compensation is meant to affect workplace safety is pricing. This section will provide a brief description of the pricing process. A central premise in this discussion is that pricing is not particularly effective in promoting safety or a healthy workplace for several reasons. One is that at the market level, the basic pricing unit, the manual rate, tends to underreward safe employers and overreward more dangerous employers. Second, at the individual employer level, the pricing process is complicated and does not provide clear incentives, particularly to smaller employers. Third, the mechanism is based on inaccurate assumptions about what aspects of losses are within employer control. And finally, although the employer may write the check to pay the premium, evidence suggests that the actual incidence of the cost falls on employees.

Pricing is one area of workers' compensation where the presence of an external stakeholder, the insurance carrier, is especially strong. The determination of the price of workers' compensation, or premium, is a multistep process. The basic formula is derived from insurance underwriting principles, which are unfamiliar to most HR professionals. It would be an exaggeration to say that insurance carriers use complexity to obscure pricing and thus keep control; however, that is often the effect. Because pricing is an area that HR professionals seem to find most obscure and difficult to understand and thus manage, it remains underdeveloped as a tool to promote safety and health.

Marketwide Manual Rates

One precept in determining the cost of workers' compensation insurance is that some jobs are inherently more dangerous than others, so firms with employees in more dangerous occupations should pay more for their insurance. To implement this concept, all of the workforce is categorized using a classification scheme based on a mix of industry and occupation where each category is thought to be homogeneous with respect to risk. A price is associated with each category, or class code, referred to as a manual rate, expressed in dollars per $100 of payroll. The more dangerous the job is, the higher the manual rate.

A variety of factors go into determining manual rates, but the basic input is the level of previous claims costs (or losses, to use the insurance term) associated with each class code in the state.[5] As the employers in any given state improve their safety records, the basic pricing unit they face, the manual rate, decreases (Burton & Chelius, 1997). This is meant to be part of the safety incentives to employers: improve safety in your workplace and your base price will decrease. There are two drawbacks to this approach as a safety incentive. First, most employers do not understand this part of insurance pricing and do not see this linkage between their behavior and manual rates. Second, improved safety lowers the manual rates for all employers in the same industry, not just those who undertook the costs of improving safety. Employers that have not improved safety enjoy the benefit of lower manual rates resulting from the improved safety measures of their competitors. Thus, those employers that do understand the incentives may rationally decide not to incur the costs of improving safety if it benefits their competitors, who may or may not make comparable safety investments.

Complex Incentives for the Individual Employer

Other than the feedback of improved safety into marketwide manual rates, the primary mechanism for safety in workers' compensation is that employers are rated by experience. Conceptually, experience rating refers to incorporating the employer's safety record into the insurance premium. The logic is economic: efficient firms need to minimize costs to stay competitive. If the costs of the failure to provide a safe workplace are accurately passed back to employers, they will respond by improving safety (Chelius, 1977). Operationally, a multiplier, referred to as the experience modification factor, is calculated and applied to the employer's basic premium. The experience modification factor, or experience mod, is the ratio of the employer's actual losses to expected losses calculated over a three-year period. If an employer operates a workplace that is as safe as the average (hypothetical) employer with the same mix of workers, then the mod factor equals one. If the employer has had a worse loss history—either more or longer claims—the mod factor will exceed one. If the safety record is better than that of the average comparable employer, then the mod will be less than one.

Conceptually, the mod factor as the ratio of actual to expected losses is straightforward. However, the calculation itself is more complicated and not likely to be transparent to employer staff who do not have an insurance background. Its complexity is evident at three points in the formula. First, the calculation begins with a determination of losses, which are meant to be equal to the total costs of benefits and other costs associated with each injury. Because workers' compensation claims may last a long time, the total costs for a claim are not necessarily known at the end of a policy year. When this is the case, an estimate of total costs, referred to as the reserve, is used to calculate the premium (McGavin, 2001).[6] Typically, this estimate does not include employer input, and few employers are sufficiently knowledgeable to challenge the underlying assumptions and logic of the reserve estimate.

Second, the formula used to rate firms by experience is designed to more accurately incorporate the losses of larger firms than of smaller. The basis for this is the assumption that losses in smaller firms occur too infrequently to be representative of the underlying probability of a loss occurring (Ruser, 1985). Thus, the formula deliberately dilutes the safety incentives for smaller employers. This might not be a problem if all employers purchased insurance from insurance carriers. However, as noted earlier, over one-fifth of workers' compensation benefits paid are through self-insurance, and the self-insured are the largest employers. Therefore, the portion of the market for which the traditional experience-rating formula would provide the most direct incentives is unlikely to be subject to the formula.

The third characteristic of experience rating that dilutes its value as a safety incentive is that the ratio of actual to expected losses is calculated as a three-year rolling average. Thus, an employer that institutes an effective safety program will not fully see the benefit of lower insurance costs for three years (Dembe, 1995; Spieler, 1994).[7]

Flawed Assumptions About Employer Control

Aside from the complexity of the formula, there is another important flaw in the experience-rating formula as a safety incentive. The formula is designed to give more weight to frequency of claim than to severity, the logic being that the employers can institute safety measures to reduce frequency but not necessarily severity. This rationale is sensible in a workplace where most injuries are traumatic and could have been prevented with safety equipment and good housekeeping. For example, in a firm where cuts are the most frequent injury type, the employer probably could lower injury rates with machine guards or other safety devices but could not control the severity of a cut once it occurs. However, the profile of injuries involving lost work time shows that 46.1% of injuries in 1999 were strains, sprains, or repetitive motion injuries (U.S. Bureau of Labor Statistics, 2001). The logic of experience rating misses the point that employers can reduce the severity of those injuries through job rotation, ergonomic interventions, and aggressive return-to-work programs. By giving far more weight to the frequency of injuries as opposed to their severity, the experience-rating process fails to provide an incentive to employers to use some of the more important tools at their disposal to promote workplace health.

The empirical evidence supporting the hypothesized effect of experience rating on safety is inconclusive (Thomason, Schmidle, & Burton, 2001). Using the feature of the experience-rating formula that makes it more accurate for larger than smaller firms, several studies found a positive effect of experience rating on larger firms; however, whether the relationship was statistically significant depended on what measure of safety was used (Durbin & Butler, 1998; Chelius & Smith, 1983; Ruser, 1985). Studies that were able to employ a pre-post experimental design rather than rely on statistical controls have found stronger support for the safety effects of experience rating (Durbin & Butler, 1998; Kralj, 1994). However, some researchers challenge the causality between experience rating and safety on the grounds that there are other confounding factors, such as the implementation of disability management programs among larger employers, that explain improvements in safety records (Spieler, 1994; Hunt, Habeck, VanTol, & Scully, 1993).

The Incidence of Workers' Compensation Premium

In addition to the weaknesses noted earlier in relying on the price of workers' compensation to promote safety is the more difficult problem: some empirical evidence suggests that the actual cost of workers' compensation insurance is not necessarily borne by employers but rather is paid for by employees in the form of lower wages (Moore & Viscusi, 1990). The framework for understanding how workers themselves pay the premium cost comes out of labor economics and is referred to as the theory of compensating wage differentials (CWD).

CWD assumes that employees are averse to risk, that is, they do not want to become injured. Holding constant other factors that might affect earnings, such as the skills or experience of the individual, CWD hypothesizes that people who assume more dangerous jobs will be paid more; they will be paid a wage premium to compensate them for taking a greater risk of injury. Two central underlying assumptions to CWD are the following: (1) employees have full information about the degree of risk they are facing in the job, and (2) they are free to choose and move among jobs of varying degrees of risk of injury and associated wage premiums (Ehrenberg, 1988).

Employers are thought to be able to provide safety in the workplace but at a cost (Reville, Bhattacharya, & Sager, 1999). Employers are seen as having to choose between two costs: (1) paying a wage premium to get workers to take a dangerous job or (2) providing a safe workplace. CWD theory hypothesizes that a matching process takes place between workers and employers; workers accept job offers from employers based on their own risk preferences and the extent to which the wage offer compensates them for taking the job.

Workers' compensation changes this equation because it guarantees income replacement and medical care. In theory, employees should be willing to accept a lower wage than otherwise for any given level of risk because the cost to them of getting hurt has dropped because of the certain availability of workers' compensation benefits. What this then means is that employees are subsidizing their own workers' compensation premiums by accepting a lower wage than they ordinarily would for a job of comparable risk (Moore & Viscusi, 1990). Whether or not employees actually do pay for their own premiums depends in part on the bargaining power of workers relative to employers. If employees are in a strong bargaining position (whether due to market or contractual conditions), they may be able to demand the higher wages, forcing the employer to pay both the wage premium for the dangerous job and the workers' compensation bill. On the other hand, if employees are in a weak bargaining position, the employer may be able to force the employees to accept wages low enough to cover the workers' compensation premium.

There is empirical support for the existence of compensating wage differentials; that is, employees in more dangerous industries do appear to receive a wage premium (Moore & Viscusi, 1990). However, the evidence is strongest for fatalities (Moore & Viscusi, 1988), suggesting that employees either do not know about the nonfatal risks they face on the job or do not have the bargaining power to secure the wage premium. Some evidence suggests that employee knowledge of job risk increases with job tenure (Viscusi & Moore, 1991). However, studies that control for various firm characteristics have found that unionized workers are more likely than nonunionized workers to be able to actually capture the wage premium associated with job risk (Dorman, 1996; Leigh, 1989).

The research that explicitly includes workers' compensation suggests that there is a trade-off between workers' compensation benefits and wages and that the decrease in wages is sufficiently large that workers in effect pay for their own workers' compensation benefits in the form of reduced wages.[8] However, the wage decrease is smaller for unionized workers, indicating that workers and employers share the cost of workers' compensation in unionized workplaces (Moore & Viscusi, 1990).

These results have great significance for the value of workers' compensation pricing as an incentive for workplace safety. It appears that to some extent employees rather than employers are bearing the cost of workers' compensation in the form of lower wages. Thus, the incentive for employers to improve workplace safety in order to avoid high costs is diluted once more. Again, this points to the conclusion that employers provide a safe workplace for reasons other than workers' compensation costs, such as a belief that it is a good investment (Spieler, 1994).

Statutory Linkages Between Safety and Benefits

One of the distinguishing features of how work-related disability is conceptualized is the sharp split between, on the one side, the objectives of safety, injury prevention, and workplace health promotion and, on the other, compensating and managing injuries after they occur (Chelius, 1991). Whether statutory structure created this split or just mirrors it, legislative mandates for safety and health promotion are typically quite separate from those that compensate the injured. In employer organizations the split commonly persists with risk management and safety being a separate unit from HR, which manages workers' compensation. One of the reforms that has taken place beginning in the mid-1980s in workers' compensation is the addition of safety and prevention provisions to state statutes. Sixteen states have a provision for mandatory safety and health programs within their state workers' compensation statute, and 10 have voluntary program provisions (Conway & Svenson, 1998). In addition, 19 states have provisions that workers' compensation insurance carriers must offer loss control services to their customers (Dembe, 1995).

Examination of the effectiveness of these sorts of statutory requirements on statewide injury rates and days of lost time suggests that, so far, these changes in statute have not been effective. State differences in changes in injury and illness rates or in changes in cases of lost workdays could not be explained by the presence or absence of a statutorily required or voluntary safety program provision in the workers' compensation statute (Conway & Svenson, 1998). This is not to say that safety programs are ineffective, because there is substantial evidence that they can be effective (Spieler, 1994). Rather, it does imply that there are limitations to the power of this type of regulation and that mandating (or even suggesting) a safety program is not the same as mandating its quality. Further, a statutory requirement is probably too blunt an instrument to be effective. Although it is broadly accepted that the basic components of traditional safety programs include hazard identification, risk control, and safety training, implementation can and should be varied to fit the idiosyncrasies of the worksite (Dembe, 1995). Monitoring employer compliance with state safety program requirements has been a challenge for the Occupational Safety and Health Administration and is unlikely to be done more effectively by state workers' compensation agencies.

Using Workers' Compensation to Maintain a Healthy Workplace

Construing workers' compensation, as it is currently configured, as an affirmative health and wellness program is difficult. This chapter's opening section discussed how workers' compensation continues to be a marginalized HR function despite its high costs to employers. To use the framework that guides this volume, the reasons for that can be summarized as the outcome of a greater presence of external stakeholders in workers' compensation than in most other HR spheres. To an unfortunate extent, organizations have taken the path of least resistance when faced with extrinsically imposed goals that do not coincide with internally held values; they abdicated responsibility for how workers' compensation is implemented and how it can be used to promote a healthy workplace. In part, they have allowed the extrinsic objectives of insurance carriers to dominate the programs. For example, safety objectives are always stated in defensive terms—lowering the costs of lost productivity and wage loss—and not in terms of proactive promotion of health. This reactive stance is more consistent with an insurance approach that values cost minimization than an employer approach that values a healthy workplace.

Allowing insurance goals to direct the implementation of workers' compensation manifests itself in the employment relationship in the form of the environment of distrust discussed earlier. Evidence that claims rates may increase or that time away from work may lengthen as benefits increase is interpreted as evidence of the presence of moral hazard, the implication being that the employer cannot trust injured workers. Once they become injured, many employees have felt that their value as employees has become diminished in their employer's eyes. They are transformed from internal stakeholders with many goals in common with the organization to claimants with extrinsic values and demands. Common sense suggests that this is not a fertile environment for planting the seeds of healthy work. One concrete step to improve the trust level might be to rethink the prevailing assumptions about who bears responsibility for disability in the workplace.

During the 1980s, workers' compensation costs were rising rapidly (Thomason, Schmidle, & Burton, 2001). Employers took a series of measures to decrease costs, one of which was instituting return-to-work programs (Spieler & Burton, 1998). The implementation of return-to-work programs represents a shift in employer mindset that may have been inspired in part by the passage of the Americans with Disabilities Act (ADA) in 1990, as well as a response to costs. The ADA comes from a philosophical position that is vastly different from workers' compensation but provides a model of understanding disability that may be of use to workers' compensation (Roberts, 1996).

First, the ADA's premise is that people want to work but are prevented from doing so because of discriminatory behavior on the part of employers. As discussed earlier, workers' compensation begins with the assumption that people do not want to work. Second, the ADA defines disability in terms of a pairing of the individual's personal characteristics (described in terms of limitations on major life activities) and features of the environment. This conceptualization of disability is based on one used by the World Health Organization that distinguishes between four sequential levels that define the consequences of disease: (1) a disease or disorder that results in (2) an impairment (functional loss at the organ level), leading to (3) a disability (activity limitation operating at the personal level), resulting in (4) a handicap (social disadvantage evident when the individual moves in society) (Johnston & Pollard, 2001). Although this sequence has been revised both by the World Health Organization and various health outcomes researchers, the basic idea is clear: a disease or disorder may result in a particular impairment or functional limitation, but the full extent of disability can only be understood within the context of where the individual lives and works (Roberts, 1996). Workers' compensation programs have traditionally been based on a conceptualization of disability as residing entirely within the physical and emotional characteristics of the individual.

The significance of these differences in beliefs about human motivation to work and causes of disability is that the acknowledgment, even if only implicit, that individuals with disabilities do face discrimination and that disability is a function of both the individual's health status and the work environment shifts the responsibility for the employee's being healthy enough to work partially onto the employer. At this point, most workers' compensation practitioners are far from embracing an ADA-like philosophy that would place the burden of a person with a disability not working on the employer rather than the employee. However, the implementation of return-to-work programs that offer transitional employment opportunities is evidence that the concept is working its way into the management of workers' compensation. The offer of light duty or transitional employment indicates a tacit acknowledgment that conditions of work may create a barrier to work.

At this point, however, two features of most return-to-work programs signal that the traditional workers' compensation philosophy dominates. First is the reliance on the medical model, which views the physician as the expert on any given individual's ability to work. Recommendations to employers that they share information about their worksite with physicians or encourage site visits is an indication of recognition that the workplace is a contributor to disability, but in many claims, the physician's report is the deciding piece of data in how a claim is handled, regardless of whether the physician is familiar with the demands of the workplace.

Second is the transitional nature of whatever reassignment or job restructuring takes place to bring a person back to work. The usual rationale for the short-term nature of transitional employment is that it is important not to confer (or appear to confer) some advantage on disability, a version of the belief mentioned earlier that some people might self-identify as being sick in order to avoid work (Stone, 1984). In this case they would be avoiding hard or less pleasant work. Implied in this imposition of time limits on altered work is an unwillingness to wholly acknowledge that the structure of ordinary work processes contributes to disability by creating barriers to working—the ADA approach. In summary, return-to-work programs represent a positive but small step in workers' compensation toward creating a healthy workplace because they represent a shift in employers' attitude about how the workplace creates barriers to work. The logic that the structure of work is integral to the creation of both disability and health needs to be extended beyond the creation of transitional work assignment to an examination of the entire work process.

Although it is now common practice for organizations to engineer safety into their new equipment purchases or at the point of new plant design, an incipient literature on the relationship between workplace health and lean production suggests that organizations continue to uncouple safety and health decisions from production design decisions to the detriment of employees' health. Two studies examine the relationship between the introduction of high performance work practices and various work-related health outcomes (Brenner, Fairris, & Ruser, 2000; Markel & Roberts, in press).

The motivation behind these two studies was the intuition that several features of the high performance workplace (such as the pace of work, team-based production and compensation, and job rotation) might both contribute to certain types of injuries and pose challenges for staying on the job or returning to work for injured employees. Rather than providing for the individualized work environment that matches the demands of the work to the abilities of the employee, as is implied in the ADA and is a basic principle of disability management (Shrey, 1995), the high performance workplace is fast-paced with little variation. Further, team production, especially when coupled with team-based rewards, is likely to pose problems when a team member has physical limitations and cannot fully contribute to the team, even if temporarily. And job rotation for the purpose of cross training, although good from a developmental perspective, may create a barrier to work, depending on the demands of the jobs in the rotation.

Both studies looked at repetitive trauma injuries and the incidence of claims. Although the two studies differed somewhat in their findings, both drew the overall conclusion that there are statistically significant and negative relationships between features of high performance work systems and health outcomes for those with work-related injuries. These studies are particularly interesting because unlike some other studies that examine relationships between psychosocial job characteristics and various health outcomes (see Chapter Four of this volume for a detailed discussion of many of these), these two studies raise a challenge to the safety implications of implementing what are thought to be positive work systems from the HR perspective. In general, the implementation of a high performance work system is associated with greater developmental opportunities for employees through cross training; for the individual worker, it implies more autonomy in decision making and opportunity to innovate; and team-based production and compensation encourage cooperation (Appelbaum & Batt, 1994). What these two studies suggest is that the health implications of these sorts of production arrangements may not be positive (Brenner et al., 2000; Markel & Roberts, in press).

The central question is really whether promoting a healthy workplace and designing work to optimize profitability are compatible goals. The incipient research on high performance workplaces suggests that they are not. Carefully designed research on how job redesign affects broader financial goals may give some insight into this key question of the feasibility of jointly considering financial objectives and health—to not have to see them as part of a trade-off that sacrifices one for the benefit of the other.

Research Directions

This chapter has provided several possible explanations for why workers' compensation has not been an important program at the organizational level in promoting a healthy workplace. Among those is the underlying philosophical underpinning of workers' compensation that people do not want to work and have to be forced to do so. Another is the question of how disability is understood, whether it resides entirely within the characteristics of the individual or is the conjunction of individual and environmental factors. Yet another is that workers' compensation safety incentives by themselves are complex and couched in insurance language and a conceptual framework that most HR professionals do not readily understand. Related to this is the frequent failure on the part of organizations to come to terms with the extrinsically generated goals and fully integrate workers' compensation into the HR domain.

Clearly, organizations vary to the extent to which they fit these descriptions. Employer trust levels of employee motivations differ. Some organizations are more willing than others to accept responsibility for creating disability and its mirror image, health. Also, organizations vary in the extent to which they have in-house expertise that can decipher and respond positively to the safety incentives contained in workers' compensation pricing. And organizations differ as to whether they see workers' compensation as a full-fledged HR function.

Some evidence suggests that employers that trust and treat their employees with respect improve workers' compensation outcomes. One study of employers, although primarily focused on the relationship between safety and health policies and practices and workers' compensation outcomes, also examines the effect of overall workplace climate, in particular the extent to which the organization sustains a people-oriented culture (Hunt et al., 1993). The study measured people-oriented culture with a 12-item scale that included questions about employee opportunity to provide input into decision making, organizational policies on information sharing, the quality of work relationships, and trust levels. The results indicated that the more people-oriented the organizational culture was, the more the total level of workers' compensation benefit payments declined. Another study found that if management is interactionally fair to employees, they are less likely to file a workers' compensation claim if injured on the job (Roberts & Markel, 2001). Both of these studies frame outcomes in terms of the reduction of costs or the absence of claims, not in terms of a healthy workplace. This suggests a need for research in which outcomes are measures of health rather than the absence of injury or cost of injury.

The workers' compensation experience with return-to-work programs may offer a research opportunity to examine the effect of varying employer willingness to take responsibility for the creation of disability. Implementation of return-to-work programs varies across organizations. One source of variation is the extent to which the organization is willing to redesign work to accommodate individuals who have been injured at work. Implementation of return-to-work programs can be seen as indicators of employer acceptance that individuals and their environment jointly create disability and health. Research could then examine the relationship between these indicators and overall health outcomes.

Research has not examined the extent to which the complexity of the workers' compensation pricing formula has deterred the use of workers' compensation for health promotion. One area of workers' compensation in which there has been essentially no primary research has been that of the relationship between employers and insurance carriers. Anecdotal evidence suggests that these relationships can be very difficult and may involve a power struggle over claims management and information flows. Because this area is effectively uninvestigated, the most basic questions about this relationship and how it affects the integration of workers' compensation into the broader HR landscape and, more importantly, health outcomes remain to be addressed.

Finally, this chapter concludes with the observation with which it began: that workers' compensation remains outside the scope of mainstream HR. I have argued that the basic reason for this revolves around the presence of a third party in the employment relationship, which confuses the question of who is in control and who has authority and creates a conflict between intrinsic and extrinsic goals that many organizations have not effectively resolved. A further reason may be another point raised earlier about employers' willingness to accept responsibility for the health of their employees. Rethinking where the responsibility for both disability and health resides is fundamental to any transformation of workers' compensation into an affirmative program for promoting workplace health.

Notes

1   The early laws did not provide for occupational illnesses, but in response to recommendations from the National Commission on Workmen's Compensation in 1972, statutory protections were extended to illnesses and repetitive injuries.

2   There is also a moral hazard risk with workers' compensation medical benefits for those employees without other medical insurance or who have insurance with a copayment or deductible.

3   Scholars distinguish between risk-bearing moral hazard, in which individuals are more careless because they know they have income insurance, and claims-reporting moral hazard, in which they exaggerate injury severity in response to benefit generosity.

4   Twenty-one states also have what are referred to as competitive state funds, which are essentially state-run and administered insurance carriers that must compete with private carriers for employers' business (U.S. Department of Labor, 2001).

5   This description omits a few steps that precede the calculation of the manual rates. Further, note that in states with administered pricing, the manual rate for any given class code will be the same for all carriers. However, in states with open competition, each carrier will set its own manual rate.

6   The term used in insurance is *actual incurred losses*. These include both the amount paid out to date on a claim and the case reserves, the estimated total value of the case.

7   Once the modified premium has been calculated, other adjustments may be applied to generate the final premium to mitigate the three-year lag in the safety incentive within experience rating. Insurers may offer what are called scheduled credits or debits that are a percentage deviation applied to reward or penalize employer safety or lack of it more immediately than experience rating allows. This is offered entirely at the insurer's discretion and is sometimes used more to secure the sale than to reward safety (Dembe, 1995).

Also, some insurers offer safety credits that reduce the premium if an employer institutes a safety program. Again, although insurers may see these measures as improving the functioning of pricing as a safety incentive, most employers are more mystified than motivated by the complexity of the pricing structure.

8   Related research also finds evidence of employees paying for their own job accommodations in the form of reduced wages, suggesting that despite the complaints employers make over the high cost of disability, their employees actually bear a significant share of the costs (Gunderson & Hyatt, 1996).

References

Appelbaum, R., & Batt, E. (1994). *The new American workplace: Transforming work in the United States.* Ithaca, NY: ILR Press.

Atkinson, W. (2000). Is workers' compensation changing? *HRMagazine, 45,* 50–58.

Berkowitz, E., & Berkowitz, M. (1985). Challenges to workers' compensation: An historical analysis. In J. D. Worrall & D. Appel (Eds.), *Workers' compensation benefits: Adequacy, equity, and efficiency* (pp. 217–239). Ithaca, NY: ILR Press.

Berkowitz, M., & Burton, J., Jr. (1987). *Permanent disability benefits in workers' compensation.* Kalamazoo, MI: W. E. Upjohn Institute for Employment Research.

Biddle, J., & Roberts, K. (in press). Evidence on the level and determinants of workers' compensation claims. *Journal of Risk and Insurance.*

Brenner, M., Fairris, D., & Ruser, J. (2000). *"Flexible" work practices and occupational safety and health.* Unpublished manuscript.

Burton, J. F., Jr., & Chelius, J. R. (1997). Workplace safety and health regulations: Rationale and results. In B. E. Kaufman (Ed.), *Government regulation of the employment relationship* (pp. 158–180). Madison, WI: Industrial Relations Research Association.

Butler, R. J., & Worrall, J. D. (1985). Work injury compensation and the duration of non-work spells. *Economic Journal, 95*(1), 714–724.

Butler, R. J., & Worrall, J. D. (1991). Claims reporting and risk bearing moral hazard in workers' compensation. *Journal of Risk and Insurance, 58*(2), 191–204.

Chamber of Commerce. (2000). *2000 Analysis of workers' compensation laws.* Washington, DC: Author.

Chelius, J. R. (1977). *Workplace safety and health.* Washington, DC: American Enterprise Institute.

Chelius, J. R. (1991, September). The use of workers' compensation to encourage occupational health and safety. *Proceedings of a conference celebrating the seventy-fifth anniversary of the Federal Employees' Compensation Act.* Edison, NJ.

Chelius, J. R., & Smith, R. (1983). Experience rating and injury prevention. In J. D. Worrall (Ed.), *Safety and the workplace.* Ithaca, NY: ILR Press.

Conway, H., & Svenson, J. (1998). Occupational injury and illness rates, 1992–1996: Why they fell. *Monthly Labor Review, 121*(11), 36–58.

Dembe, A. (1995). Alternative approaches for incorporating safety into state workers' compensation reform. *Journal of Insurance Regulation, 13*(4), 445–461.

Dionne, G. (1982). Moral hazard and state-dependent utility function. *Journal of Risk and Insurance, 49*(2), 405–423.

Dorman, P. (1996). *Markets and mortality: Economics, dangerous work, and the value of human life.* Cambridge: Cambridge University Press.

Durbin, D., & Butler, R. (1998). Prevention of disability for work-related sources: The roles of risk management, government intervention, and insurance. In T. Thomason, J. F. Burton Jr., and D. E. Hyatt (Eds.), *New approaches to disability in the workplace* (pp. 63–86). Madison, WI: Industrial Relations Research Association.

Ehrenberg, R. G. (1988). Workers' compensation, wages, and the risk of injury. In J. F. Burton Jr. (Ed.), *New perspectives in workers' compensation* (pp. 71–96). Ithaca, NY: ILR Press.

Ehrenberg, R. G., & Smith, R. S. (1996). *Modern labor economics.* Reading, MA: Addison-Wesley.

Ellenberger, J. N. (1998). Fraud: Another view. *On Workers' Compensation, 8*(8), 7–8.

Gunderson, M., & Hyatt, D. E. (1996). Do injured workers pay for reasonable accommodation? *Industrial and Labor Relations Review, 50*(1), 92–104.

Hirsch, B. T., MacPherson, D., & DuMond, J. M. (1997). Workers' compensation recipiency in union and nonunion workplaces. *Industrial and Labor Relations Review, 50*(2), 213–236.

Hunt, H. A., Habeck, R., VanTol, B., & Scully, S. (1993). *Disability prevention among Michigan employers* (Upjohn Institute Technical Report No. 93–004). Kalamazoo, MI: W. E. Upjohn Institute for Employment Research.

Hyatt, D. E. (1996). Work disincentives of workers' compensation permanent partial disability benefits: Evidence for Canada. *Canadian Journal of Economics, 29*(2), 289–308.

Hyatt, D. E., & Law, D. K. (2000). Should work-injury compensation continue to imbibe at the tort bar? In M. Gunderson & D. E. Hyatt (Eds.), *Workers' compensation: Foundation for reform* (pp. 238–259). Toronto, Ontario: University of Toronto Press.

Johnston, M., & Pollard, B. (2001). Consequences of disease: Testing the WHO International Classification Of Impairments, Disabilities, and Handicaps (ICIDH) Model. *Social Science and Medicine, 53*(10), 1261–1273.

Kralj, B. (1994). Employer responses to workers' compensation rating. *Industrial Relations, 49*(2), 41–61.

Kreider, B. (1999). Latent work disability and reporting bias. *Journal of Human Resources, 34*(4), 734–769.

Krueger, A. (1990). Incentive effects of workers' compensation insurance. *Journal of Public Economics, 41*(1), 73–99.

Larson, A. (1953). *The laws of workmen's compensation.* New York: Matthew Bender.

Larson, A. (1988). Tensions of the next decade. In J. F. Burton Jr. (Ed.), *New perspectives in workers' compensation* (pp. 21–44). Ithaca, NY: ILR Press.

Leigh, J. P. (1989). Compensating wage differentials for job-related death: The opposing arguments. *Journal of Economic Issues, 23*(3), 823–842.

Leonard, J. S. (1986). Labor supply incentives and disincentives for disabled persons. In M. Berkowitz & M. A. Hill (Eds.), *Disability and the labor market* (pp. 61–94). Ithaca, NY: ILR Press.

Markel, K., & Roberts, K. (2002). Are injured workers in organizations with high involvement workplace practices more likely to lose time from work? In P. Voos (Ed.), *Proceedings of the 54th annual meeting of the Industrial Relations Research Association, Atlanta, GA* (pp. 32–39). Champaign, IL: Industrial Relations Research Association.

Markiewicz, B., & Roberts, K. (2001, October). *The Worker Injury National Survey Project (WINS): Exploring the feasibility of building a national information resource.* Paper presented at the meeting of the International Association of Industrial Accident Board Commissioners, Portland, ME.

McGavin, M. F. (2001). *Blueprint for workers comp cost containment.* Dallas, TX: International Risk Management Institute.

Meyer, B., Viscusi, W. K., & Durbin, D. (1995). Workers' compensation and injury duration: Evidence from a natural experiment. *American Economic Review, 85*(3), 322–340.

Mont, D., Burton, J. F., Jr., Reno, V., & Thompson, C. (2001). *Workers' compensation: Benefits, coverage, and costs.* Washington, DC: National Academy of Social Insurance.

Moore, M. J., & Viscusi, W. K. (1990). *Compensation mechanisms for job risks.* Princeton, NJ: Princeton University Press.

Moore, M. J., & Viscusi, W. K. (1998). Doubling the estimated value of life: Results using new occupational fatality data. *Journal of Policy Analysis and Management, 7*(3), 476–490.

Nackley, J. V. (1984). *Primer on workers' compensation.* Washington, DC: Bureau of National Affairs.

National Academy of Social Insurance. (2001). *Workers' compensation: Benefits, coverage, and costs.* Washington, DC: Author.

National Safety Council. (1998). *Accident facts.* Itasca, IL: National Safety Council.

Plumb, J. M., & Cowell, J. W. (1998). An overview of workers' compensation. In T. Guidotti & J. W. Cowell (Eds.), *Occupational medicine: Workers' compensation, 13*(2), 241–272.

Ray, M. (1986). *Work-related injury: Workers' assessment of its consequences for work, family, mental, and physical health.* Pullman: Washington State University, Injured Workers Project.

Reville, R. T., Bhattacharya, J., & Sager, L. (1999). *Measuring the economic consequences of workplace injuries, functional, economic, and social outcomes of occupational injuries and illnesses.* Washington, DC: U.S. Department of Health and Human Services.

Roberts, K. (1996). Managing disability-based diversity. In E. Kossek & S. Lobel (Eds.), *Managing diversity* (pp. 310–331). Cambridge, MA: Blackwell.

Roberts, K., & Gleason, S. (1991). What employees want from workers' comp. *HRMagazine, 36*(12), 49–54.

Roberts, K., & Gleason, S. (1994). Procedural justice in the workers' compensation claims process. In J. Rojot & H. Wheeler (Eds.), *Employee rights and industrial justice* (pp. 77–88). Norwell, MA: Kluwer.

Roberts, K., & Markel, K. (2001). Claiming in the name of fairness: Organizational justice and the decision to file for workplace injury compensation. *Journal of Occupational Health Psychology, 6*(4), 332–347.

Rosenman, K. D., Gardiner, J. C., Wang, J., Biddle, J., Hogan, A., Reilly, M. J., Roberts, K., & Welch, E. (2000). Why most workers with occupational repetitive trauma do not file for workers' compensation. *Journal of Occupational and Environmental Medicine, 42*(1), 25–34.

Ruser, J. W. (1985). Workers' compensation insurance, experience rating, and occupational injuries. *Rand Journal of Economics, 16*(4), 487–503.

Shainblum, E., Sullivan, T., & Frank, J. (2000). Multicausality, non-traditional injury, and the future of workers' compensation. In M. Gunderson & D. E. Hyatt (Eds.), *Workers' compensation: Foundations for reform* (pp. 174–211). Toronto, Ontario: University of Toronto Press.

Shavell, S. (1992). On moral hazard and insurance. In G. Dionne & S. E. Harrington (Eds.), *Foundations of insurance economics.* Norwell, MA: Kluwer.

Shrey, D. (1995). Worksite disability management and industrial rehabilitation. In D. Shrey (Ed.), *Principles and practices of disability management in industry* (pp. 3–54). Winter Park, FL: GR Press.

Spieler, E. A. (1994). Perpetuating risk? Workers' compensation and the persistence of occupational injuries. *Houston Law Review, 31,* 119–264.

Spieler, E. A., & Burton, J. F., Jr. (1998). Compensation for disabled workers: Workers' compensation. In T. Thomason, J. F. Burton Jr., and D. E. Hyatt (Eds.), *New approaches to disability in the workplace* (pp. 205–244). Madison, WI: Industrial Relations Research Association.

Stone, D. (1984). *The disabled state.* Philadelphia: Temple University Press.

Sum, J. (1996). *Navigating the California workers' compensation system: The injured worker's experience.* San Francisco: California Commission on Health and Safety.

Thomason, T., Hyatt, D. E., & Roberts, K. (1998). Disputes and dispute resolution. In T. Thomason, J. F. Burton Jr., and D. E. Hyatt (Eds.), *New approaches to disability in the workplace* (pp. 269–298). Madison, WI: Industrial Relations Research Association.

Thomason, T., Schmidle, T. P., & Burton, J. F., Jr. (2001). *Workers' compensation: Benefits, costs, and safety under alternative insurance arrangements.* Kalamazoo, MI: W. E. Upjohn Institute for Employment Research.

U.S. Bureau of Labor Statistics. (2001, March 28). *Lost-worktime injuries and illnesses: Characteristics and resulting time away from work* (U.S. Department of Labor News Release 01-71) [On-line]. Available: http://stats.bls.gov/special.requests/ocwc/oshwc/osh/case/osnr00 12.pdf

U.S. Department of Labor. (2001). *State workers' compensation laws.* Washington, DC: Employment Standards Administration.

Viscusi, W. K., &. Moore, M. J. (1991). Worker learning and compensating differentials. *Industrial and Labor Relations Review, 45*(4), 80–96.

Williams, C. A. (1986). Workers' compensation insurance rates. In J. Chelius (Ed.), *Current issues in workers' compensation* (pp. 209–236). Kalamazoo, MI: W. E. Upjohn Institute for Employment Research.

Worrall, J. D., & Appel, D. (1985). Some benefit issues in workers' compensation. In J. D. Worrall & D. Appel (Eds.), *Workers' compensation benefits: Adequacy, equity, and efficiency* (pp. 1–18). Ithaca, NY: ILR Press.

CHAPTER 13

The Role of External Policies in Shaping Organizational Health and Safety

Chris Brotherton

This is an exciting time in which to be involved in health and safety. In the United Kingdom in June 2000, the government published a strategy statement, *Revitalising Health and Safety* (Health and Safety Commission; Department of Environment, Transport, and the Regions, 2000). New legislation is being drafted and is soon to be presented to Parliament. Organizations are reviewing their policies, and trade unions are becoming increasingly active in health and safety matters.

Across Europe, nations differ widely, despite the directives issued by the European Parliament, in the way in which they address Community Law. In the United States, progress seems less vigorously involved in seeking to protect workers' health through the use of legislation, but each year sees an increase in the advancement of health standards in work. This chapter will review some of the main developments in health and safety in a range of countries. It will examine some of the underlying philosophical approaches to health and safety and explore some challenges to health and safety at work nationally and internationally. In the terms used in Hofmann and Tetrick (Chapter One of this volume), this chapter positions its arguments at the macro-level and in the section of their model concerned with the long-term and extrinsic effects on organizational health. Primarily, this chapter takes the legal framework as the prime indicator of the social forces operating in organizations. It is the law, shaped by government's overall responses to the labor market, industrial relations, and general views of risk that provides the general framework for health and safety.

The United Kingdom

Occupational health and safety law is founded on both statute law and the common law. Breach of a statute (an Act of Parliament) and regulations made under a statute generally gives rise to criminal liability. This means that an offender can be brought before the criminal courts, for example, the Magistrate's Court, and if found guilty of the offense, can be fined, imprisoned, or both. The principal U.K. health and safety legislation is the Health and Safety at Work etc. Act (1974) and the Management of Health and Safety at Work Regulations (1992). A person may also be in breach of a common-law duty. The common law is the unwritten law and is based on the decisions of the courts that have been bound by the doctrine of precedent into a body of authoritative principles and rules. Common law is fundamentally "judge-made" law and forms the basis for the law of tort. The torts of negligence and breach of statutory duty have played an important part in the development of civil liability with respect to occupational health and safety in the United Kingdom.

The principal contribution of common law to health and safety at work involves the rights of employees, their dependents, and other persons to sue an employer for damages for death, personal injury, or disease. An employer is expected to take reasonable care for the safety of the employees and other people affected by the employer's operations. So if an employer knows, or ought to have known, of a risk to the health and safety of the employees, the employer may be liable if (1) an employee dies, is injured, or suffers disease as a result of exposure to that risk or (2) the employer failed to take reasonable care.

The Health and Safety at Work etc. Act followed from the 1972 Committee of Inquiry on Safety and Health at Work, the Robens

Committee, which produced a report that heralded a significant change of approach in British health and safety regulation. It recommended the introduction of measures that would

- Provide a more self-regulating system for health and safety
- Ensure wider coverage of those affected by the risks associated with work
- Clarify duties for health and safety in a single comprehensive framework
- Enable a greater degree of involvement of employers, workers, and their organizations in health and safety
- Create a national authority for health and safety
- Provide new enforcement powers to health and safety inspectors

Prior to these recommendations, the system for health and safety regulation in force in the United Kingdom was piecemeal and prescriptive. The traditional measures available were complex and sometimes incomprehensible to the people they affected, often marked by incomplete and overlapping coverage, produced with little involvement from either those they were intended to protect or those whose activities they were meant to regulate, and provided limited procedures for their enforcement. The Robens recommendations were greatly praised as a radical departure from traditional approaches to health and safety regulation.

The Health and Safety at Work etc. Act was intended to provide a framework through which the Robens Committee's ideas of preventing injuries and ill health through self-regulation could flourish. Its aim was to facilitate a shift of emphasis in British legislative provisions away from prescriptive standards toward a goal-setting approach and to create greater participation of representatives of employers and employees in making and maintaining preventive health and safety standards.

Under the Act a tripartite national authority was created, the Health and Safety Commission. Subsequently, a large number of industry and subject-based joint advisory committees were established, and a philosophy of policymaking was advanced under which the people who created risks and those who worked with them would be involved in decision-making about what level of risk was acceptable.

The Act also established the Health and Safety Executive (HSE) as the executive arm of the Health and Safety Commission, with the responsibility for achieving compliance with its provisions.

The Act enables the Secretary of State to approve regulations that can spell out the details of the specific legislative standards. At the time of its introduction, it did not replace the existing provisions made under previous statutes or indeed the statutes themselves. Rather, it was envisaged that these earlier provisions would be replaced gradually by regulations made under the Act. By 1999 the Health and Safety Commission had introduced over 80 sets of such regulations.

An innovation of the HSW Act was its provision for the use of Approved Codes of Practice (ACOPs), instruments that do not impose legal duties but set out the means by which a legal duty can be accomplished. ACOPs generally accompany new sets of regulations. In the event of a breach of one of the Act's regulations accompanied by an approved code of practice, the normal direction of the burden of proof is reversed, and the court requires the defendant to show that the means used to discharge the relevant duty were equivalent to those laid down in the ACOP. The Act sets out a general set of duties for employers, employees, the self-employed, controllers of the premises, and manufacturers and suppliers of articles and substances used at the workplace. These duties were intended to give everyone concerned with health and safety at work clear, concise, and accessible notions of their basic legal obligations in order to remedy the criticism of the previous legislative system— that it was too complex, unwieldy, and alienating to those people whose activities it was intended to regulate. The duties were also intended to facilitate greater attention to the management of health and safety.

The Act requires employers to ensure the health, safety, and welfare of their employees. The Act makes it clear that this duty of care extends to include the provision of plant, systems of work, information, training and supervision, means of access and egress, the working environment, and the use of articles and substances. These general duties of employers, which draw on previously established common law principles, are qualified by the concept of reasonable practicability. This phrase draws its legal definition from case law in which it is established that the duty holder must take into account the hazard, danger, and injury that may occur and balance it against the cost, inconvenience, time, and trouble needed to counter it.

The judgment of Asquith in *Edwards v. National Coal Board,* 1949, is often quoted for this legal definition (cited in Tolley, E15015):

> . . . it seems to me to imply that a computation must be made by the owner in which the quantum of risk is placed on one scale and the sacrifice involved in the measures necessary for averting the risk (whether in money, time or trouble) is placed in the other. If it be shown that there is a gross disproportion between them—the risk being insignificant in relation to the sacrifice—the defendants discharge the onus on them.

This qualification on the duties under the Act poses several problems. It contradicts the intention of the Act to make everyone's duties clear in relation to the prevention of occupational injury and ill health, because

> . . . ultimately the extent of their duty can only be determined after the event, when it has been tested in court. It is also subject to differences in perception—differences which are affected by the passage of time, the extent and development of knowledge and experience and changes in societal expectation in relation to risk. (James & Walters, 1999, p. 5)

The Act contains a statutory framework for worker representation and consultation in the workplace. Trade-union safety representatives and safety committees are the means by which these aims are achieved. About 120,000 safety representatives receive training in health and safety at work in any one year in the United Kingdom. James and Walters (1999) comment that the development of the regulatory system following the introduction of the HSW Act occurred rather slowly for the remainder of the 1970s, with little sign of the major revision of previous legislation advocated by the Robens Committee.

The regulators faced a large task, but in 1979 there was a change of government.

For the next 18 years a succession of Conservative governments with fundamentally different policies to those prevalent at the time of Robens and the passing of the HSW Act (indeed, different to those prevalent since the end of the Second World War) governed the United Kingdom. As well as pursuing a political agenda in which corporatism was rejected, trade unions attached, institutions of the welfare state dismantled, and public expenditure massively reduced, these governments were committed to deregulatory economic strategies under which monetarism, market forces, and privatization dominated the political economy of the country.

This was indeed an extraordinary situation: on the one hand, development was rooted in the final stages of the tradition of British corporatism, and on the other, a government with overall responsibility for its implementation and operation, whose basic political and economic strategy could not have been more opposed to corporatism and its manifestations.

At a crucial stage of its development, the HSW was faced with a government that based its policy on deregulation and cuts in finance that lead to decreases in front-line services. (James & Walters, 1999, pp. 7–8)

Simultaneously, the HSW gradually acquired new and onerous areas of enforcement responsibility as new legislation on asbestos licensing, genetic modification, and pesticides was introduced and responsibilities were transferred from the police, railways, nuclear safety research, and off-shore safety.

The new Labour government has introduced a strategy statement titled *Revitalising Health and Safety* (Health and Safety Commission, 2000) with the aim of injecting new impetus into the health and safety agenda; identifying new approaches to further rates of accidents and ill health caused by work, especially approaches relevant to small firms; ensuring relevance for the Health and Safety Commission and HSW's work over the next 25 years; and gaining maximum benefits from links between occupational health and safety and other government programs. The strategy document has set the first ever targets for the United Kingdom's health and safety system:

- Reduce the number of working days lost per 100,000 workers from work-related injury and ill health by 30% by the year 2010

- Reduce the incidence rate of fatal and major injury accidents by 10% by 2010
- Reduce the incidence rate of cases of work-related ill health by 20% by 2010
- Achieve half the improvement under each target by 2004

The delivery of the new targets will depend "crucially on the commitment of stakeholders to pioneer new action" (Health and Safety Commission, 2000, p. 1). The government aims to motivate employers to see the benefits of a good health and safety regime and to engage more small firms more effectively in a program of tailored sector-specific guidance. Another aim is to put the government's own house in order by removing Crown immunity, taking action on procurement, and promoting educational coverage of health and safety.

The strategy document explores the responsibilities of boards of directors, including specific reporting duties. More controversially, *Revitalising Health and Safety* (Health and Safety Commission, 2000) draws attention to the weaknesses in current legislation as it applies to company directors, which make it difficult to set the law in motion. A new proposal from the Law Commission, cited in the same source, states "that a special offence of corporate killing should be created" (p. 25).

> In cases where management arrangements had failed to ensure the health and safety of workers or the public, a death would be regarded as having been caused by the conduct of the corporation. Individuals within a company could still be liable for the offences of reckless killing and killing by gross carelessness, as well as the company being liable for the offence of corporate killing. (Health and Safety Commission, 2000, p. 25)

The United Kingdom and Europe

In recent years, the European Union (EU) has provided the principal motor for changes in U.K. health and safety legislation. Article 118A in the foundation treaty (the Treaty of Rome) gives health and safety prominence in the EU's objectives. The EU's Social Charter also contains a declaration on health and safety, although this has no legal force. The EU is multilayered in its political and administrative arrangements and multichanneled in its forms of communication.

Writing of the transformation of Western Europe since 1945, Nugent (1999, pp. 21–22) tells us

> . . . although the logic of circumstances and of political and economic changes have brought the [nations] much more closely together, there can hardly be said to have been a common and coherent integrationist force at work in Western Europe in the post-war years. Far from the states being bound together in the pursuit of a shared visionary mission, relations between them have often been extremely uncomfortable and uneasy, based as they have been on a host of different needs and different perceptions of what is possible and necessary. In consequence the processes of co-operation and integration have operated in many different forums and at many different levels, in many different ways, and at many different speeds. Even within the EE/EU, which has been at the integrationist core, the course of the integration process has varied considerably, with the mid-1970s until the early 1980s being the slowest integrationist advance, and the mid-1980s until the early 1990s being the fastest.

It is not surprising then that the underlying philosophies of health and safety are diverse. The United Kingdom has implemented EU directives in health and safety management provision and use of work equipment, physical working conditions, personal protective equipment, manual handling operations, and display screen equipment. In addition, the United Kingdom has enacted all the following as a result of directives: construction regulations, control of substances hazardous to health, electricity, consultation with all employees—whether or not represented by trade unions, first aid, training, noise, and reporting of injuries and dangerous occurrences.

The European Perspective

The European Community Occupational Health Directives adopted by the then-twelve member states between 1989 and 1994 are now law virtually throughout Western Europe. The political and administrative reasons for this are that the 1992 agreement establishing the European Economic Area between the twelve Community States and the European Free Trade Area (EFTA) provided for the latter to transpose the Directives into their own national law. Three EFTA states—Austria, Finland and Sweden—subsequently joined the European Union and so naturally had to incorporate the established body of Community laws and regulations. Switzerland too, although democratically opting to stay outside the European Economic Area, has selectively transposed parts of EU occupational health. Each country has a different set of characteristics for its industrial relations and its occupational health laws. The EFTA organizations had no remit to harmonize regulations concerning health at work. The countries have different arrangements for welfare and for insurance. All have different histories politically and economically. Detailed comparisons and contrasts are difficult to make rigorously in a short space, but broad generalizations are possible.

The Nordic countries—Sweden, Finland, and Norway—have been key inspirations for Community legislation. The Nordic model's central features are its long tradition of government regulation of labor market conflicts in cooperation between the employers and trade unions; a unified trade-union movement, significantly better organized that in other industrialized countries; a long tradition of collective bargaining; a largely common law system that regulates conflicts according to the principle of labor peace and the right of collective action; cooperation between the state, trade unions, and employers on economic policy and information supplemented by consultation at different levels. The Nordic model is a strong form of social democracy. The basis of consultation found in Sweden, Norway. and Finland is enshrined in these countries' constitutions. For many years their geopolitical situation made them actually and potentially vulnerable to invasion from other countries. The avoidance of internal conflict was set as a high priority lest it provided an opportunity for attack by a neighbor.

The Nordic occupational health model grew out of the questioning of Tayloristic work organization in the sixties. Prior to that, employers' rights to organize production had seldom been challenged. In the early 1970s, a groundswell of general worker discontent appeared. Sweden experienced a series of wildcat strikes, and in all of the Nordic countries, a series of surveys gave evidence of dissatisfaction with unhealthy and alienating working conditions. Fierce employers' opposition to the linked issues of working conditions and industrial democracy raised further questions about how far industrial democracy and the quality of working life were compatible with the ability of privately owned business to flourish.

Today, the Nordic countries' core objectives of occupational health are much more the focus of political debate than anywhere else in the world and have produced legislation, guidance, and programs that have transformed working conditions. Adequate resource transfers are made through mutualization of costs or other public funding arrangements. The national level of industrial relations (between employers and unions) is significantly engaged in debate about the general working environment. At plant level, the forum for co-decision is seen as the decisive link in developing prevention policies. Vogel (1998, pp. 24–25) writes

> In this way, the political issues are not masked by political intervention on technical issues. This is a fundamental sea-change. Since the idea of work-related risks first appeared, the entire development of occupational health has been increasingly shaped by the influence of specialists—be it occupational health doctors, industrial hygienists, safety engineers or ergonomists and psychologists. Their job was to say within a fairly broad legal framework—what prevention measures were required against what related risks. Obviously the Nordic approach never denied the role of specialists. Without them, many issues would not have been brought to light, and many complaints would not have led to practical changes. The difference lay in integrating their activities into an overall program of political priorities and loosening the constraints which contractual relations between preventive services and companies may imply. . . . Attempts to open up the workplace or bring employers' occupational health policy under public and workers' control are supported by various instruments: a highly-qualified, better staffed labour inspectorate than elsewhere in Europe . . ., information and research systems which ensure regular contacts between company preventive staff and public and union agencies with occupational health responsibilities, and public-funding arrangements.

We can see that the Nordic model builds in a high level of visibility of occupationally related diseases. It also maintains a gender perspective in its analysis of work and preventive policies. As Vogel

(1998, p. 29) puts it: "The general impression is that excellent research has led to specific measures where the conditions are right (trade-union pressure and employers' acceptance)." The analysis shows us that the Nordic countries have had a consistent policy of market control. Generally, the import and marketing of work equipment and chemical substances have been controlled by the labor inspectorate, sometimes assisted by other specialized agencies, in order to compel manufacturers to integrate prevention requirements at the design stage. For their part the trade unions have sought to turn market controls from a purely public responsibility into a form of workers' control grounded in workers' direct workplace experience. The Swedish Confederation of white-collar employees has, for example, devised voluntary standards for display screen equipment based on a comprehensive survey of trade-union members, partly for negotiators to ensure that employers purchase only compliant equipment. The EU Directives on occupational health and safety add little to the Nordic countries' ability to manage the issues that arise. Rather, these countries set the baseline for the rest of the European Community, if not the rest of the world.

Austria

The present boundaries of Austria form the heart of what was once the Austro-Hungarian Empire, at its height taking in lands now part of the Czech Republic as well as Silesia (Poland) and Bohemia (Hungary and Slovakia). Vogel (1998, p. 125) tells us that "the first short-lived labour organisations appeared in Austria in the first third of the nineteenth century. Worker participation in the brief 1894 revolution in Vienna was through newly founded political organizations and trade unions, and a workers' press with titles like *Workers News*. The crushing of the revolution in October 1846 and the outer-strike legislation of 1852 were to deny a voice for the labor movement for two decades."

Military defeats for Austria led to a break in civil society that could be filled by the trade-union movement, which eventually formed the Social Democratic Party. The Christian Socialist movement, rooted in the social doctrines of the Catholic Church and conservatism, was formed in the early part of the twentieth century, as was a pan-Germanic liberal movement. The Christian Socialist movement advocated national unity and was deeply anti-Semitic. A century on, these parties are still the main political forces in Austria. Austria's political system can perhaps be characterized as being centralized and bureaucratic but also socially polarized. The days of empire concentrated power in the hands of those at the center, but involvement of the surrounding areas and the groups based there was far less direct.

Vogel (1998, p. 130) sums up the effects of the fractionated geopolitics of Austria's historical development on the present state's employment relations in the following section of his analysis:

> Collective bargaining is governed by the Work Organisation Act *(Arbeits Verfassung Gesetz, ArbVG)* which prescribes the hierarchy of the sources of law and frames company agreements between employers and works councils on statutory matters or those laid by collective agreements from a higher source. Collective agreements include a collective industrial peace clause imposing compulsory mediation and arbitration to avert disputes. Nationally, the joint wages and prices board plays an important role in framing economic policy. Workers and employers' representative organisations play a no less important, although primary consultative role in other fields.

Vogel (1999, pp. 130–131) contrasts Austria with the Nordic countries by telling us that

> . . . if the main differences between Austrian neo-corporation and Scandinavian system had to be summed up in a nutshell, the decisive criterion would have to be the maintenance of a wide social divide. In a country where the labor movement never exercised real leadership, the "Social partnership" failed to establish significant mechanisms of redistribution. It has the highest wage dispersion indexes in Europe. The bases of compromise were less ambitious than in the Scandinavian countries and have essentially been limited to as near full employment as possible and social policies that avert extreme poverty.

Within this context Austrian occupational health legislation comprises three broad sets of rules. First, public policy rules backed by criminal penalties set limits on freedom of contract in the employment relationship. They lay down safety obligations, dating back to the 1850s. Austrian legal theory classes these obligations into occupational safety rules, regulation of working time, and the protection of specific groups such as young people, women, and pregnant women. Factory and labor inspectors were instituted in 1883 to police compliance with the very few provisions for the protection of workers. German law applied in Austria under the "invading" Nazi regime. The insurance system was expanded at this time and remains a central feature of Austrian social provision. European directives lead to major legislative reforms including the Working Time Act (1972), which reduced the working week to 40 hours. The Workplace Protection Act (1972) produced the first self-contained, systematic legislation.

Second, the employer has a civil-law "duty to render aid" to workers under the provisions governing labor contracts. Vogel (1998, p. 135) reports that actions invoking the employer's civil liability are rare.

Third, in 1887 the Compulsory Industrial Accident Insurance Act, modeled largely on its German forerunner, was passed. This Act as variously amended has determined for a long time the shape of preventive practices and, by restricting employers' liability to cases of willful negligence, was largely responsible for impeding the development on the civil obligation to render aid.

The first occupational health services in Austria were generally instituted in the late nineteenth century by large companies and financed by the insurance companies. Prevention was seldom the priority for these services; rather, they encouraged employers' control of absenteeism and of keeping health-care spending in check. Employers often used the medical expertise of these services to oppose recognition of occupational diseases. So silicosis, generally diagnosed as bronchitis or tuberculosis, was not accepted as an industrial disease until 1935.

The Occupational Safety and Health Act of 1994 established that employers must develop a prevention policy based on a hierarchy of priorities; identify and evaluate risks and make the evaluation available to workers; communicate with other employers when they share a workplace; appoint safety officers; inform, consult workers, and involve them in all matters with a decisive effect on safety and health. Employers must provide medical checks for workers subject to specific exposures. Employers have to provide ergonomically designed workstations if they include visual display units and provide eye tests. As well, they must systematically introduce a safety service and occupational health service for all workers.

The United States

Any synopsis of employment relations in the United States is bound to be partial and perhaps inaccurate, particularly when applied to occupational health and safety. Cornfield (1987, p. 331) writes: "Throughout the twentieth century in the United States, managements have deployed their new production technologies in order to maintain enterprise profitability, while workers have attempted to protect their job security." In this, U.S. workers and their representative organizations are not unlike others in the world. Cornfield reviewed case studies of 14 major U.S. industrial sectors and discerned two patterns of labor-management relations. The first and more prevalent pattern is increasing managerial control in the shop or office. Managerial control strategies have taken such forms as job de-skilling, technical control, and bureaucratic control of internal labor markets. The second pattern is an increase in formal labor-management cooperation at the company level and industry-wide management decision making. Recent developments in formal cooperation include such voluntary established measures as joint labor-management productivity, technology monitoring, research and long-term industry development committees, joint lobbying efforts for the enactment of protective legislation of products and government policies, and union representation on company boards. Cornfield suggests that this divergence in labor relations trends has resulted from the uneven development of unionization and uneven occurrence of adverse macroeconomic conditions among U.S. industries since 1945.

More recently, Castells (1997) suggests that the United States is an example par excellence of the service economy model, characterized by a rapid phasing out of manufacturing employment after 1970. The model emphasizes an entirely new employment structure where the differentiation among various service activities becomes the main element in creating the employment structure of society. The model emphasizes capital management services over producer services and keeps expanding the social service sector because of a dramatic rise in health-care jobs and to a lesser extent in education employment. The model also indicates an expansion in the managerial category, which includes a considerable number of middle managers.

Insofar as it is possible to characterize the United States it is an example of a "free-market" philosophy, controlled only by those interest groups that seek to influence decision making in the political institutions and as a federal system of government with inevitable tensions between the central and more local elements. In the United States, as elsewhere, voluntary efforts in the free market have not succeeded historically in reducing the incidence of diseases and injuries; workers have demanded government intervention into the activities of the private sector. This intervention takes the form of standard setting, enforcement, and the transfer of information.

Three federal laws—the Mine Safety and Health Act of 1969, the Occupational Safety and Health (OSH) Act of 1970, and the Toxic Substances Control Act of 1976—are at the core legislation concerning health at work.

The OSH Act established the Occupational Safety and Health Administration (OSHA) in the Department of Labor to enforce compliance with the Act; the National Institute for Occupational Safety and Health in the Department of Health and Human Services (under the Centers for Disease Control and Prevention) to perform research and conduct health hazard evaluations; and the independent, quasi-judicial Occupational Safety and Health Review Commission to hear employer contests of OSHA citations. The Office of Pollution and Toxic Substances in the Environmental Protection Agency administers the Toxic Substances Control Act.

The OSH Act requires OSHA to (1) encourage employers and employees to reduce hazards in the workplace and to implement new or improved safety and health programs; (2) develop mandatory job safety and health standards and enforce them effectively; (3) establish "separate but dependent responsibilities and rights" (OSH Act, 1970, Introduction) for employers and employees for the achievement of better safety and health conditions; (4) establish reporting and record-keeping procedures to monitor job-related injuries and illnesses; and (5) encourage states to assume the fullest responsibility for establishing and administering their own occupational safety and health programs, which must be at least as effective as the federal program. The OSH Act originally covered all employers and their employees, except self-employed people; family-owned and -operated firms; state, county, and municipal workers; and workplaces already protected by other federal agencies or other federal statutes. In 1979 Congress exempted from routine OSHA safety inspections approximately 1.5 million businesses with fewer than 11 people. Because federal agencies (except the U.S. Postal Service) are not subject to OSHA regulations and enforcement provisions, each agency is required to establish and maintain its own effective and comprehensive job safety and health program. OSHA provisions do not apply to state and local governments in their role as employers. OSHA requires, however, that any state desiring to gain OSHA support or funding for its own occupational safety and health program must provide a program to cover its state and local government workers that is at least as effective as the OSHA program for private employees.

OSHA can bring its own standard-setting procedures either on its own or on petitions from other parties, including the secretary of state for health and human services, the National Institute for Occupational Safety and Health, state and local governments, any nationally recognized standards-producing organization, employer or labor representatives, or any other interested person. The standard-setting process involves input from advisory committees and from the National Institute for Occupational Safety and Health. When OSHA develops plans to propose, amend, or delete a standard, it publishes its intentions in the *Federal Register*. Subsequently, interested parties have opportunities to present arguments and pertinent evidence in writing or at public hearings. OSHA is authorized to set emergency temporary standards, which take effect immediately but are supposed to be followed up by the establishment of permanent standards within six months. OSHA must first determine that workers are in grave danger to exposure to toxic substances or new hazards and are not adequately protected by existing standards. The employer may challenge the setting of new standards in court.

Under OSHA a worker can request a workplace inspection by writing and signing the request. When an inspector visits a workplace, a representative of the workers can accompany the inspector on the "walk-around." If specific requests for inspections are not made, OSHA can make random inspections, on an infrequent basis, to those workplaces with a lower than average safety record. OSHA covers about 5 million workplaces, and it conducts about 90,000 inspections a year. OSHA employees number about 1,000 federal inspectors and a further 2,000 within state agencies. OSHA can fine employers up to $7,000 for each violation of the OSH Act that is discovered in a workplace inspection and up to $70,000, or up to six months' imprisonment, if the violation is willful or is repeated. The failure to abate hazards can lead to a $7,000 fine per day. Management can appeal through the Occupational Safety and Review Commission, which was established under the OSH Act.

The OSH Act provides a right to know for workers and regulatory agencies when exposure to hazards is taking place. The disclosure can take a number of forms. The democratization of the workplace is the key to the success of disclosure. In particular, a duty to disclose only such information as has been requested can provide a narrower flow of information than a duty to disclose everything. Under OSHA's Hazard Communication Standard, employers have a duty to inform workers of the identity of substances with which they work through labeling the product container and disclosing the purchase. Employers appear to be under no legal obligation to amend inadequate, insufficient, or incorrect information provided by the manufacturer. Employers must transmit information on the standards and their requirements and on operations as well as their locations.

Workers have the right to refuse hazardous work, and this should carry with it the right to access all information on the process. However, there appears to be no adequate way of enforcing this right under the provisions.

The now defunct Congressional Office of Technology Assessment reported:

> OSHA's current economic and technological feasibility analyses devote little attention to the potential of advanced or emerging technologies to yield technically and economically superior meth-

ods for achieving reductions in workplace hazards. . . . Opportunities are missed to harness leading-edge or innovative production technologies to society's collective advantage, and to achieve greater worker protection with technologically superior means.

Intelligently directed effort can yield hazard control options—attributes that would no doubt, enhance the "win-win" character of OSHA's compliance requirements in many cases and support the achievement of greater hazard reduction. (cited in Ashford, 2000, pp. 211–236)

Of all of the examples of national policy for occupational health and safety, the United States appears to have the lightest legal armory at its disposal. This is perhaps not a surprise given the characterization of the U.S. philosophy of employment relations set out earlier in this section. The real test comes when one examines the effectiveness of the work organization's approach to health and safety. There is scope for good research here. Anecdotally, several people working in the offshore oil industry tell me that they feel that U.S. rigs are safer than U.K. rigs because procedures and techniques are much more direct in the U.S. rigs. Of course, even within a nation's facilities, one can find considerable variation in behavior and attitudes toward safety, which in turn will affect individual perceptions of safety (Mearns, Flin, Fleming, & Gordon, 1997). At the time of the Mearns et al research, health and safety considerations were at the core of operations. On the other hand, companies based in the United Kingdom report difficulty in developing corporate health and safety policies and procedures that equally encompass their U.S. operations because approaches vary in each of the states.

Overview

This chapter has sought to illustrate a range of international examples and to root each within the social context that shapes its employment relationship. As might be expected, there are similarities and differences. In terms of the overall coherence of health and safety, the Nordic countries offer a complete and apparently integrated approach, whereas the United States appears, at least in its policies, to be partial and minimal. Austria comes between these examples, as does the United Kingdom. In Austria the role played by insurance is much greater than in the other countries. Each country is clearly learning from the other. Certainly, the United Kingdom's strategy document, *Revitalising Health and Safety* (Health and Safety Commission, 2000), is providing lessons drawn from other countries as a way forward in particularly difficult areas for health and safety development. All countries will use the accident and injury rates that workers experience as a means of evaluating the effectiveness of health and safety provision. This, viewed in terms of Hofmann and Tetrick's model (see Chapter One), is extremely external and very long-term!

According to the International Labour Office (Feyer, 2001) about 250 million accidents that result in injury occur annually in workplaces worldwide. Of these, 335,000 are fatal accidents, an estimated rate of 14 per 100,000 workers annually. Of course, the rates are different for individual countries and regions. The International Labour Office has estimated the annual fatal occupational injury rate for established market economies as 5 per 100,000 workers, whereas the rate for Asia, which has had a relatively rapid and unregulated pace of industrialization is much higher, at 23 fatal occupational injuries per 100,000 workers.

The cost to society of occupational injuries is also clear when one considers the problem relative to other health problems. Injuries in general are a leading cause of hospitalization and premature death in most countries. Occupational injuries represent a substantial part of the entire injury problem. The United States has some 5,350 fatal and 5,650,000 nonfatal injuries and illnesses a year, of which 1,700,000 require recuperation away from work (U.S. Department of Labor, Bureau of Labor Statistics, 2002).

Each year in the EU, about 8,000 people die because of injuries they receive at work. A further 10 million have their health affected to one degree or another by accidents or work-related illness (Mason, 1992).

The economic burden imposed by occupational injury on nations, companies, and individuals is high. In the United States, the National Safety Council (1996) has estimated the total economic burden imposed by occupational injuries at $120 billion per year. Beneath the economic costs lies a huge amount of human suffering and wasted resources such as the amount of health care and lost production. To reduce this burden will take a huge amount of effort by all of the parties—governments, trade unions, government agencies, and so on. Just as the creation of health and safety policies and procedures was due to the originating activities of the main stakeholders in the employment relationship, so too the maintenance and revitalization of health and safety will rely on the efforts of all of the partners.

At an organizational level, an increased awareness of health and safety is discernible among company managers and workers. Europe has hugely increased its awareness of risk. Several large-scale accidents in Britain and in Europe have produced a massive shift and awareness toward the risks associated with living in contemporary society (Beck, 1992; Giddens, 1998). The raising of consciousness of risk may yet still have to work through to the operations of organizations, but clearly the political agenda is now more focused on health and safety than it was even a decade ago. Trade unions too have placed health and safety high in their priorities so as to reengage government in dialogue over the employment relationship in general (Brotherton, 2003).

There are interesting health and safety developments at an organizational level. For example, in the United Kingdom, the Trades Union Congress, the Confederation of British Industry (representing employers), and the government departments concerned are working on partnership initiatives. These partnership schemes involve such developments as management and unions working together in producing health and safety standards in the paper-making industry, developing new forms of representation and training in the media and communications industry, and redesigning jobs in refuse collection, together with joint problem solving and risk management (Trades Union Congress, 2001). In the United States, the Oregon State Safety Employment Act (1973) has led to a wide number of initiatives. Particularly impressive, since OSHA's federal attempt to set ergonomics standards faced considerable opposition, at least initially, are the worksite redesign programs. The program enables Oregon OSHA to offer research and development grants to employers, employer groups, employee groups, and educational institutions to address workplace health and safety and ergonomics problems. Oregon OSHA has a wide range of projects; it also includes prototypes to assist manual handling of large sheets of plyboard, replace floors in a transport company to help prevent slips and falls, develop an automated packing system for a coffee company so as to cut out movement that induces repetitive strain injury, reduce noise from vibrators used in casting operations, and so on (Oregon OSHA Web site). Creative and innovative health and safety projects are taking place in organizations on both sides of the Atlantic. We need to see whether these initiatives have a real impact on the way in which all the stakeholders within organizations actually manage health and safety.

Challenges for the Health and Safety Agenda

Researchers need to evaluate health and safety work, not only to ensure value for money but to begin a process in which training and development is made a live issue in the management of organizations. At present, little evaluation seems to be taking place (Rabinowitz, Bridger & Lambert, 1998; Bridger & Friedberg, 1999). Despite the fact that many health and safety developments are transcending national boundaries, there is not yet a clear-cut and precise definitional and measurement approval to the analysis of problems identified in the discipline. Cooper (2000) makes this point about the term *safety culture*, but it applies to most in the disciplines that contribute to health and safety. Most frustrating of all when attempting an international review of health and safety is the fact that fundamental policy indicators, such as accident reporting, are compiled and calculated on a different basis in each country. Given the costs of injury and accidents, to say nothing of the costs of research and care to remedy problems, a more broad ranging, if not a fully common, means of data gathering and monitoring might produce considerable savings.

Finally, the greatest challenge at all levels—whether national policy, local organization, or individual—is to make health and safety much more a core and strategic consideration in the way in which work is conducted. The benefits of doing this can be seen by examining the Nordic countries model. Too often, health safety is seen as an add-on, an additional set of constraints that organizations have comply with. In many countries health and safety is poised to move closer to the strategic position. As it does so, the pace of technical change, globalization, and the changing composition of the labor market will all offer new challenges. High on the risk-exposure rankings are white-collar jobs, because traditional manufacturing has at least to some extent dealt with exposures to hazard. New approaches have then to be found for old problems in new settings.

Research Challenges for Psychology

The world of work is changing rapidly in response to new technology and new business conditions shaped by global markets. General trends include a large increase in the number of small businesses, changes in management methods, increased use of subcontracting and temporary staff, changes in working hours and in patterns of the working day, and the ability to work away from home. All of these issues can have implications for occupational safety and health. This chapter has demonstrated that the legislative framework is very slow to respond to rapid change. It is also evident that the general model of enforcement and compliance on which most countries rely is becoming increasingly less appropriate to the occupational health and safety of workers. That model sets the baseline but does not move the frontier. In the modern world of work, occupational health and safety has to be every worker's and employer's active responsibility. Healthy and safe behaviors have to become established as prime values and motives for each worker, lest the injury and ill health rates soar. Clearly, the developing and changing world stretches any conceivable system of inspection beyond its practical limits, let alone makes it difficult to operate in a sustained way and so change values and motives. New psychological research at the individual and the organizational level could focus on how to raise health and safety to the status of prime values and motives. Within organizations, health and safety is increasingly developed within partnership initiatives. The move to partnerships is just one indication of a new mode of industrial relations, at least in the United Kingdom. Partnership, rather than conflict, is itself an indication of a new psychological state. From the emerging case studies of partnership initiatives comes evidence that the pattern of social relations necessary to sustain these developments could be facilitated and enhanced by psychologists working with the partners in action research (Brotherton,

2003). Where the organizational framework is less clearly articulated and work is conducted in isolation (by people working from home or by the self-employed), occupational health has to be promoted by the kinds of initiatives—such as training, advice, and social support—that have formed the staple diet of practitioner psychologists for a century or more. Research as to how to bring the benefits of these skills into the changing world of work with all its fragmentation and isolation seems an urgent requirement.

Occupational stress, bullying, and harassment, as well as violence at work are becoming important issues for the health and safety agendas of most countries. These are vital areas in which psychologists have already made significant research contributions. The research of psychologists in these vital areas of health and safety seems likely to be in increasing demand. Psychologists have also begun to play a role in understanding diversity in the workplace. Diversity raises important considerations for health and safety; women's work, an aging workforce, and disability all raise important issues for the way in which ergonomics tackles job design and sets standards for safe work. As people from different cultures come together in the workplace, new challenges arise for communications related to health and safety; research on how to manage health and safety matters with a diverse workforce seems essential.

This chapter has concerned itself with macro-level issues for health and safety. Here lies the greatest challenge to research: To what extent can health and safety issues be researched globally and across national boundaries? The legal and institutional framework for health and safety that has shaped the world of work hitherto has led to different, often irreconcilable, bases for recording, testing, and monitoring. Surveying the detailed literatures, one is left with an awareness that the various governments involved are coming to see these differences as undesirable. Can the physical sciences that contribute to health and safety cope with cross-national cooperation? The task for the physical sciences is relatively straightforward compared with that for psychology. But the overriding need is to see research on a comparative and international scale into the various ways in which the legal framework affects the real behaviors involved in safe and healthy work.

Is this an exciting time to be involved in health and safety? Certainly, it is a challenging time.

References

Ashford, N. (2000). Government regulation of occupational health and safety. In B. S. Levy & D. H. Wegman (Eds.), *Occupational health: Recognising and preventing work-related diseases and injury* (pp. 211–236). Philadelphia: Lippincott.

Beck, U. (1992). *Risk society.* Thousand Oaks, CA: Sage.

Bridger, R. S., & Friedberg, S. S. (1999). Managers' estimates of safe loads for manual handling: Evidence of risk compensation. *Safety Science, 32,* 103–111.

Brotherton, C. (2003). Reworking industrial relations and social psychology. In P. Ackers & A. Wilkinson (Eds.), *Reworking industrial relations: New perspectives on employment and society* (Chapter 5). Oxford: Oxford University Press.

Castells, M. (1997). *The Information Age: Economy, society, and culture. Vol. 2: The power of identity.* Oxford: Blackwells.

Cooper, M. D. (2000). Safety culture. *Safety Science, 33,* 105–120.

Cornfield, D. B. (1987). Labor-management co-operation or managerial control? Emerging patterns of labor relations in the United States. In D. B. Cornfield (Ed.), *Workers, managers, and technological change: Emerging patterns of labor relations* (pp. 24–26). New York: Plenum Press.

Feyer, A. M. (2001). Occupational injury. In W. Carwowski (Ed.), *International encyclopedia of ergonomics and human factors,* (Vol. 3, pp. 1569–1572). London: Taylor and Francis.

Giddens, A. (1998). Risk society: The context of British politics. In Jane Franklin (Ed.), *The politics of risk society* (pp. 1–12). London: Polity Press.

Health and Safety Commission; Department of Environment, Transport, and the Regions. (2000). *Revitalising health and safety.* (Strategy statement.) London: Department of Environment, Transport, and the Regions.

James, P., & Walters, D. (1999). *Regulating health and safety at work: The way forward.* London: Institute of Employment Rights.

Loche, J. (1981). The politics of health and safety. Sir Alexander Redgrave Memorial Lecture, IOSH.

Mason, D. (1992). Health and safety across Europe. *Environmental Health,* May.

Mearns, K., Flin, R., Fleming, M., & Gordon, R. (1997). *Human and organisational factors in offshore safety.* Norwich, England: Health and Safety Executive.

National Safety Council. (1996). *Accident facts.* Itasca, IL: National Safety Council.

Nugent, N. (1999). *The government and politics of the European Union.* (4th ed.). London: Macmillan.

Occupational Safety and Health Act (1970). *Journal of Physiotherapy, 55*(2), 12–15.

Oregon State OSHA Web site. http://www.cbs.state.or.us/external/osha/about.htm

Rabinowitz, D., Bridger, R. S., & Lambert, M. I. (1998). Lifting technique and abdominal belt usage: A biomechanical physiological and subjective investigation. *Safety Science 28,* 155–164.

Lord Robens. *Safety and health at work.* Report of the Robens Committee, 1970–1972. Cmnd 5034. London: HMSO.

Tolley's health and safety at work handbook. (12th ed.). London: Tolley.

Trades Union Congress. (2001). *Partners in prevention: Revitalising health and safety in the workplace.* London: Trade Union Congress.

U.S. Department of Labor, Bureau of Labor Statistics. (2002). [On-line]. Available: http://www.bls.gov.

Vogel, L. (1998). *Prevention at the workplace: The impact of community directives on preventive systems in Sweden, Finland, Norway, Austria, and Switzerland.* Brussels: European Trade Union Technical Bureau for Health and Safety.

CHAPTER 14

Concluding Comments
Integration and Future Directions

Lois E. Tetrick
David A. Hofmann

In the process of editing this volume, we have gained multiple insights into health and safety at both the individual and organizational levels. The purpose of this concluding chapter is to integrate some of these insights and challenge researchers to address the needs for theory development and empirical testing of the theories that result. We will begin by presenting a summary of the conceptualization of health and its implications for occupational health and safety based on the generalized framework we developed in Chapter One. Then we expand that framework to more explicitly incorporate various levels of analysis and describe how the chapters in this volume contribute to our understanding of individual and organizational health across levels. Last, we briefly discuss a couple of future research directions that will further our understanding of individual and organizational health and safety.

Conceptualization of Health

In Chapter One of this volume, we reviewed a number of different ways in which health has been conceptualized. These various conceptualizations of health ranged from discussions centered on the absence of illness to more recent discussions expanding the construct to include notions of flourishing and optimal functioning. Within organizational health, we found an equally broad and wide-ranging discussion of health including such things as the lack of accidents, employee morale, survival, financial performance, growth, and flexibility.

Based on this review, we developed a framework for understanding health and safety of individuals and organizations, which we believe generalizes across levels of analysis. The framework drew on the idea of balancing competing goals falling along two continua: (1) a short-term versus long-term continuum and (2) an intrinsic versus extrinsic continuum (see Figure 1.1). Each of the contributors to this volume made reference to this framework—either implicitly or explicitly—in terms of both where the existing literature resides as well as how future research could address parts of the framework that the literature has not yet addressed. Most of the topics covered in this volume have traditionally been considered from only one quadrant in our framework. For example, Burke and Sarpy (Chapter Three) and Zohar (Chapter Seven) characterized the current state of safety research as focusing primarily on short-term, extrinsic goals. Other chapters in this volume that discussed the legal environment of organizations—such as Perrewé, Treadway, and Hall (Chapter Ten) with their discussion of the Federal Medical Leave Act, Roberts (Chapter Twelve) on workers' compensation, and Brotherton (Chapter Thirteen) on the legal and political environment in the European Union and the United States—tend to describe the current state of the research as falling into the short-term, extrinsic quadrant.

At the risk of overgeneralizing somewhat, we might say that the current state of the literature regarding many of the topics covered in this volume typically has adopted a relatively short-term focus as well as a definition of health emphasizing the removal of negative aspects of health and safety. A couple of exceptions are Shaw and Delery (Chapter Eight) on strategic human resource management and Griffiths and Munir (Chapter Eleven) on health promotion programs. We believe, however, that the authors of the various contributions have done an excellent job of covering new ground by considering the implications of a broader conceptualization of health for their chosen domain of interest.

Levels of Analysis

In Chapter One we acknowledged the need for taking a multilevel approach in understanding health and safety and provided a conceptualization of health that we believed generalized across these various levels of analysis. Looking across the various contributions, however, we are now in a better position to explicitly recognize the various levels of analysis considered throughout this volume. Figure 14.1 depicts our original framework now with individual, group or team, organizational, and societal levels of analysis explicitly recognized.

The individual level of analysis is likely the most familiar for occupational health and safety research and focuses on developing

**Figure 14.1. Schematic Representation
of the Concept of Health.**

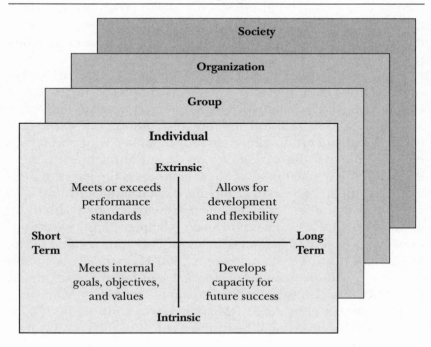

and maintaining individual health and safety. This level incorporates individual differences (see Spector, Chapter Two) as well as many interventions in occupational safety such as training (see Burke and Sarpy, Chapter Three). Parker, Turner, and Griffin (Chapter Four) augment the individual level of analysis on job design by considering potential group and organizational effects. Similarly, Smith, Sulsky, and Ormond (Chapter Nine) incorporate organizational concerns as well as individual concerns with shiftwork and alternative work arrangements like telework.

The group level of analysis focuses on understanding group- or team-level effects on health and safety including, but not limited to, team effectiveness. Tesluk and Quigley (Chapter Five) present a discussion of how group phenomena (e.g., group process) can affect health and safety. Taking a slightly different approach, Zohar (Chapter Seven) focuses more explicitly on safety, considering the role of safety climate within groups as well as its relationship to leadership. Although we are positioning both of these chapters at the group or team level, clearly both group norms and climate have cross-level influences on individual health and safety. Although some recent research has investigated these cross-level relationships (see Chapters Five and Seven), more work remains to be done on this front.

Duffy, O'Leary-Kelly, and Ganster (Chapter Six) also bridge between individual and group levels of analysis. The focus of this chapter is on antisocial work behavior and its effects on individual and organizational health. Therefore, the authors acknowledge organizational and group effects on individual behavior and recognize the role of individual effects on organizational health. Also implicit if not explicit in all of these chapters is the effect of individual health and safety on group and organizational health.

The next level in our schematic representation of health is the organizational level. Shaw and Delery (Chapter Eight) clearly address the organizational level of analysis. They indicate that strategic human resources management (SHRM) typically has been concerned with the financial performance of organizations, which they describe as being a distal reflection of organizational health. This chapter extends the SHRM approach to consider both individual and organizational health as a result of the employment relationship.

Interestingly, the remaining chapters of this volume might be categorized as focusing on a more macrosocietal level of analysis. Certainly, issues of work-family balance (Chapter Ten), health promotion (Chapter Eleven), and workers' compensation (Chapter Twelve) have implications for the other levels of analysis, but they also extend beyond the organizational "fence." Each of these chapters recognizes the role of external factors (public health, economic development, legal constraints) on individual and organizational health. In addition, Brotherton (Chapter Thirteen) summarizes the development of external constraints through legislation in the United States and Europe in affecting organizations' policies and practices concerning employees' health and safety. These chapters suggest the need to incorporate a historic perspective in understanding individual and organizational health, whether at the societal, organizational, group, or individual level of analysis.

Future Research Directions

Each chapter in this volume has suggested future research directions, and we will not repeat these here. However, a few research needs appear to result from the consideration of all the chapters as a whole.

Further Conceptualization of the Construct of Health

We have attempted to integrate the concepts of individual and organizational health along two dimensions: short-term versus long-term goals and intrinsic versus extrinsic goals. This framework seemed to move away from focusing on health as the absence of illness or positive health; we believe that one might be able to identify criteria in each of the four quadrants that represent the presence of positive health as well as the absence of illness. However, as one reads the chapters in this book, it is clear that much of the existing literature still focuses on health as the absence of illness.

Many of the chapters in this volume focus either on health or on safety although the authors generally tried to incorporate both health and safety in their extensions of the literature. At the individual level of analysis, we surmise that most people would subsume safety within the umbrella concept of health although we are un-aware of a direct treatment of the relation between health and safety. However, it occurs to us that health also could affect safety. Therefore, considering the theoretical relationship between the two concepts is important. Also, as we described health and safety at the various levels of analysis, it was not clear whether safety has group, organization, and society analogues as health appears to have. Perhaps at the higher levels safety becomes more synonymous with health. Again, further theoretical development of both health and safety is needed with the accompanying empirical work to advance our understanding of individual and organizational health and safety.

Consideration of Cross-Level Effects

The chapter authors discuss cross-level effects and the implications for understanding individual and organizational health. Many of the authors have recognized potential group effects on individual health and safety. Fewer have recognized potential organizational and societal effects on individual health and safety. Generally speaking, little empirical research has actually investigated these potential cross-level effects. The chapter authors offer suggestions for future research directions to extend our theory and research to incorporate these effects. However, the reader should not construe our call for more cross-level research to suggest that there is no further need for within-level research.

Incorporation of Time Within Health and Safety

Much of the literature presented in the chapters either has taken a short-term perspective or has been mute in considering short-term versus long-term goals. This is not unique to research on health and safety; much of the literature on organizational behavior has not considered temporal issues (see Goodman, Ancona, Lawrence, & Tushman, 2001; also see Humphrey, Moon, Conlon, & Hofmann, manuscript submitted for publication but not yet accepted, for an exception). We encourage theory building and empirical research that explicitly considers time to further the understanding of the temporal aspects of health and safety.

In Conclusion

This volume has presented many of the core topics within the area of occupational safety and health. We have attempted to give readers a view of existing research on individual and organizational health and safety as well as to identify some of the theoretical and research gaps. The authors of the individual chapters have presented future directions, and we have raised some additional directions for future research here. A considerable body of knowledge on many occupational health and safety topics exists; however, the works tend to address only health (and mostly ill health) or safety; focus on only one level of analysis, typically the individual level; and not integrate across various literatures. We hope that readers might use the contributions in this volume as a starting point in efforts to rectify these shortcomings.

References

Goodman, P. S., Ancona, D. G., Lawrence, B. S., & Tushman, M. L. (Eds.). (2001). Forum on time and organizational research [Special issue]. *Academy of Management Review, 26*(4).

Humphrey, S. E., Moon, H., Conlon, D. E., & Hofmann, D. A. (2002). Making progress in a progress decision: The role of level of completion and schedule on safety. Manuscript submitted for publication but not yet accepted.

Name Index

Subject Index

A

Abuse, workplace, 178–181

Abusive supervision. *See* Antisocial work behavior

Academy of Management Journal, 342

Active mental health, 106–110; and aspiration and positive self-regard, 109–110; and autonomy, independence, and productivity, 108–109; and competence, 107–108

Air National Guard, 181

American Management Association, 182

Americans with Disabilities Act (ADA), 47, 360, 361, 363

Antisocial work behavior: and abuse, bullying, and undermining, 178–181; challenges for researchers on, 190–192; and environmental model of organizational health, 189–190; and individual outcomes, 175–181; and organizational outcomes, 181–182; research on relationship between, and health, 192–195; and sexual harassment, 176–178; toward theoretical and empirical progress in, 182–190; and wellness model of organizational health, 184–188

Anxiety, definition of, 36

Anxiety-comfort, 93

Approved Codes of Practice (ACOPs), 375

Asia, 390

Aspiration, 109–110

Australia, 326

Austria, 380, 382–385, 389–390

Autonomy, 108–109

B

Behavioral safety performance: questions concerning, 77–79; questions concerning nature of worker characteristics and, 79; and safety-related work behaviors, 110–111

Brabantia (Netherlands), 333

British Standards Institute, 222, 224

Bullying, 178–181. *See also* Antisocial work behavior

Burnout, 93

C

Canada, 65, 326

Cardiovascular disease (CVD), 92, 103–104, 117

Caregiver, role of, 294–295

Centers for Disease Control and Prevention, 81, 386

Chamber of Commerce, 350

Chernobyl disaster, 201

Christian Socialist movement (Austria), 382, 383

Climate, safety: antecedents of, 208–210; cognitive challenges in formation of, 205–206; conceptual description of relationships between leadership dimensions and, 211; leadership and, injury mediation model,